Understanding Pulmonary Pathology

Understanding Pulmonary Pathology
Applying Pathological Findings in Therapeutic Decision-Making

Richard L. Kradin, M.D., D.T.M. & H.
Associate Physician (Pulmonary/Critical Care Unit) and Associate Pathologist,
Massachusetts General Hospital, Boston, MA, United States
Associate Professor of Medicine and Pathology,
Harvard Medical School, Boston, MA, United States

AMSTERDAM • BOSTON • HEIDELBERG • LONDON • NEW YORK • OXFORD • PARIS
SAN DIEGO • SAN FRANCISCO • SINGAPORE • SYDNEY • TOKYO

Academic Press is an imprint of Elsevier

Academic Press is an imprint of Elsevier
125 London Wall, London EC2Y 5AS, United Kingdom
525 B Street, Suite 1800, San Diego, CA 92101-4495, United States
50 Hampshire Street, 5th Floor, Cambridge, MA 02139, United States
The Boulevard, Langford Lane, Kidlington, Oxford OX5 1GB, United Kingdom

Notices
Knowledge and best practice in this field are constantly changing. As new research and experience broaden our understanding, changes in research methods, professional practices, or medical treatment may become necessary.

Practitioners and researchers must always rely on their own experience and knowledge in evaluating and using any information, methods, compounds, or experiments described herein. In using such information or methods they should be mindful of their own safety and the safety of others, including parties for whom they have a professional responsibility.

To the fullest extent of the law, neither the Publisher nor the authors, contributors, or editors, assume any liability for any injury and/or damage to persons or property as a matter of products liability, negligence or otherwise, or from any use or operation of any methods, products, instructions, or ideas contained in the material herein.

Library of Congress Cataloging-in-Publication Data
A catalog record for this book is available from the Library of Congress

British Library Cataloguing-in-Publication Data
A catalogue record for this book is available from the British Library

ISBN: 978-0-12-801304-5

For information on all Academic Press publications
visit our website at https://www.elsevier.com/

  Working together
to grow libraries in
developing countries

www.elsevier.com • www.bookaid.org

Publisher: Mica Haley
Acquisition Editor: Stacy Masucci
Editorial Project Manager: Samuel Young
Production Project Manager: Edward Taylor
Designer: Matthew Limbert

Typeset by TNQ Books and Journals

Contents

Acknowledgments ix
Introduction xi

1. Approach to the Sampling of the Lung

Handling of the Samples 3
Histochemical Stains 4
Immunostains 4
Further Reading 4

2. Anatomy of the Lung

Gross Pulmonary Anatomy 5
Airway Anatomy 5
Pulmonary Vessels 9
Bronchial Circulation 13
Pulmonary Lymphatics 13
The Elastic Structure of the Lungs 13
Pleura 15
Pulmonary Vascular Elastic Fibers 16
Immune Anatomy 17
Further Reading 19

3. Pulmonary Responses to Injury

Acute Inflammation 21
Chronic Inflammation 23
Fibrosis 23
Histiocytic (Macrophage) Reactions 23
The Airway Wall 27
The Alveolar Interstitium 29
Pulmonary Vessels 31
Pleura 34
Further Reading 36

4. The Radiographic Patterns of Common Lung Diseases

Subba R. Digumarthy

Introduction 37
Airway Diseases 37
Small Airways Disease 41
Direct Signs 41
Indirect Signs 42

Infection 42
Diffuse Lung Disease 46
Lung Neoplasia 54
Further Reading 58

5. Diseases of the Airways

Asthma 63
Chronic Obstructive Pulmonary Disease 67
Chronic Bronchitis 67
Emphysema 69
Exacerbations of COPD 75
Bronchiectasis 76
Cystic Fibrosis 77
The Small Airways 81
Bronchiolitis 81
Obliterative Bronchiolitis 82
Bronchiolitis Obliterans Organizing Pneumonia 84
Follicular Bronchiolitis 87
Diseases of the Upper Airways 90
Tracheal Stenosis 90
Other Tracheobronchial Disorders 91
Amyloidosis 91
Relapsing Polychondritis 93
Tracheopathia Osteoplastica 93
Tracheobronchomalacia 93
Further Reading 94

6. Interstitial Lung Disease

Usual Interstitial Pneumonitis 95
Management 102
Nonspecific Interstitial Pneumonitis 102
Lymphocytic Interstitial Pneumonitis 110
Hypersensitivity Pneumonitis 111
Diffuse Alveolar Damage/Acute Interstitial Pneumonia 114
Acute Interstitial Pneumonia 119
Diffuse Alveolar Injury With Delayed Alveolar Re-Epithelialization 120
Other Poorly Defined Entities 120
Sarcoidosis 122
Granulolymphocytic Interstitial Lung Disease 126
Pleuro-Pulmonary Fibroelastosis 127

IgG4-Mediated Lung Disease   127
Further Reading   128

7. Nonneoplastic Smoking-Related Disorders

Respiratory Bronchiolitis (Interstitial Lung Disease)   129
Desquamative Interstitial Pneumonitis   130
LCH (or Eosinophilic Granuloma)   131
Smoking-Related Interstitial Fibrosis   134
Further Reading   134

8. Pulmonary Vascular Disorders

Pulmonary Arterial Hypertension   137
Idiopathic Pulmonary Arterial Hypertension 138
Pulmonary Veno-Occlusive Disease   139
Pulmonary Capillary Hemangiomatosis   140
Other Causes of Pulmonary Vascular Disease 141
Pulmonary Thromboembolic Disease   141
Chronic Thromboembolic Pulmonary Hypertension   144
Other Forms of Pulmonary Thromboembolic Disease   145
Tumor Emboli   149
Pulmonary Angiitis   149
GPA—Wegener's Granulomatosis   150
Pulmonary Alveolar Hemorrhage Syndromes   152
Antialveolar Basement Membrane Disease or Goodpasture's Disease   153
Drug-Induced Angiitis   155
Arteriovenous Malformations   156
Further Reading   156

9. Pulmonary Infection

The Approach to Sampling for Infection   157
Transbronchial Biopsy   157
Fine Needle Aspiration Biopsy   157
Transbronchial Needle Aspiration Biopsy   157
Video-Assisted and Open Thoracoscopic Biopsy   158
Handling Lung Biopsies   158
Pulmonary Injury in Infection   158
Pulmonary Host Response   158
Pulmonary Defenses   159
Patterns of Lung Injury due to Infection 160
Tracheobronchitis/Bronchiolitis and Miliary Infection   160
Diffuse Alveolar Damage   160
RNA Viruses   162
Influenza   162
Serious Acute Respiratory Syndrome   163
Middle Eastern Respiratory Syndrome   164

Respiratory Syncytial Virus   165
Parainfluenza   166
Measles   166
DNA Viruses   167
Adenovirus   167
Cytomegalovirus   168
Herpesvirus   168
Varicella Zoster   170
Hantavirus   171
Other "Atypical Pneumonias"   172
Mycoplasma Pneumonia   172
Epstein—Barr Virus   172
Pneumocystis jirovecii   174
Bronchiectasis   175
Acute Bronchopneumonia   177
Bacterial Infections   178
Pneumococcal Pneumonia   178
Group A Streptococci   179
Staphylococcal aureus   181
Gram-negative Bacteria   182
Klebsiella   182
Lung Abscess due to Oropharyngeal Aspiration 183
Actinomycosis   183
Nocardia   184
Legionella   188
Rhodococcus equi   188
Tropheryma whippelii (Whipple's Disease) 189
Granulomatous Pneumonia   190
Mycobacterial Infection   190
Spectrum of Pulmonary Tuberculous Infection 193
Reactivation Tuberculosis   195
Atypical Mycobacteria   196
Melioidosis   199
Fungal Infection due to Yeasts   200
Histoplasmosis   201
Blastomyces   204
Cryptococcus   205
Coccidioides immitis   206
Paracoccidioides   207
Candida spp.   207
Hyphate Fungi   208
Aspergillus spp.   209
Immune Disorders due to Aspergillus Infection 209
Allergic Bronchopulmonary Aspergillosis   212
Bronchocentric Granulomatosis   213
Hypersensitivity Pneumonitis   214
Aspergillus bronchitis and Chronic Necrotizing Aspergillosis   215
Fungus Balls   215
Angioinvasive Aspergillosis   216
Other Aspergillus Species   217
Other Hyphate Fungi   219
Pseudoallescheria (Scedosporium)   220
Fusarium   221

Differential Diagnosis 221
Dematiaceous (Pigmented) Fungi 222
Other Fungi 222
Parasites 223
Protozoa 224
Toxoplasma 225
Cryptosporidium 225
Microsporidia 226
Nematodes (Round Worms) 227
Dirofilaria 229
Trematodes (Flukes) 230
Paragonimiasis 232
Cestodes (Tapeworms) 234
Microbes Associated With Bioterrorism 235
Anthrax 236
Yersinia pestis (Plague Pneumonia) 238
Francisella tularensis (Tularemia
   Pneumonia) 241
Pleural Infection 242
Further Reading 242

10. Lung Cancer

Squamous Cell Carcinoma 245
Small-Cell and Large-Cell (Neuroendocrine)
   Carcinoma 247
Adenocarcinoma 249
Multiple Tumors 251
Molecular Analysis 252
Immunohistology 253
Large-Cell Undifferentiated Carcinoma 253
Pleomorphic Carcinomas 254
Malignant Melanoma 254
Metastatic Carcinoma 256
Sarcomas of the Lung 257
Tumors of Intermediate Malignant Potential 260
Salivary Gland Tumors 261
Mucoepidermoid Tumors 261
Adenoid Cystic Carcinoma 261
Mixed Tumors 263
Benign Tumors 263
Hamartoma 263
Sclerosing Hemangioma 263
Benign Clear Cell "Sugar" Tumor
   (Perivascular Cell Tumor) 263
Glomus Tumors 263
Squamous Papillomas 266
Alveolar and Papillary Adenoma 267
Benign Tumorlets 269
Further Reading 269

11. Lymphoid Lesions

Benign Intrapulmonary Lymph Nodes 271
Nodular Lymphoid Hyperplasia 271

Lymphoid Interstitial Pneumonitis 271
Lymphomatoid Granulomatosis 272
Extranodal Marginal Zone B-Cell Lymphoma
   (MALT Lymphoma) 273
Primary B-Cell Lymphoma 274
Hodgkin's Lymphoma 274
Intravascular Large B-Cell Lymphoma 275
T-Cell Lymphoma 275
Leukemia 276
Further Reading 277

12. Iatrogenic Lung Diseases

Drug-Induced Pneumonitis 279
Opiate-Induced Changes 279
Amiodarone 279
Methotrexate 279
Mesalamine 281
Sulfonamides 282
Cytotoxic Drugs 283
Radiation 283
Nitrofurantoin 284
Transfusion-Related Acute Lung Injury 284
Newer Drugs 285
Further Reading 286

13. Toxic Exposures

Toxic Gases 287
Asbestos 287
Asbestos-Related Cancer 289
Talcosis 290
Silicosis 291
Silicatosis 291
Coal Worker's Pneumoconiosis 292
Hard Metal Pneumoconiosis 293
Beryllium Disease 294
Siderosis 295
Aluminosis 295
Further Reading 295

14. Sundry Disorders

Pulmonary Edema 297
Dendriform Ossification 297
Alveolar Microlithiasis 298
Dystrophic Calcification 298
Aspiration Pneumonia 299
Food 299
Lipid Aspiration 300
Cocaine Inhalation ("Crack") 300
Amyloidosis 300
Hyalinizing Granuloma 300
Pulmonary Alveolar Lipoproteinosis 303
Pulmonary Scars 305

Lymphangioleiomyomatosis   305
Chronic Eosinophilic Pneumonia   307

15. Pulmonary Transplant Pathology

Acute Allograft Rejection   309
Airways Inflammation   311
Chronic Graft Airway Rejection   311
Chronic Vascular Rejection   312
Humoral Rejection   313
Other Forms of Graft Injury   313
Postimplantation Response   313
Aspiration   313
Posttransplant Lymphoproliferative Disorder 314
Recurrence of Disease in an Allograft   316
Further Reading   316

16. Pediatric Lung Disease

Hyaline Membrane Disease   317
Sequestrations   317
Intralobar Sequestration   318
Bronchial Atresia and Congenital Cystic
   Adenomatoid Malformation   318
Neuroendocrine Hyperplasia of Childhood   320

Interstitial Lung Diseases   320
Surfactant Protein Deficiency   320
ABCA3 Deficiency   320
Surfactant-Protein C Deficiency   320
Malignant Tumors   323
Further Reading   324

17. Pleural Diseases

Reactive Fibrinous Pleuritis   325
Pleural Plaques   325
Benign Tumors of the Pleura   326
Desmoid Tumors   326
Neural Tumors   326
Metastatic Pleural Tumors   327
Malignant Mesothelioma   327
Rare Pleural Malignancies   330
Further Reading   332

Glossary of Terms   333
Index   337

Acknowledgments

I wish to acknowledge the people who have helped in the production of this textbook. The medical artwork was done by S. Cashman, M. Forrestall-Lee, and S. Conley. L. Arini assisted in the compilation of the text.

My career could not have taken its unique path without the support of my early mentors, including the late R.T. McCluskey, a former Chairman of MGH Pathology, H. Kazemi, the former Chairman of the MGH Pulmonary and Critical Care Unit, and R.B. Colvin, a former Chairman of MGH Pathology and my research mentor in immunopathology.

Introduction

This text is an attempt to bridge the practices of pulmonary medicine and pulmonary pathology. Although in the past, it was common for clinicians to be engaged directly in the study of pathology, this is no longer the case. For many reasons, the practices of clinical medicine and pathology have diverged. But clinical medicine and pathology provide cogent and complementary perspectives on disease. The information gleaned by clinicians is based on empirical observation of the patient at the macroscopic level, whereas the surgical pathologist primarily examines disease at the microscopic scale.

Translating microscopic images into words is not a simple exercise, especially in the practice of medical pathology. A detailed description of a lung biopsy would require pages of words, far more information than the clinician wants or needs to know. Detailed descriptive diagnoses tend to confuse clinicians who are largely unacquainted with the jargon of pathology. How to formulate a pathological diagnosis succinctly is an art as well as a goal that is often only achieved after a long career.

When formulating diagnoses, it is important to recognize what may be lost in translation. Consider the following example: Two people are attempting to have a conversation when neither understands the other's language. In this case, sophisticated discourse is impossible, and the conversationalists might have to revert to signing to make their point.

Next, let's assume that both individuals speak similar languages, e.g., Spanish and Italian, but neither is fluent in the other. In this scenario, elements of the conversation will be understood, whereas others will not be. In fact, this is comparable to what occurs when pathologists and clinicians converse. Although they share a basic medical education, their respective languages have diverged sufficiently due to their specific training that the details of the conversation are no longer fully comprehended.

Based on my experience as both a pulmonologist and pulmonary pathologist, I am convinced that the difficulties involved in conveying detailed information from one discipline to another have currently become great enough to detract from patient care. Although a standard medical school education continues to include introductions to pathology, the time allotted to its teaching has diminished. Medical subspecialties do not routinely attend autopsy conferences or study specimens through the microscope. In the United States, pathologists-in-training are no longer required to participate in direct patient care beyond medical school. Consequently, there has been an emergence of "medical dialects" that interfere with communication.

As a medical resident, I thought of diseases as distinct entities as they are presented in most medical textbooks. However, after training in pathology, I realized that my conceptions of disease had been artificial. Although certain diseases, like cancer, can be diagnosed by widely accepted criteria, the nonneoplastic medical diseases, which make up the bulk of disorders addressed by medical subspecialists, are difficult to define with precision. Their pathologies are frequently nonspecific, so that clinicopathological correlation is often required, if an accurate diagnosis is to be established, and even then, the diagnosis may remain in doubt.

In addition, there are many pathological findings that I had never encountered in clinical training. These represent the *esoterica* of pathology, i.e., entities recognized by pathologists, but unknown to clinicians. Pathologists may choose not to reveal these observations as efforts at communicating them to clinicians tend to produce undesirable confusion. Clinicians may experience the same frustration in describing certain clinical facets of disease to pathologists, who for the most part are no longer conversant with the subtleties of physiology and therapeutics.

The unintended result is the insularity of medical specialties. For pathologists, this has resulted in an increased inclination to focus primarily on the diagnosis of neoplastic diseases, as this is the pathologist's area of greatest expertise vis-à-vis medical internists. As pathologists may have difficulties in interpreting biopsies of nonneoplastic diseases and in recognizing the implications of their diagnoses, they frequently choose to defer to the clinical impressions of the medical subspecialist in interpreting their pathological findings.

Furthermore, pathological observations are too often ignored by medical researchers who lack training in pathology. Creative researchers can develop some very interesting theories concerning the pathogenesis of disease, but which have little bearing on disease as it appears under the microscope. I have reviewed research articles for prestigious medical research journals in which the published images of an animal model did not actually show what they were purported to, or lacked any resemblance to the disease being investigated. A substantial amount of funded research is ill conceived for this reason.

So how does one convey information concerning the pathology of lung disease to pulmonologists and vice versa? Routinely attending mandatory clinico-pathological conferences is a good first step. I have personally come to know my medical colleagues' understanding of pathology. I make a concerted effort at these conferences to be certain that pathological diagnoses are clearly explained, with respect to their specificities. When I sense that a pathological diagnosis is perhaps being misconstrued, I will guide my colleagues back to the reality at hand. Although it may be difficult to rein in a clinician's diagnosis, optimal patient care may require it.

There tends to be a dichotomous appreciation of pathologists among clinicians. Some expect that pathologists know all, others that they know next to nothing. As in most things, the correct answer is somewhere in between.

Another challenge is a widespread confusion among pulmonologists concerning who, when, and how to harvest lung samples. There are multiple reasons for this problem. Pulmonologists do not understand the challenges that face pathologists in the diagnosis of medical lung disease. The findings in the lung may be patchy, as in pulmonary vascular disease, or complex and multifocal, as in autoimmune disorders. In such cases, a transbronchial biopsy is likely to be a low yield procedure with respect to establishing an accurate diagnosis. Some pulmonologists are routinely hesitant to pursue larger thoracoscopic biopsies and tend to default to less invasive procedures that they can control. However, routinely opting for a low diagnostic yield procedure with the idea that a larger procedure can subsequently be pursued can be inappropriate and put the patient through unnecessary discomfort.

It is likely that the nuances of both pulmonary medicine and pathology can only be optimally synthesized by a single mind. Understandably, this is rarely possible, yet a deep knowledge of lung anatomy and microscopic anatomy is an absolute requirement for understanding pulmonary disease. I often wish that pulmonologists would spend more time formally learning lung pathology, but it is unreasonable to expect them to become experts. Instead, it is my hope that they will use this text to foster diagnostic acumen. They should also use this text to learn which questions are important to pose to their colleagues in pathology, to achieve sophistication in diagnosis, therapy, and research.

Although this project was originally aimed at educating pulmonologists, pathologists will find that this text will enhance their appreciation of lung disease, a field that many find challenging.

A chapter on the radiology of common lung diseases authored by Dr. S. Digumarthy has been included as an integral part of the text as it will certainly help both clinicians and pathologists approach lung pathology in practice.

As access to references is widely available on the Internet, I have chosen not to provide a detailed bibliography. Instead, I have cited a number of articles that should provide the reader with a broad overview of the topics presented here. Some of the ideas presented in this text represent my own perspectives on lung pathology and have never been published. However, as they represent the observations made over a long career engaged in the study of lung disease, I am confident that they will be confirmed by others in practice.

Finally, I have included a glossary of terms widely used by pulmonary pathologists that lead to considerable confusion among pulmonologists. This was at the suggestion of the editors, and I fully agree that it will be of substantial value to the reader. Most of these terms that appear in the text are in italics.

Richard L. Kradin

Chapter 1

Approach to the Sampling of the Lung

The basics of lung pathology include how to optimally harvest and process lung samples. There are currently a variety of options available for obtaining lung cells, fluids, and tissue, and they differ with respect to their potential diagnostic yield. Specialty training and individual can bias which approach is taken, but this may not always be optimal for establishing an accurate diagnosis.

Determining which approach to take depends on a variety of factors. Most important is whether the patient can tolerate the procedure. Seriously hypoxemic patients in respiratory distress are not able to tolerate bronchoscopic procedures. Patients with tenuous cardiovascular status may not be candidates for anesthesia. Severe pulmonary hypertension is a contraindication for most invasive procedures. Coagulopathies must be corrected before considering either a lung biopsy or a bronchoscopic brushing. Each patient presents a different decision matrix.

If the patient is judged sufficiently healthy to undergo either an endoscopic or a surgical biopsy, the next concern should be the diagnostic yield. A variety of disorders can potentially be diagnosed via bronchoalveolar lavage (Table 1.1). This includes most infections, as well as malignancy, the presence of diffuse pulmonary hemorrhage, alveolar proteinosis, eosinophilic pneumonia, and Langerhans' cell histiocytosis.

If one opts for a bronchoalveolar lavage (BAL), the question should arise as to whether to perform a concomitant endoscopic biopsy. In a diffuse pulmonary disease, transbronchial biopsy generally adds little morbidity to the procedure and tends to enhance the diagnostic yield. Central endobronchial lesions can readily be visualized and biopsied. Endoscopic ultrasound-guided biopsies and transpulmonary needle biopsies increase the diagnostic yield for sarcoidosis and malignancy and can be helpful in staging lung cancers as potentially inoperable.

Malignant pulmonary nodules that are peripherally located can be accurately sampled via fine needle aspirate biopsies and this may be the optimal approach especially if the likelihood of infection and additional information from a BAL is low. However, nonmalignant nodules are rarely definitively diagnosed by minimal sampling, with the exception of infectious granulomatous disease. Severe emphysema is a contraindication for fine needle aspiration (FNA) due to the increased risk of pneumothorax, and certain nodules cannot be sampled due to their location.

Thoracoscopic biopsies are the gold standard for the diagnosis of diffuse interstitial pneumonias and benign nodular diseases. The diagnosis of interstitial disease requires that the pathologist be provided sufficient tissue with which to achieve a diagnosis. Many disorders are patchy and cannot be reliably sampled by a transbronchial approach. Ideally, in diffuse interstitial disease wedge biopsies should be harvested via the thoracoscope from all lobes of the lung and for benign nodular diseases, from areas showing radiographic abnormalities.

TABLE 1.1 Diagnosis by Bronchoalveolar Lavage (BAL)

Infection (viral, bacterial, mycobacterial, helminths)	Culture, PCR, electron microscopy (EM)
Malignancy	Cytology, EM
Pulmonary alveolar proteinosis	Gross inspection, EM
Langerhans' cell histiocytosis	Immunostains for CD1a, and S-100, EM
Eosinophilic pneumonia (PNA)	Eosinophil count
Diffuse alveolar hemorrhage	Blood, hemosiderin-laden macrophages
Lipid aspiration	Oil-Red O stain
Asbestosis	Asbestos bodies

Understanding Pulmonary Pathology. http://dx.doi.org/10.1016/B978-0-12-801304-5.00001-0

TABLE 1.2 Diagnoses that Generally Require a Video-Assisted Thoracoscopic (VATS) or Open Lung Biopsy Procedure

Usual interstitial pneumonia/idiopathic pulmonary fibrosis

Nonspecific interstitial pneumonia

Hypersensitivity pneumonitis (subacute/chronic)

Organizing pneumonia

Rheumatoid nodules

Langerhans' cell histiocytosis

Erdheim—Chester disease

Rosai—Dorfman disease

Drug-induced pneumonitis

Vasculitis

Nodular infections

Pulmonary veno-occlusive disease

Pulmonary capillary hemangiomatosis

Sclerosing hemangioma

IgG4 disease

Amyloidosis

Isolated scars

The tip of the right middle lobe and the lingula can show nonspecific chronic changes and these areas should not be sampled. The surgeon should avoid areas of lung with features of end-stage lung, as these may not be diagnostic (Table 1.2).

Cryobiopsies increase the amount of lung tissue that can be sampled and may in the future decrease the need for thoracoscopic biopsies. But at present, there have been untoward hemorrhagic complications in patients and this approach requires further evaluation.

Perhaps the most critical question is whether an invasive sampling approach will influence treatment. This is a difficult subject to discuss out of context. For example, there are many inflammatory medical disorders that respond well to corticosteroids and it can be argued that a biopsy is unlikely to add much to an empiric therapeutic approach. Although there is merit to this argument, especially if a patient is not a good candidate for a procedure or is reticent to undergo one, obtaining a biopsy can foster confidence in adopting a therapeutic approach and may ultimately reassure patients. As will be discussed, characterizing the histological features of the inflammatory response may assist in guiding therapy. Finally, research into disease must be based in part on pathological observations. For all of these reasons, biopsies should be pursued, ideally in the early stage of the disease, whenever there is doubt concerning the diagnosis, assuming that the benefits outweigh the actual risks.

How best to approach the diagnosis of noninfectious pulmonary disorders merits a detailed explaination. Endobronchial biopsies will generally suffice for the diagnosis of intraluminal neoplasia, both benign and malignant, when sampling is adequate. But the ability of transbronchial biopsies to yield an accurate diagnosis is limited to diffuse diseases with a lymphangitic pattern of spread, including lymphangitic carcinoma, lymphoma, and sarcoidosis, and they can generally be diagnosed by this approach if approximately five samples are retrieved. The pulmonary lymphatics course adjacent to the small airways and therefore are readily sampled. Peribronchiolar and diffuse diseases can also be diagnosed in some cases. Transbronchial biopsy (TBB) is a reasonable first choice when the differential diagnosis, based on an in-depth appreciation of the clinical and radiographic findings, includes eosinophilic pneumonia, organizing pneumonia, hypersensitivity pneumonitis, Langerhans' cell histiocytosis, desquamative interstitial pneumonitis (DIP), and respiratory bronchiolitis. All published series on the role of TBB tend to include a variety of diagnoses but the level of reproducibility for some is poor (Table 1.3).

Finally, pulmonologists must be cautious in their interpretation of the results of certain procedures. For example, the finding of *Aspergillus* spp. in a BAL specimen does not mean that a peripheral nodule in the lung is necessarily caused

TABLE 1.3 Biopsies Established by Endobronchial/Transbronchial Biopsy

Malignant
Endobronchial and lymphangitic malignant tumor
Lepidic adenocarcinoma
Carcinoid
Granular cell myoblastoma
Adenoid cystic carcinoma
Granular cell tumor
Squamous papilloma
Benign
Sarcoidosis
Amyloidosis (airway)
Granulomatous polyangiitis (airway)
Eosinophilic pneumonia
Diffuse alveolar damage/acute interstitial pneumonia
Acute panbronchiolitis
Infection (if nodular or localized[a])
Hypersensitivity pneumonitis
Organizing pneumonia[a]
Drug-induced pneumonia
Obliterative bronchiolitis[a]
Langerhans' cell histiocytosis[a]

[a]*Diagnosis most often requires a VATS or an open biopsy.*

by the fungus. In such a case, and in the absence of immunosuppression, it would be wise to obtain a tissue biopsy. Certain findings, for example, organizing pneumonia in a small biopsy, may reflect a nonspecific change adjacent to a tumor or abscess. In the same vein, a diagnosis of organizing pneumonia does not mean that the disease is cryptogenic and the pulmonologist must consider a list of possible causes. The diagnostic process should not stop after a biopsy diagnosis has been rendered until all etiologies have been excluded. On the other hand, a diagnosis of malignancy made with a high degree of confidence, even in a small sample, is rarely an error, and additional diagnostic approaches are rarely required.

Ultimately, the choice of biopsy depends on a thoughtfully considered differential diagnosis. If there is doubt concerning which approach to pursue, the question should be discussed with the diagnosing pathologist who ultimately will have to make the diagnosis and may have a clearer idea as to what type of sampling is optimal.

HANDLING OF THE SAMPLES

Bronchoalveolar lavage fluids should be divided into samples for cytological, hematologic, biochemical, and microbiological examination. Cell counts and differential counts can be extremely helpful in assessing infection. In autoimmune disorders, serological testing can help establish the correct diagnosis. In cases in which a diagnosis of alveolar proteinosis is a consideration the milky opaque appearance of the fluid can be diagnostic and a small aliquot can be centrifuged and processed for electron microscopy. If lymphoproliferative disease is being considered, fluids should be examined by cytofluorimetry.

Specific questions should be posed directly to the hospital cytologist who may be inclined to comment solely on the presence or absence of malignancy. If there is a question of Langerhans' cell histiocytosis, the cytologist should be asked to apply appropriate immunostains and/or process the fluid for electron microscopy. If there is a question of pulmonary

TABLE 1.4 Commonly Histochemical Stains Applied in the Diagnosis of Pulmonary Disease

Stain	Purpose
Hematoxylin and eosin	Standard stain for assessing histopathology. Most forms of inflammation, microorganisms, and tumors can be identified accurately with this single combination stain
Elastic stains	Determine underlying architecture, evaluate vascular disease, identify obliterated airways, and determine invasion by tumor or fungi
Trichrome stains	Detects and distinguishes new (gray) from established collagen (dark blue) deposition. Distinguishes muscle (red) from collagen. Distinguishes fibrin (red) from collagen
Periodic acid Schiff	Detects glycogen and glycogenated proteins. Helpful in distinguishing alveolar lipoproteinosis from pulmonary edema and in the diagnosis of certain tumors
Iron stain	Detects hemosiderin, a breakdown product of red cell hemoglobin, and can be used to diagnose early and chronic pulmonary hemorrhage. Asbestos bodies are easily detected with iron stains
Congo red	Used to detect amyloid fibers. True amyloid stains red and is apple green when examined under polarizing light

hemorrhage, which may be seen in diffuse alveolar hemorrhage (DAH), lymphangioleiomyomatosis (LAM), or pulmonary veno-occlusive disease (PVOD), the cytologist should be asked to note the presence of hemosiderin-laden macrophages.

The pulmonologist should be aware that in most centers, the standard approach to handling tissue by surgeons and surgical pathologists is to place it directly into buffered formalin prior to processing it for microscope slide production. However, this approach may be suboptimal. When infectious disease is a consideration, tissue should first be cultured and saved or frozen for the possible application of PCR. Frozen tissue is also required if direct immunofluorescence is to be applied, if there is a question of immune complex disease, or for research purposes.

In some cases, especially certain unusual malignancies, infections, or storage diseases, tissue should be harvested for ultrastructural examination, which requires glutaraldehyde fixation and appropriate buffering, as formaldehyde is a suboptimal fixative for ultrastructural examination.

HISTOCHEMICAL STAINS

The standard histochemical stain applied to biopsies in pathology is the hematoxylin and eosin stain. It allows the pathologist to diagnose most disorders. However, a host of other stains may be required to optimize the diagnosis of lung diseases (Table 1.4).

IMMUNOSTAINS

Various immunological antibodies have been commercially developed for the detection of cell-associated antigens. These are important in the diagnosis and subclassification of malignancies but they also play a role in the diagnosis of nonmalignant disorders. Other techniques including in situ hybridization and molecular phenotyping play an important role in the diagnosis of malignant and benign lung disorders.

FURTHER READING

Jones, K.D., Urisman, A., 2012. Histopathologic approach to the surgical lung biopsy in interstitial lung disease. Clin Chest Med 33 (1), 27–40.
A brief review of the interpretation of lung biopsies in interstitial disease.
Herth, F.J., Eberhardt, R., Becker, H.D., et al., 2006. Endobronchial ultrasound-guided transbronchial lung biopsy in fluoroscopically invisible solitary pulmonary nodules: a prospective trial. Chest 129 (1), 147–150.
A discussion of the endoscopic ultrasound-guided biopsies approach to the biopsy of small lung nodules.
Poletti, V., Casoni, G.L., Gurioli, C., et al., 2014. Lung cryobiopsies: a paradigm shift in diagnostic bronchoscopy? Respirology 19 (5), 645–654.
A discussion of the merits and concerns surrounding cryobiopsies.
Poletti, V., Chilosi, M., Olivieri, D., 2004. Diagnostic invasive procedures in diffuse infiltrative lung diseases. Respiration 71 (2), 107–119.
A study that argues the merits of BAL in the diagnosis of lung disease.

Chapter 2

Anatomy of the Lung

The practice of surgical pathology is based on "morbid anatomy," i.e., the interpretation of changes in normal lung structure. The lung develops from the embryonic *foregut*. The developing lung in the first trimester of gestation consists of epithelial-lined tubules that course within undifferentiated *mesenchyme*. This is referred to as the *pseudoglandular* phase of development (Fig. 2.1). With increasing gestational age, branching tubular precursors of what will ultimately become the conducting airways invade the *mesenchyme* to produce the *canalicular* phase. In the late third trimester, the process of alveolarization occurs and does not fully mature until term. This *alveolar* phase includes the differentiation of the gas-exchanging units of the lung.

At birth, the fetal lung is normally filled with amniotic fluid but must inflate without collapsing as it enters the gaseous medium of the ambient air. Surfactant lipoproteins produced by mature pulmonary alveolar epithelial cells reduce surface tension according to the *Laplace equation* ($P = 2T/R$; i.e., intra-alveolar pressure = 2 × tension/radius of the alveolus) and surfactants prevent the inflated alveoli from collapsing. When surfactant production is deficient due to prematurity or congenital abnormalities, ventilation and gas exchange may be reduced leading to the neonatal *respiratory distress syndrome*. After birth, the thickened blood vessels of the fetal circulation become progressively thin walled as they mature into a high-capacity, low-pressure, system for the conduction of pulmonary blood flow.

GROSS PULMONARY ANATOMY

The lungs are located in the chest cavities on either side of the heart and other mediastinal structures. The right lung is larger and normally has three lobes, whereas the left lung has two lobes (Fig. 2.2). The right lung includes 10 segments, the left lung 8 (Fig. 2.3). These segmental bronchi can be identified by the bronchoscopist (Table 2.1). Fibrous *major fissures* separate the upper from lower lobes of both the right and left lungs and an additional *minor fissure* partitions the right middle lobe (Fig. 2.2). These fissures may be anatomically complete or incomplete. This is potentially important as air can move from one segment or lobe to another via collateral ventilation if fissures are incomplete.

AIRWAY ANATOMY

Lung structure determines lung function (Table 2.2). The airways serve different physiological activities depending on their caliber. The larger conducting airways consist of ∼15 orders of fractal-like, asymmetric, dichotomously branching, cartilaginous tubes that efficiently carry air to and from the lung (Fig. 2.4A and B). The conducting airways represent the normal anatomic *dead space* of the lung (∼150 mL), as they do not participate directly in gas exchange.

The large airways are lined by a complex pseudostratified ciliated columnar epithelium (Fig. 2.5A and B) that secretes serous and mucinous glycoproteins, which defend against microbial invasion and injury due to inhaled particulates.

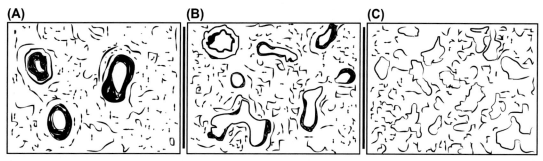

FIGURE 2.1   Development of the embryonic lung: (A) the pseudoglandular phase, (B) the canalicular phase, and (C) the alveolar phase.

Understanding Pulmonary Pathology. http://dx.doi.org/10.1016/B978-0-12-801304-5.00002-2

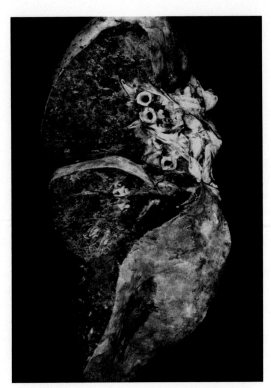

FIGURE 2.2   The adult right lung has three lobes separated by a major and minor fissure.

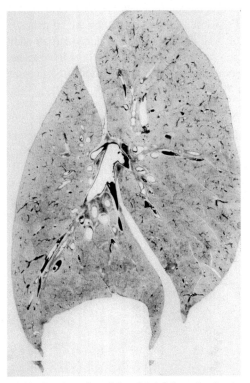

FIGURE 2.3   A thin Gough section of the adult left lung showing its major fissure.

TABLE 2.1 Gross Lung Lobar and Segmental Anatomy

Right Lung	Left Lung
• Upper lobe 　• Apical 　• Anterior 　• Posterior	• Upper lobe 　• Apical-posterior 　• Anterior 　• Superior 　• Inferior
• Middle lobe 　• Medial 　• Lateral	
• Lower lobe 　• Superior 　• Anterior basal 　• Lateral basal 　• Medial basal 　• Posterior basal	• Lower lobe 　• Superior 　• Antero-medial basal 　• Lateral basal 　• Posterior basal

TABLE 2.2 Essential Aspects of Pulmonary Microanatomy

Bronchi are cartilaginous airways, bronchioles lack cartilage

The acinus is an idealized unit of gas exchange that includes all airway structures distal to the terminal bronchiole

Airways course with pulmonary arteries to the level of the terminal bronchiole

Pulmonary veins run in the interlobular septa

Lymphatics run adjacent to bronchovascular septa, interlobular septa, and both visceral and parietal pleura

The visceral pleura reflects along the chest wall as the parietal pleura

Secreted mucus mobilized by ciliated epithelial cells acts as a *mucociliary escalator* that sweeps inhaled particulates adorally toward the mouth to be expectorated or swallowed.

The airways course together with the pulmonary arteries within fibrous bronchovascular septa to the level of the terminal bronchiole (Fig. 2.6). The terminal bronchioles enter the lung at the center of the secondary *pulmonary lobule*, a hexagonal lung tissue unit subtended by adjacent interlobular septa. The pulmonary veins course within these septa. The interlobular septa can be visualized by computed tomography scanning and with the naked eye (Fig. 2.7).

The location of disease within the pulmonary lobule is an important feature in diagnosis. Airway disease tends to be located at the center of lobules, whereas certain forms of pulmonary fibrosis are located peripherally. Diseases associated with the microcirculation are randomly distributed in the pulmonary lobule. For this reason, biopsy samples that do not adequately sample the pulmonary lobule are often inadequate for diagnostic purposes.

Cartilage in the trachea normally forms a continuous semicircle with a membranous posterior membrane, whereas the bronchial cartilage is segmented. The distribution of cartilage in the airways is irregularly identified in airways that are less than ~2 mm in diameter. Serial sectioning of a small caliber airway at this level can reveal cartilage in some sections but not in others. For this reason, they are best referred to generically, both anatomically and physiologically, as *small airways*.

Bronchioles by definition lack cartilaginous investment. The membranous *terminal bronchioles* ramify into three to five pulmonary acini that include the gas-exchanging surfaces of the lung (Fig. 2.8). The acinus is an idealized structure with (1) approximately three orders of branching respiratory bronchioles that are partially alveolated and contribute to gas exchange; (2) the alveolar duct that has the highest O_2 tension within the acinus; and (3) the terminal alveolar sacs (Fig. 2.9). However, what can actually be seen radiographically or with the naked eye is the *pulmonary lobule* and not the acinus.

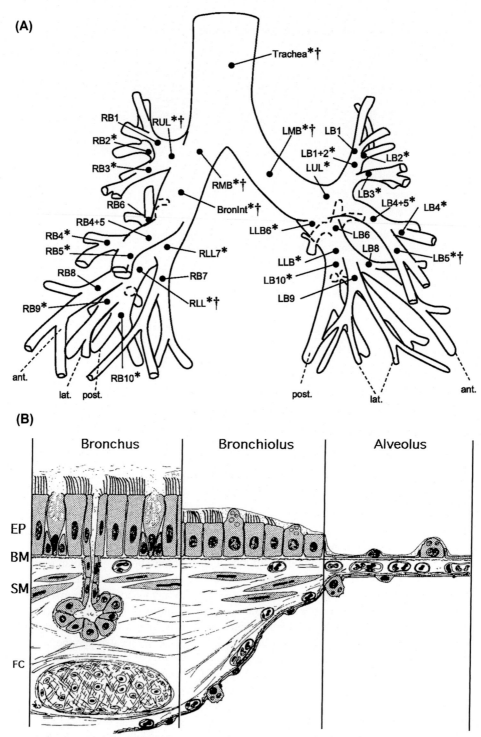

FIGURE 2.4   (A) The branching segmental anatomy of the normal airways, (B) diagram of airway structure moving from proximal to distal. *BM*, basement membrane; *EP*, epithelium; *F*, fibrocartilage; *SM*, smooth muscle.

The alveolar wall is lined by a flattened squamous alveolar type I cell that can only be visualized with the electron microscope (Fig. 2.10) and by alveolar type II cells that secrete surfactant (Fig. 2.11). The latter proliferate nonspecifically during inflammation, and an increase in the number of alveolar epithelial cells is a nonspecific indicator of alveolar injury.

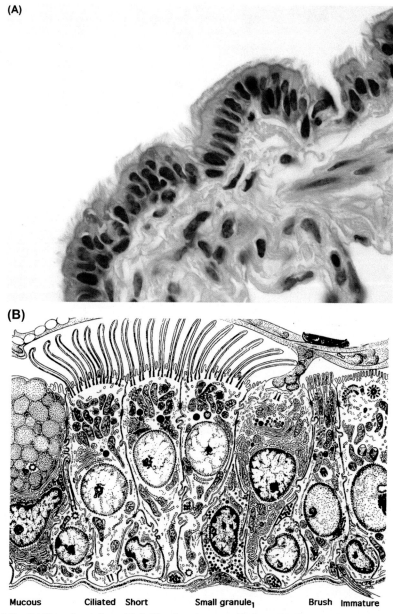

FIGURE 2.5   (A) Normal pseudostratified ciliated respiratory epithelium, (B) diagram of respiratory epithelium.

Gas exchange takes place across the basement membrane shared by the alveolar type I cell and the alveolar endothelial cell (Fig. 2.10). The nongas exchanging surface of the alveolus includes matrix-producing cells and elastin fibers, and it serves as a potential space for the accumulation of fluid and inflammatory cells.

PULMONARY VESSELS

The main pulmonary artery arises from the right cardiac ventricle and immediately branches into two trunks that carry deoxygenated blood to each lung. As noted, the pulmonary arteries course in fibrous septa with their accompanying airways. This allows pathologists to landmark these structures with the light microscope. The pulmonary artery is generally ~15% smaller in luminal diameter than its accompanying airway. Changes in their relative size suggest underlying pathology due to either airway or vascular disease. The pulmonary arteries distal to the terminal bronchiole diverge as they enter the pulmonary lobule to supply the pulmonary alveolar capillaries, where the bulk of the gas exchange occurs between inhaled O_2 and circulating CO_2.

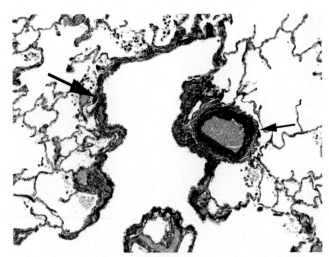

FIGURE 2.6   A pulmonary airway (*large arrow*) and artery (*small arrow*) course together in the fibrous bronchovascular septa (elastic stain).

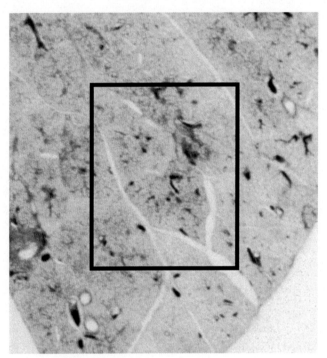

FIGURE 2.7   A secondary pulmonary lobule (box) is delineated by the interlobular septae.

Hypoxemia, i.e., decreased partial pressures of O_2 (pO_2) in the systemic arterial blood is a cardinal feature of lung disease (Table 2.3). The physiological causes of hypoxemia include (1) hypoventilation, (2) V/Q abnormalities in which there is suboptimal matching of ventilation and pulmonary blood flow, (3) the shunting of deoxygenated blood, which can reflect (a) an intracardiac shunt of deoxygenated blood from the right into the left ventricle, (b) shunting by blood vessels in the lung, or (c) areas of lung that are receiving blood but not ventilated (atelectatic lung), and (4) a block to the normal diffusion of O_2 across the alveolar wall. Carbon monoxide reduces O_2 saturation by competing with O_2 for binding sites on hemoglobin. Methemoglobin induced by certain drugs and foods (fava beans) in patients with glucose 6-phosphate dehydrogenase (G6PDH) deficiency and can also produce hypoxemia.

An elevated pCO_2 in the arterial blood (Table 2.4) *always* indicates hypoventilation due to (1) decreased pulmonary bellows function, (2) decreased central ventilatory drive, or (3) increased peripheral CO_2 production due to hypermetabolism, e.g., fever, when unmatched by an adaptive increase in ventilation. The level of arterial pCO_2 is a function of the

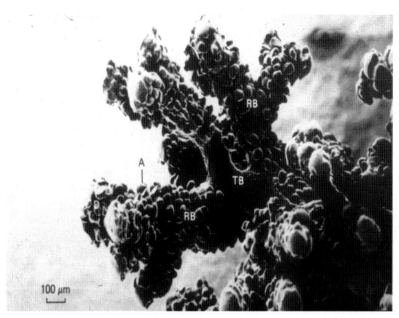

FIGURE 2.8   The pulmonary acinus seen in three dimensions via the scanning electron microscope. *A*, alveolus; *RB*, respiratory bronchiole; *TB*, terminal bronchiole.

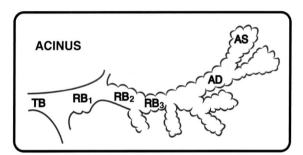

FIGURE 2.9   Diagram of the pulmonary acinus. *A*, alveolus; *AD*, alveolar duct; *RB*, respiratory bronchiole; *TB*, terminal bronchiole.

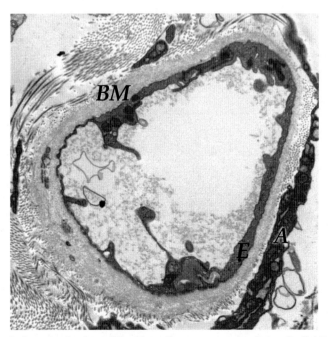

FIGURE 2.10   An electron micrograph shows the alveolar wall lined in part by a squamous alveolar type I cell, (A) immediately adjacent to the alveolar capillary endothelial cell (E) and sharing a basement membrane (BM).

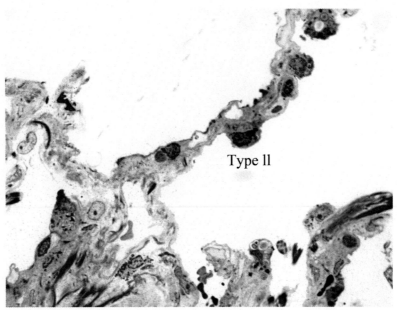

Type II

FIGURE 2.11 A cuboidal alveolar type II epithelial cell shown in a 1-μm section.

TABLE 2.3 Causes of Hypoxemia

Hypoventilation
Ventilation/perfusion abnormality
Pulmonary shunting (V/Q ratio = 0)
Abnormal diffusion of O_2 across alveolar wall $\left(DL_{O_2} = V_{O_2}/PA_{O_2} - P_{capO_2} \right)$
$P_{O_{2_{alv}}} \simeq P_{O_{2_{insp}}} - PA_{CO_2}\Big/\text{respiratory quotient}$

TABLE 2.4 Causes of Increased pCO_2 and Respiratory Acidosis

Decreased VA
Decreased respiratory drive from central nervous system (CNS).
Increased metabolic demand with CO_2 production and nonadaptive ventilation
Increased dead space

metabolic production of CO_2 and inversely proportional to alveolar ventilation (VA). Elevated pCO_2 decreases the plasma pH leading to respiratory acidosis, whereas hyperventilation results in respiratory alkalosis. The final pH of the blood is determined according to the Henderson—Hasselbalch equation by the actions of both the lungs and renal tubules. Their activities compensate for each other, so that, e.g., a patient with a markedly but chronically elevated pCO_2 may have a near normal plasma pH due to renal tubular retention of HCO_3^-, and vice versa. Critical physiological parameters can be expressed as follows:

1. Minute ventilation, $V_E = TV \times RR$ (tidal volume, the amount of air in a single breath × respiratory rate)
2. $pCO_2 = VCO_{2x}$ k/VA, i.e., pCO_2 = metabolic rate of production of $CO_2 \times 0.863$/alveolar ventilation
3. $pH = pk + k\left[HCO_3^-\right]/[CO_2]$, where k is an empirically determined constant

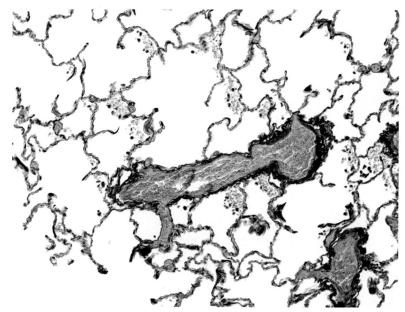

FIGURE 2.12   A pulmonary vein courses within an interlobular septum (elastic stain).

BRONCHIAL CIRCULATION

The blood supplies to the airways and pleura are derived from the systemic bronchial arteries, which arise from the aorta or the intercostal arteries, and carry fully oxygenated blood. Anatomic shunts between the pulmonary and bronchiolar circulation account for ~3% of the normal *shunt fraction* observed in the arterial blood of a normal subject breathing room air. It is necessary to recognize anatomic variations in the bronchial circulation, as interventions aimed at controlling pulmonary hemoptysis via bronchial arterial coiling can potentially occlude aberrant vessels that supply the spinal cord.

It is a generic rule that all "new growth" within the lung, including neoplasia, remodeled airways, and lung fibrosis, derives its blood supply from systemic bronchial arteries. New monoclonal antibody therapies, e.g., bevacizumab, aimed at inhibiting the growth of tumors, e.g., antivascular endothelial growth factor, utilize this principle but have not been studied in bronchiectasis or interstitial lung disease.

The pulmonary veins carry oxygenated blood from the postcapillary venules to the left atrium. Small veins course through interlobular septae (Fig. 2.12) and drain into either the superior or inferior pulmonary veins immediately prior to entering the left cardiac atrium. As a consequence, laser ablation procedures that target the atrial conducting system, sarcoid, histoplasma infections, and mediastinal tumors can cause large vessel pulmonary veno-occlusive disease.

PULMONARY LYMPHATICS

Lymphatic drainage from the lung is complex. There are two pulmonary lymphatic systems. One drains the lung distal to the terminal bronchiole and forms a superficial plexus that courses over the convexity of the visceral pleura to drain into the thoracic duct in the mediastinum. A second, deep lymphatic plexus begins at the terminal bronchiole and drains centripetally to the hilum and the thoracic duct. Lymphatic vessels are widely distributed in the lung in the bronchovascular septa, interlobular septa, and pleura.

THE ELASTIC STRUCTURE OF THE LUNGS

The lung is an *elastic* organ (Fig. 2.13; Table 2.5). Its elements are invested with specialized elastic fibers that expand with respiratory effort and relax spontaneously without during exhalation. The elastic fibers in the lung are variably arranged. Axial airways show dense elastic fibers in their walls, whereas the elastic structures of the alveoli are thin and poorly developed (Fig. 2.14A, *arrows*).

In emphysema, there is a loss of the alveolar elastic fibers that normally tether the small airways (Fig. 2.14B). This leads to a narrowing of the lumen and increased lung compliance ($\Delta V/\Delta P$) during inhalation. It also results in

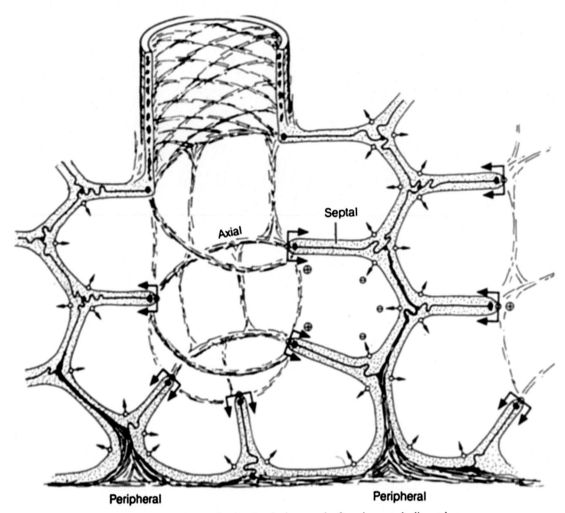

FIGURE 2.13 Diagram showing the elastic network of an airway and adjacent lung.

TABLE 2.5 The Lung as an Elastic Organ

Elastic fibers are found in virtually all parts of the lung

The pulmonary airways show a dense distribution of elastic fibers

Pulmonary arteries have poorly developed internal and external elastic lamina as compared to systemic arteries and are invested with differing amounts of elastic fibers based on their size (i.e., elastic versus muscular pulmonary arteries)

Pulmonary veins and venules show a poorly defined internal elastic lamina

The alveoli contain elastic fibers

The visceral pleura shows a high degree of variability in its investment of elastic fibers, which can pose difficulties when determining whether the visceral pleura has been "invaded" by tumor cells in cases of peripheral malignancies

energy-dependent work during exhalation. The amount of work required for ventilation depends on the underlying state of the lung. In emphysema, large tidal volumes and slow respiratory rates (pink puffer) tend to decrease the work of breathing, whereas in diseases characterized by interstitial fibrosis, the elastic fibers in the lung are increased, resulting in "stiff" lungs and decreased compliance. In restrictive interstitial diseases, the work of breathing is decreased by smaller but more frequent tidal breaths (tachypnea).

The sensation of *dyspnea* may be caused by an increase in (1) the work of breathing, (2) increased stiffness due to scarring and infiltration, (3) hypoxemia, and (4) decreased arterial blood pH in carotid chemoreceptors. However, dyspnea

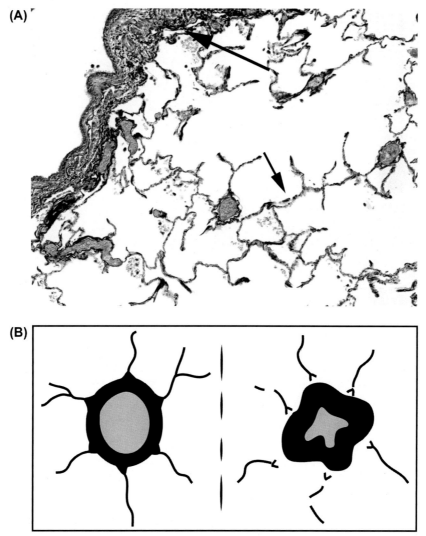

FIGURE 2.14   (**A**) Elastic stain showing thick elastin fibers of an airway (*large arrow*) and the adjacent alveoli (*small arrow*). (**B**) Left panel shows normal elastic fiber tethering of small airways; right panel shows disrupted elastic fiber tethering in emphysema.

is by definition a subjective symptom, so that it is difficult to evaluate it objectively. It should be noted that hypoxemia can be observed in the absence of dyspnea and vice versa. However, patients with severe objective pulmonary dysfunction will generally experience dyspnea.

A significant number of patients seen in pulmonary clinical practice will have nonpulmonary causes for their dyspnea. Cardiac disease must be excluded. Anxiety accounts for approximately one-third of patients who complain of dyspnea even when they have underlying cardiopulmonary deficits. Gastroesophageal reflux can produce dyspnea presumably by stimulating vagal afferents, as can deconditioning. Cardiopulmonary exercise testing can determine whether the sensation of dyspnea has a ventilatory limit, demonstrate the presence of a cardiac limitation, uncover anxiety-driven chronic hyperventilation, or unmask deconditioning as the cause of dyspnea.

PLEURA

Both lungs are invested with a thin membrane lined by mesothelial cells, the *visceral pleura*. The visceral pleura is apposed to the lung and reflects onto the chest wall to form the *parietal pleura*. A fine elastic network is present at the limiting membrane of the alveolated lung, whereas a denser elastic layer is seen deeper within the pleura (Fig. 2.15, *arrows*).

The space between the visceral and parietal pleural surfaces normally contains several milliliters of transudate fluid, which reduces the friction between the lung and chest wall during ventilation. In lung pleural disease, or extra-thoracic disorders, this potential space can be filled by air (pneumothorax), transudate cardiac edema fluid, lymph (chylothorax),

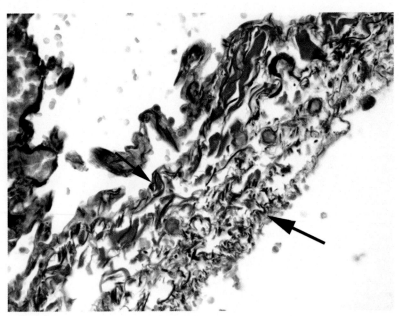

FIGURE 2.15 The elastic layers of the visceral pleura. *Small arrow* shows limiting elastic membrane. Dense elastic fibers are present deeper in the pleura (*large arrow*).

blood (hemothorax), or inflammatory exudative effusions. The pleura communicates with the peritoneal lining of the abdominal viscera, so that the intra-abdominal fluid, i.e., ascites, pancreatitis, and subdiaphragmatic infection, can also produce pleural effusions. Patients with Meigs syndrome and ovarian fibromas may develop pleural effusions.

PULMONARY VASCULAR ELASTIC FIBERS

In addition to the changes that occur with ventilation, the pulmonary vessels must respond to the pulsating pressures of the entire cardiac output at pressures generated by the right ventricle (normal = 25/10 mm Hg) for the pulmonary arteries or left cardiac ventricle (normal = 120/80 mm Hg) for the bronchial arteries. Large elastic pulmonary arteries show well-defined internal and external elastic laminae, as well as elastic fibers that course within the muscular tunica media at a lower density than in systemic arteries (Fig. 2.16). The number of elastic fibers in the media decreases with the caliber of the vessels so that *muscular* pulmonary arteries include only an internal and external elastic lamina. The pulmonary veins

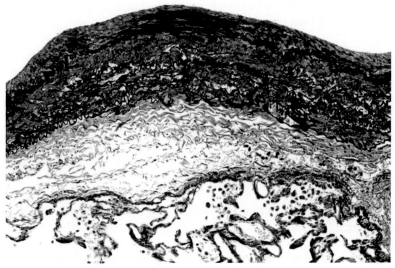

FIGURE 2.16 The elastic lamina of a large caliber pulmonary artery.

generally exhibit an obvious external elastic lamina with a poorly defined internal elastic lamina (Fig. 2.12). Bronchial arteries that carry blood at systemic pressures show well-developed internal and external elastic laminae, as well as dense elastic fibers within their tunica media.

The distribution of elastic fibers in the lung allows pathologists to define anatomic structures in health and disease. For example, when a small airway has been obliterated by scarring, its former presence can be discerned by virtue of residual elastic fibers. The staging of a lung cancer is influenced by whether tumor cells invade the elastic lamina of the pleura. Elastic fibers also increase or decrease in pulmonary fibrosis and emphysema, respectively, and pulmonary vascular diseases invariably show an increased density of elastic fibers within vessels.

IMMUNE ANATOMY

Lymph nodes are abundantly situated around the major airways, in the pulmonary hilum, and in the mediastinum. Surgical oncologists label the regional lymph nodes as *stations* for the purpose of tumor staging. A simple way to remember their relative placement is that lymph node stations with two digits, i.e., ≥10 are *hilar* nodes, whereas single digit stations designate *mediastinal* nodes (Fig. 2.17).

Drainage to the regional lymph nodes shows substantial variation that is not predictable. It is not uncommon to have a malignant tumor metastasize to one nodal group, skip an adjacent nodal group, only to be found again in a more distant lymph node. Positron emission tomography (PET) scanning and mediastinoscopy are generally used for determining the presence of nodal metastases, but findings at surgery can show disease undetected by either method. It is also important to recognize that PET scanning makes assessments based on radioactive fluoro-deoxyglucose (FDG) uptake and that increased FDG uptake can be seen when lymph nodes are inflamed and enlarged in the absence of metastasis. Consequently, patients with lung carcinoma should not be excluded as surgical candidates without the benefit of sampling the mediastinal lymph nodes.

Bronchial-associated lymphoid tissue is not well developed in humans. Inhaled antigens absorbed from the airway lumen are transferred to mucosal lymphoid tissue and then to hilar nodes. Immune cells are located throughout the large

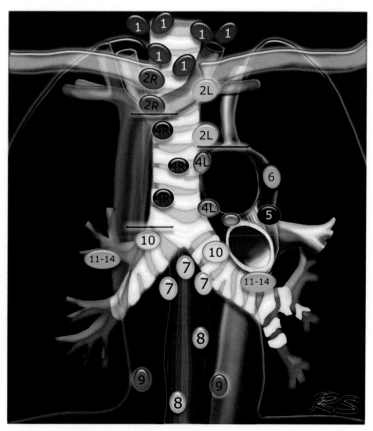

FIGURE 2.17 Diagram of the thoracic lymph node stations.

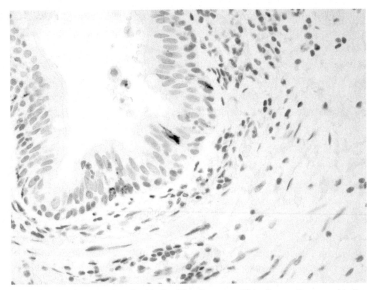

FIGURE 2.18   A CD1a + Langerhans' cell with dendritic features within the respiratory epithelium (peroxidase).

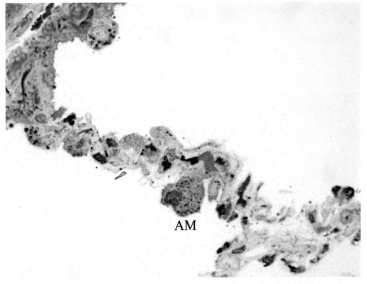

FIGURE 2.19   Normal alveolar macrophage (AM) attached to an alveolar wall in a thin section.

and small airways. Resident Langerhans' cells in the airway epithelium play an important role in the initial processing of soluble and particulate antigens (Fig. 2.18).

Resident alveolar macrophages scavenge particulates deposited within the alveolar sacs. Under the light microscope following standard formalin fixation, they are seen within the alveolar lumen, but this is in fact an artifact, as they are normally located along the surfaces of the alveolar walls in vivo (Fig. 2.19).

Particulate antigens catabolized by alveolar macrophages are released to local dendritic cells for processing. Most antigen presentation occurs in the regional lymph nodes, and there is active trafficking of lung macrophages and dendritic cells to hilar lymph nodes. It is noteworthy that other types of cells can also translocate to regional lymph nodes during inflammation, including benign mesothelial cells, and they should not be taken as evidence of metastatic disease.

Pulmonary inflammation is mediated primarily by circulating granulocytes, lymphocytes, and monocytes recruited to areas of antigen challenge or lung injury. Overall, the normal milieu of the lung is immunosuppressive, which allows the vulnerable gas—exchange surface to remain uninflamed.

FURTHER READING

Taylor, C.R., Weibel, E.R., 1981. Design of the mammalian respiratory system. I. problem and strategy. Respir Physiol 44 (1), 1—10.
This paper introduces a series of reports on the structure and function of the respiratory system of mammals.
Haefeli-Bleuer, B., Weibel, E.R., 1988. Morphometry of the human pulmonary acinus. Anat Rec 220 (4), 401—414.
The paper that defined the pulmonary acinus.

Chapter 3

Pulmonary Responses to Injury

A critical factor in pulmonary medical pathology is recognizing how the lung reacts to injury. Without an in-depth knowledge of this subject, diagnoses are necessarily based primarily on "pattern recognition" and will tend to lack interpretive depth.

Each compartment in the lung, i.e., the airways, alveolar wall, and vessels, has a limited yet characteristic number of responses to injury. Furthermore, the patterns of lung injury are redundant, i.e., they have multiple etiologies. Both the pathologist and clinician must be cognizant of this fact and develop their differential diagnosis based on these patterns (Table 3.1).

Knowing the pathways of inflammation can help to predict responses to therapy. For example, lung lesions rich in eosinophils are likely to be responsive to the administration of corticosteroids, whereas pauci-inflamed fibrotic lesions are unlikely to respond to anti-inflammatory medications, but may be responsive to drugs that antagonize fibrosis. Pathologists should be queried as to the degree of inflammation and the type of inflammatory cells within a biopsy, as this may be important in formulating therapy as the "diagnosis." In this chapter, the pathways of pulmonary inflammation are briefly reviewed. Detailed descriptions of inflammatory changes in specific pulmonary diseases are addressed under specific chapter headings.

ACUTE INFLAMMATION

A number of pathways contribute to *acute inflammation*. The primary event in acute pulmonary inflammation is injury to the microvasculature. This leads to the exudation of edema fluid, plasma proteins, and clotting factors that foster the extravascular generation of fibrin. If one immunostains an acutely inflamed lung, one sees the diffuse deposition of extravascular fibrin that may not be obvious in routine hematoxylin and eosin (H&E)-stained sections (Fig. 3.1). Fibrin forms a gel that entraps free water and it also serves as potential "scaffolding" for subsequent lung repair. The complex sum of humoral, cellular immune, cytokine, chemokine, complement, and coagulation pathways determines the fate of acute pulmonary injury (Fig. 3.2).

TABLE 3.1 Patterns of Pulmonary Injury in Pulmonary Disease

Organizing Pneumonia	Postinfectious, aspiration, drug injury, autoimmune disease, transplant rejection, idiopathic
Diffuse alveolar damage (DAD)	Viral infection, sepsis, shock, drug toxicity, exacerbation of chronic interstitial disease, idiopathic
Obliterative bronchiolitis	Inhaled toxins, postinfectious, lung transplant rejection, graft versus host disease
Nonspecific interstitial pneumonitis	Autoimmune disease, hypersensitivity pneumonitis, drug toxicity, post-DAD
Usual interstitial pneumonitis	Idiopathic, autoimmune disease, drug toxicity, pneumoconiosis, congenital disease
Granulomatous disease	Sarcoidosis, infection, hypersensitivity pneumonitis, immune deficiency, drug hypersensitivity, beryllium
Pulmonary edema	Left atrial hypertension, drug toxicity, intracranial disease, posttransfusion, postpleural fluid evacuation
Alveolar proteinosis	Autoimmune, immunodeficiency, pneumoconiosis, congenital surfactant defects
Desquamative interstitial pneumonitis	Smoking, drugs, fibrosis

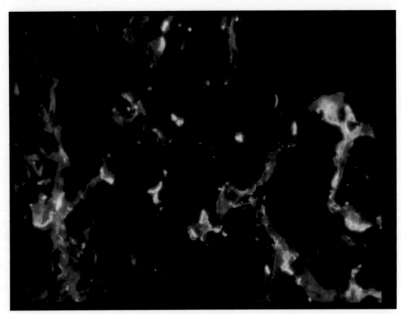

FIGURE 3.1 The diffuse deposition of fibrin in a lung acutely exposed to hyperoxia (immunofluorescence).

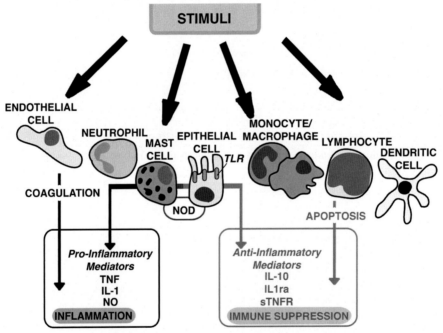

FIGURE 3.2 Regulation of acute inflammatory response.

Acute inflammation activates the complement cascade, which induces the rapid influx of neutrophils via the generation of chemoattractant complement fragments. Activated cells in the lung release a panoply of cytokines that determine the sequential changes seen in inflammation. In bacterial pneumonias, large numbers of neutrophils accumulate to produce a *pyogenic* response or *abscess* formation (Fig. 3.3). In other cases, low-level neutrophil exudation coexists with chronic inflammation. In some diseases, e.g., rheumatoid arthritis, the transit of neutrophils through the lung is rapid, so that their role may be underestimated based on the microscopic examination of a lung biopsy. This is true in emphysema and the interstitial pneumonias, as well. In these disorders, bronchial lavage specimens will tend to show increased percentages of neutrophils in the lavage fluid.

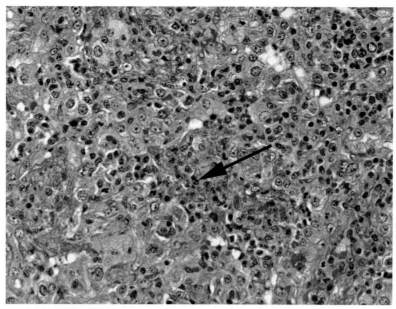

FIGURE 3.3   Acute neutrophilic pulmonary inflammation (*arrow*).

CHRONIC INFLAMMATION

Chronic inflammation generally develops as part of the sequence of cellular events following acute inflammation. The cell types that characterize what pathologists term *chronic inflammation* primarily including lymphocytes, macrophages, and plasma cells (Fig. 3.4A). These leukocytes mediate innate as well as adaptive immunity. For historical reasons, pathologists continue to refer to macrophages as *histiocytes*. The majority of chronic inflammatory lymphocytes seen in most diseases are CD3+ T lymphocytes, comprised by a majority of CD3+CD4+ lymphocytes and a smaller complement of CD3+CD8+ lymphoid cells. CD20+ B cells are rarely observed in large numbers in pulmonary chronic inflammation outside of nodular aggregates termed *follicles* (Fig. 3.4B). These follicles may develop active *germinal centers*, reminiscent of what is seen in reactive lymph nodes. CD56+ natural killer cells do not comprise numerically prominent numbers in most pulmonary chronic inflammation. PD1+ lymphocytes play a role in immunosuppression and their numbers vary with disease.

FIBROSIS

Most inflammatory reactions in the lung resolve via the development of fibrosis, which may be mild or severe. Fibrosis includes fibroblasts, macrophages, mast cells, and eosinophils (Fig. 3.5). Fibrosis can exhibit different appearances depending on its acuity versus chronicity. This can be crudely assessed histologically with the aid of trichrome stains. Active fibroplasia includes type III collagen and glycosaminoglycans and shows *amphophilic* staining when viewed under the microscope with the standard H&E stain (Fig. 3.6). Established scars tend to be rich in cross-linked type I collagen and exhibit deep pink staining. Hyalinized paucicellular fibrosis is a common feature of asbestos pleural plaques (Fig. 3.7), scarring seen in sarcoidosis, chronic histoplasmosis, and hyalinizing granuloma. As a general rule, early fibroplasia is *potentially* reversible, whereas dense established scars are not, a distinction with important therapeutic implications.

HISTIOCYTIC (MACROPHAGE) REACTIONS

A variety of inflammatory and irritant reactions are characterized by the accumulation of pulmonary macrophages. In these responses, macrophages tend to infiltrate the pulmonary interstitium and the alveolar spaces. In chronic inflammation CD68 + macrophages constitute the dominant cell type (Fig. 3.8). They are derived from circulating blood monocytes with different proinflammatory capacities than resident alveolar macrophages, the latter's primary role being to scavenge particulates and maintain a sterile and immunosuppressive environment via the release of transforming growth factor beta-β and interleukin-10.

(A)

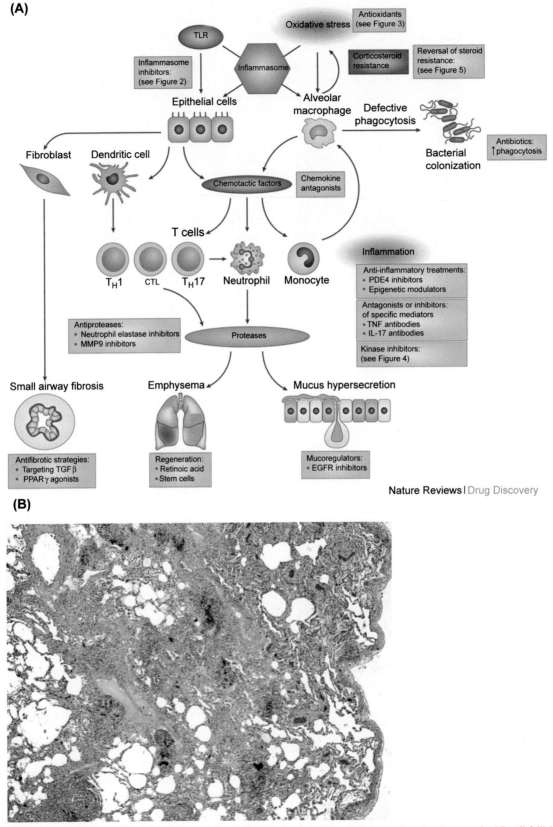

Nature Reviews | Drug Discovery

(B)

FIGURE 3.4   (A) Pathways of chronic pulmonary inflammation, (B) lung with chronic inflammation showing organized B-cell follicles.

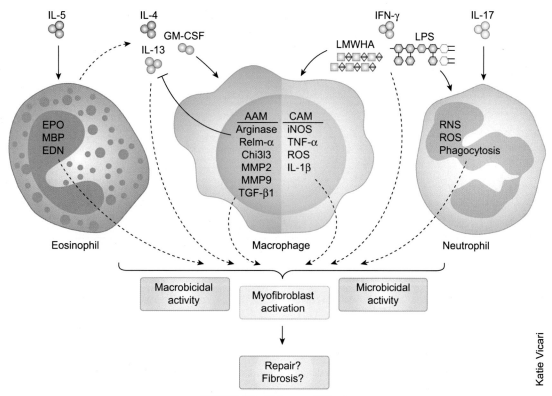

FIGURE 3.5 Pathways of fibrosis.

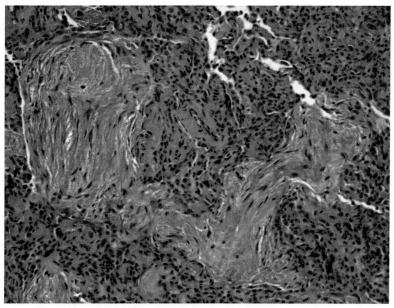

FIGURE 3.6 Intraluminal fibrosis in organizing pneumonia.

When exudate pulmonary macrophages encounter particulates that are not readily catabolized, and in the presence of interferon-γ, they develop increased intracytoplasmic lysosomes and assume an *epithelioid* appearance. Tight-knit microscopic aggregates of epithelioid macrophages are termed *granulomas* (Fig. 3.9). They are a host defense mechanism related to the primitive clotting system seen in invertebrates, e.g., the horseshoe crab (limulus), designed to limit the penetrance of foreign particulates and virulent organisms.

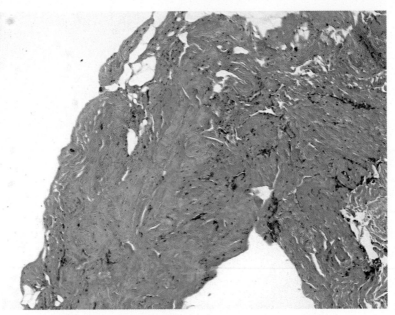

FIGURE 3.7 Established hyaline fibrosis in a pulmonary scar.

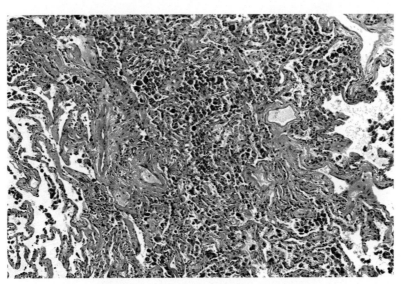

FIGURE 3.8 Pulmonary macrophages in alveoli.

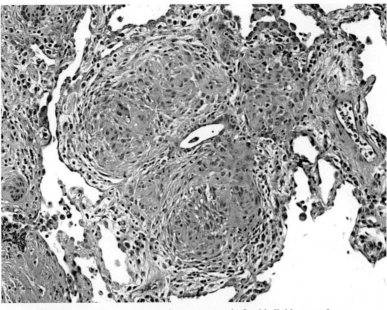

FIGURE 3.9 A compact granuloma composed of epithelioid macrophages.

The pathways noted above participate in virtually all of the inflammatory disorders of the lung. Even disorders that are generally considered to be noninflammatory, e.g., atherosclerosis, generally show some degree of chronic inflammation. However, each pulmonary compartment has its own specific set of reactions to injury.

THE AIRWAY WALL

The responses of the airway wall depend on the nature of the insult. Chronic irritant exposure produces changes in the lining respiratory epithelium. This includes an increase in mucin-secreting cells within the surface epithelium and mural minor salivary glands (Fig. 3.10). If irritation persists, the surface lining respiratory epithelium undergoes *metaplastic* change toward keratin-producing squamous cell differentiation, which mimics the protective surface of the skin (Fig. 3.11).

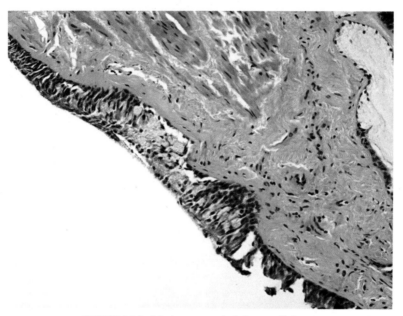

FIGURE 3.10   Mucinous metaplasia in a small airway.

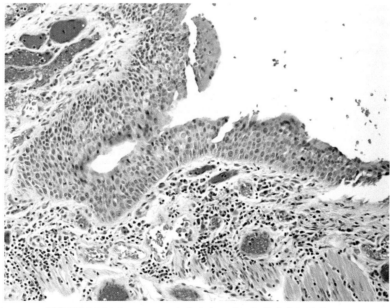

FIGURE 3.11   Squamous metaplasia in a chronically inflamed airway.

Depending on the nature of the insult there may be an acute, chronic, or mixed inflammatory response. When there is an atopic component, eosinophils are prominent.

If the airway epithelial lining ulcerates, there is an outpouring of plasma proteins from the injured microvasculature that releases fibrin. Active fibrosis fills the injured airway lumens to produce fibrous plugs or *Masson bodies* (Fig. 3.6). This is accompanied by the ingrowth of capillary-like vessels. This *granulation tissue* may include acute and chronic inflammation, and the fragility of small new vessels can lead to local hemorrhage (Fig. 3.12).

The response to particulate challenges includes the accumulation of intra-alveolar macrophages. Proximal obstruction of airway lumens evokes an accumulation of alveolar macrophages that become vacuolated due to the accumulation of intracytoplasmic lipid (Fig. 3.13). The mechanism of this change is unknown, but it is a reliable marker of small airways obstruction.

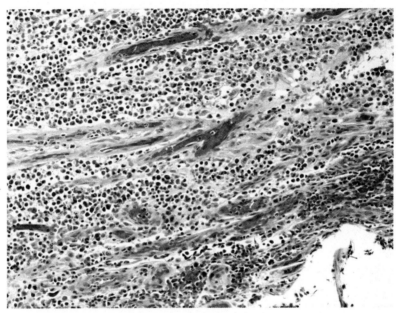

FIGURE 3.12   Granulation tissue comprised of inflamed capillary-like vessels.

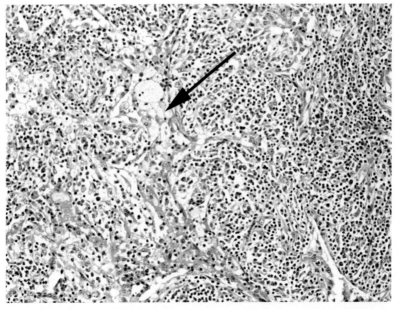

FIGURE 3.13   Lipoid pneumonia behind an area of airway obstruction by tumor. Note the presence of lipid-laden foamy macrophages (*arrow*).

THE ALVEOLAR INTERSTITIUM

Inflammation of the alveolar wall may be acute or chronic. Fibrosis expands the interstitium and injury may produce microulceration with out-pouching of active fibroplasia, lined by reactive alveolar epithelium, termed *fibroblastic foci* (Fig. 3.14). Their histology is comparable to that of *Masson bodies* in the small airways; however, the fibroblastic focus is *limited* to the alveolar interstitium.

In severe alveolar injury, the alveolar surfaces are diffusely ulcerated with loss of the normal epithelial lining. Exudate plasma proteins become entrapped within the remains of necrotic alveolar epithelial cells and the resulting admixture of fibrin and epithelial debris apposed to the alveolar wall is termed a *hyaline membrane* (Fig. 3.15). Its presence is *prima facie* evidence of acute alveolar damage.

Alveolar wall injury results in the proliferation of alveolar type II cells that re-epithelialize the denuded alveolus. Active reparative type II epithelial cells show *reactive cytological atypia*, characterized by increased cytoplasm and nuclear

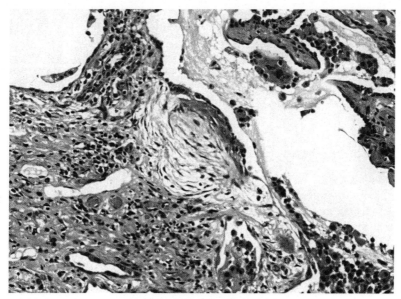

FIGURE 3.14 A fibroblastic focus.

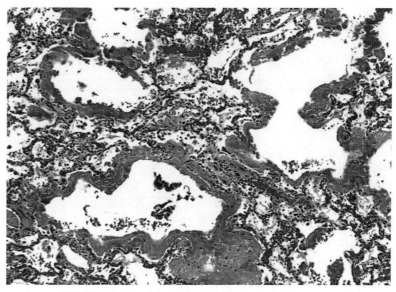

FIGURE 3.15 Hyaline membrane lining alveolar ducts and sacs.

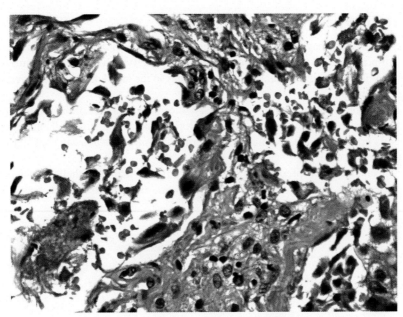

FIGURE 3.16 Reactive atypia in alveolar epithelial lining cells.

enlargement that can be severe (Fig. 3.16). Viral infection, cytotoxic drugs, and radiation also produce severe reactive alveolar type II cell atypia. It can be difficult and even impossible to distinguish reactive atypia from neoplasia, and cytologists and pathologists should be cautious about diagnosing malignancy in the setting of severe inflammation, chemotherapy, or radiation effects.

Subsequent fibrosis in alveolar injury results from a confluence of fibroblastic foci showing edema, reactive fibroblasts and myofibroblasts, macrophages, mast cells, and a smaller but definite increase in eosinophils at 2−3 weeks. Eventually, the interstitial fibrosis becomes condensed and less edematous but alveolar type II cell hyperplasia may persist.

In chronic interstitial inflammation and fibrosis there may be proliferation of smooth muscle precursors in the walls of small airways (Fig. 3.17). There may also be an accumulation of metaplastic adipose tissue that develops beneath the pleura. Metaplastic bone termed *dendriform ossification* is commonly seen as a nonspecific change in fibrotic lungs. *Corpora amylacea*, spheroid structures comprised of glycoproteins tend to develop in edematous and fibrotic lungs

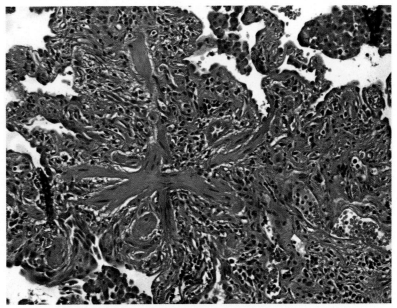

FIGURE 3.17 Metaplastic smooth muscle extending into alveolar walls.

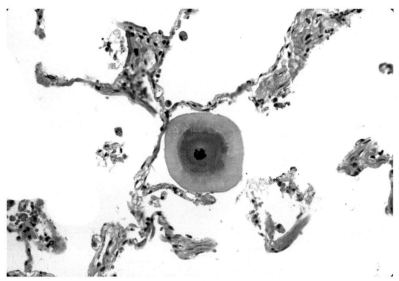

FIGURE 3.18   Corpora amylacea.

(Fig. 3.18), and focal calcified spherules so-called "blue bodies" develop in scarred lungs when intra-alveolar macrophages accumulate. All of these findings currently have no recognized clinical significance and are judged as pathologic "curiosities."

PULMONARY VESSELS

Changes in the pulmonary vessel wall generally reflect (1) chronic elevations of pulmonary arterial pressures, (2) thrombo/embolic disease, or (3) vascular wall inflammation. When chronically elevated pulmonary arterial pressure due to (a) chronic airway and interstitial diseases, (b) chronic hypoxemia, or (c) circulating vasoactive factors develop, the pulmonary arterial microvasculature undergoes a stereotypic mode of remodeling. Mildly elevated pulmonary arterial pressures lead to thickening of medial smooth muscle and extension of vascular muscle distally to the level of the pre-capillary arteriole (Fig. 3.19). The vascular intima is thickened due to the deposition of collagens and glycosaminoglycans. The vascular elastic fibers are concomitantly increased and pulmonary arterial vessels begin to assume the appearance of

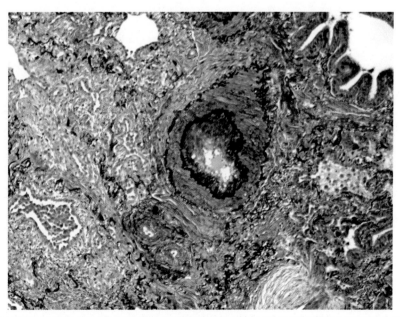

FIGURE 3.19   Vascular remodeling in pulmonary hypertension showing thickened vascular intima and tunica media (elastic stain).

FIGURE 3.20 Inflamed pulmonary artery with focal necrosis (*arrow*).

systemic arteries. Long-standing pulmonary hypertension tends to result in the development of atherosclerosis, which may vary in degree from fatty streaks to calcified plaques with aneurysmal changes.

When pulmonary arterial pressures are greatly elevated as in idiopathic pulmonary hypertension, congenital heart disease with left to right vascular shunting, or in reaction to vasoactive factors, the small muscular arteries and arterioles can undergo mural necrosis termed *fibrinoid necrosis* with luminal fibrin deposition (Fig. 3.20). With time, the thrombosed vascular lumen is revascularized by capillary-like vessels and recanalizes with multiple small caliber vascular channels termed *plexiform lesions*. Microvascular shunting around a partially occluded small vessel produces *dilatation* or *angiomatoid lesions* (Fig. 3.21). This finding is patchy and must be diligently searched for by the pathologist. As a rule, pulmonary vascular disease is focal and easily left unsampled by a small biopsy and can even go unidentified in a cursory examination of large samples of lung. Elastic stains help to highlight changes in the pulmonary vasculature.

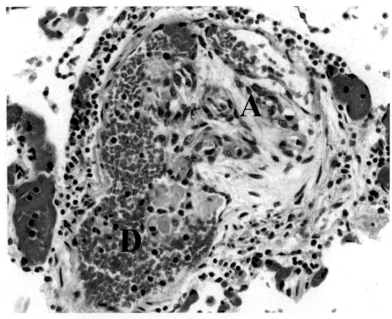

FIGURE 3.21 Angiomatoid (A) and dilatation (D) lesions.

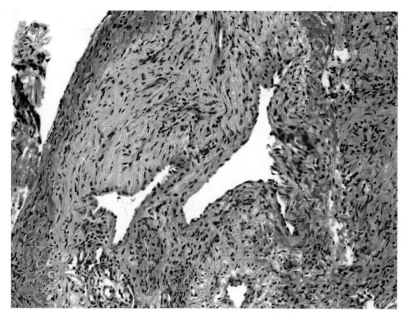

FIGURE 3.22 Recanalized vascular intima.

In cases of thromboembolic disease, the vascular wall undergoes remodeling in conjunction with the luminal thrombus to produce a remodeled *neointima* (Fig. 3.22). This often shows recanalization with new blood channel channels in larger caliber muscular and elastic arteries.

Angiitis indicates vascular wall inflammation. It may be primary or secondary to surrounding inflammation. In the presence of diffuse lung inflammation, in situ thrombi can form vessels due to the release of tissue thromboplastins and obliterate vascular lumens or lead to eccentric intimal proliferations adjacent to the focus of surrounding inflammation (*endarteritis obliterans*). It is generally impossible to distinguish an in situ thrombus from a thromboembolus, especially in small caliber arteries, so that the context of disease should be considered before searching for an extra-pulmonary source. Vasculitis indicates the presence of necrosis with fibrin deposition in the vascular wall (Fig. 3.23). Identifying the size of the involved vessel, and the type of inflammation, i.e., acute, chronic, granulomatous, is helpful in establishing a definitive diagnosis.

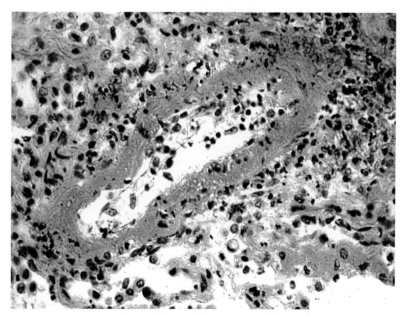

FIGURE 3.23 Pulmonary vasculitis.

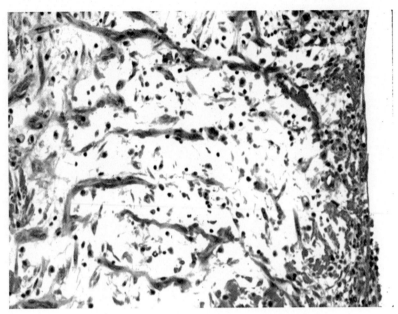

FIGURE 3.24 Vascular arrangement in inflammatory pleuritis.

PLEURA

The pleura is a mesothelial cell-lined surface whose reactions to injury are stereotypic. Most clinical pleural disease comes to the attention of clinicians via patient complaints of pleuritic chest pain or dyspnea due to the development of pleuritic or a pleural effusion. Pleural disease may be primary, due to sympathetic inflammation of the adjacent lung, or adjacent to subdiaphragmatic inflammation. Chronic chest wall pain is an ominous sign that usually indicates malignancy as the sensory nerves are located in the chest wall so that persistent pain usually indicates involvement by an infiltrative process.

When the pleural microvasculature is injured, fibrin deposits along the pleural surface. There is reactive mesothelial hyperplasia, which may be exuberant with reactive cytological atypia that can be difficult to distinguish from malignancy.

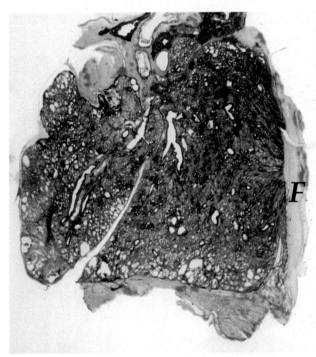

FIGURE 3.25 Diffuse pleural fibrosis (fibrothorax) (F).

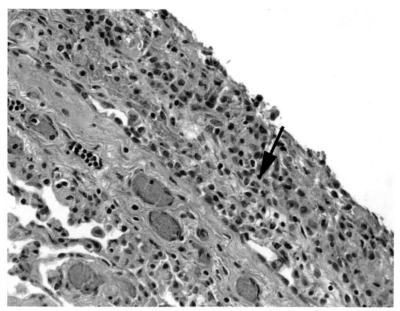

FIGURE 3.26 Eosinophilic pleuritis in patient with pneumothorax (*arrow*).

The ingrowth of capillary-like vessels in the inflamed pleura tends to be oriented perpendicular to the inflamed surface (Fig. 3.24). This constellation of changes is referred to as *reactive fibrinous pleuritis* and is often all that is found in benign pleural effusions caused by many different etiologies. In general, unless tumor and granulomatous infection are serious diagnostic considerations, pleural biopsies rarely yield specific diagnostic information.

When there are recurrent episodes of pleural effusion, pleural fibrosis tends to be layered like the rings of a tree in cross-section. Loculated effusions can entrap the underlying lung and diffuse pleural fibrosis may lead to circumferential fibrothorax that restricts the expansion of the underlying lung, requiring decortication if the lung is to re-expand (Fig. 3.25). For this reason, efforts at biochemical or surgical lysis of adhesions should be pursued early when there is a large degree of loculation.

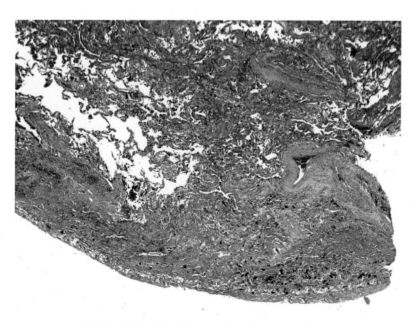

FIGURE 3.27 Pleural scarring following pneumothorax.

Air within the pleura due to pneumothorax leads to mesothelial hyperplasia, chronic inflammation, and an increase in tissue eosinophils termed *eosinophilic pleuritis* (Fig. 3.26). Although a small pneumothorax will generally resorb, especially with supplemental O_2, larger pneumothoraces require catheter drainage and may include the accumulation of pleural blood. Pleural scarring and chronic inflammation are indicators of previous pneumothoraces (Fig. 3.27).

FURTHER READING

Colten, H.R., Krause, J.E., 1997. Pulmonary inflammation—a balancing act. N Engl J Med 336 (15), 1094—1096.
A good review of the various pathways leading to pulmonary inflammation.
Matthay, M.A., Zimmerman, G.A., 2005. Acute lung injury and the acute respiratory distress syndrome. Am J Respir Cell Mol Biol 33 (4), 319—327.
A review of the mechanisms underlying acute lung injury and repair.
Abraham, E., 2000. Coagulation abnormalities in acute lung injury and sepsis. Am J Respir Cell Mol Biol 22 (4), 401—404.
A paper that focuses on the importance of the coagulation system in lung injury.
Hunninghake, G.W., Crystal, R.G., 1981. Pulmonary sarcoidosis: a disorder mediated by excess helper T-lymphocyte activity at sites of disease activity. N Engl J Med 305 (8), 429—434.
An early paper that demonstrated the role of Th1 cytokines in granuloma formation.

Chapter 4

The Radiographic Patterns of Common Lung Diseases

Subba R. Digumarthy, M.D.

[1]*Assistant Professor of Radiology, Harvard Medical School, Boston, MA, United States; Radiologist, Massachusetts General Hospital, Boston, MA, United States*

INTRODUCTION

The radiological approach to pulmonary disease requires a basic knowledge of pathology along with an understanding of anatomy, radiological signs, and pattern recognition. There can be an overlap of signs and patterns in many diseases; these can be resolved to a great extent by a review of the clinical history and other laboratory tests.

AIRWAY DISEASES

The diseases of airways can be simplistically divided into conditions affecting large and small airways. This distinction is based on the luminal diameter, an airway greater than 2 mm in diameter is considered a large airway. This approach has a dual advantage of separating airways based on histology and imaging findings.

The large airways have cartilage in their walls and include trachea, main, lobar, segmental, subsegmental, and smaller branches to the lobular bronchiole in the center of secondary pulmonary lobule. The secondary pulmonary lobule is the smallest organized unit of pulmonary parenchyma seen on high-resolution computed tomography (HRCT). Therefore, the vast majority of large airways are directly visualized on high-resolution CT. Some of the largest distal airways such as lobular bronchi, however, are not routinely visualized on HRCT. The radiological approach to these diseases is based on wall thickness, collapsibility of wall, and luminal diameter.

Wall thickness: The walls of trachea, major bronchi, and first few generations of subsegmental bronchi are well seen on routine imaging studies. The walls are of uniform thickness and smooth. The cartilaginous rings in the walls can calcify in the proximal large airways, particularly in the elderly. The walls of subsegmental bronchi are very thin and barely perceived. The most common manifestation of acute bronchitis related to infection, inhalation injury of reactive airway disease, and also chronic bronchitis related to smoking is increased airway wall thickening. There may be endobronchial filling defects due to mucus.

Collapsibility: The presence of cartilage rings in the walls allows support and prevents their collapse during expiration. If there is loss or deficiency of cartilage normal support is lost and results in the collapse of the lumen during expiration and dynamic process such as coughing. As the normal cartilage rings are incomplete, there is a luminal narrowing up to 50% in normal people, due to buckling of membranous wall unsupported by cartilage. In some young people, there can be narrowing up to 70% that is still considered normal. Demonstration of tracheomalacia and bronchomalacia from excessive collapsibility of the airway wall requires modified scanning protocols such as end -expiratory scanning or dynamic scanning during breathing and coughing (Fig. 4.1).

Luminal diameter: The normal bronchi are approximately the same size as their accompanying branches of pulmonary arteries, and they progressively taper when followed from lung hilum to the periphery. If the bronchus is larger than accompanying pulmonary artery (ratio > 1.5) or if there is loss of normal tapering, the bronchi are considered to be dilated. The normal trachea is less than 3.5 cm in diameter and there is no internal comparison. Stenosis is narrowing of the lumen of large airways and can be focal or diffuse (Fig. 4.2).

Bronchiectasis: Irreversible dilation of bronchi is termed bronchiectasis (Fig. 4.3). The descriptive terms cylindrical, varicoid, and saccular are commonly used in radiological reports. These are useful not only in conveying the severity of bronchiectasis but also in helping to determine the underlying etiology. Saccular or cystic bronchiectasis is a feature of

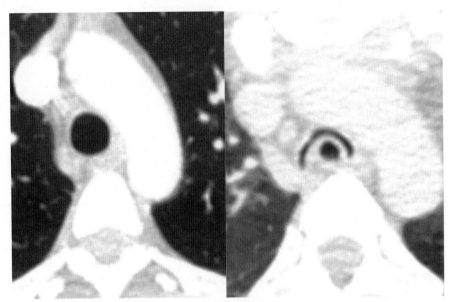

FIGURE 4.1 Collapse of the trachea due to excessive buckling of posterior wall and collapse of anterolateral cartilaginous wall. There is more than 70% luminal narrowing and this is consistent with tracheomalacia.

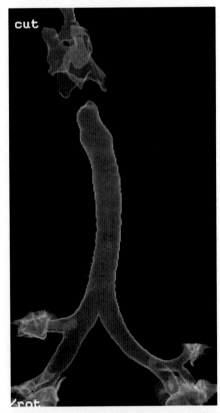

FIGURE 4.2 Normal appearance of trachea and major bronchi in reconstructed image using the three-dimensional external volume rendering technique.

cystic fibrosis, ciliary motility syndromes, and sequel of extensive scarring, after tuberculosis. Focal bronchiectasis is usually the result of scarring in areas affected by pneumonia. Diffuse bronchiectasis on the other hand implies underlying developmental and congenital causes such as cystic fibrosis, immotile cilia syndrome, and Mounier—Kuhn syndrome that result in recurrent infections. The distribution of bronchiectasis is also helpful in determining the underlying cause. In cystic fibrosis, upper lobes and upper lungs are preferentially affected, whereas the lower lungs are affected preferentially in ciliary motility syndromes (Fig. 4.4).

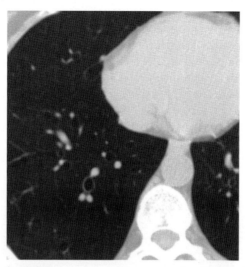

FIGURE 4.3   In bronchiectasis, the bronchi are larger than accompanying arteries and do not taper as they branch.

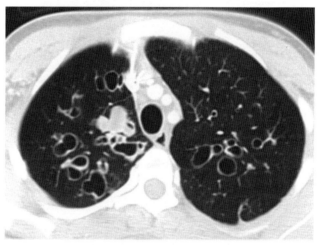

FIGURE 4.4   Cystic bronchiectasis in the upper lobes and superior lower lobes in a patient with cystic fibrosis. Noted also the bronchial impaction in medial right upper lobe and bronchial wall thickening.

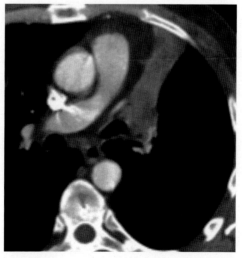

FIGURE 4.5   Left upper lobe collapse caused by squamous cell carcinoma in the left upper lobe bronchus.

Stenosis: The lumens of large airways can be narrowed by a variety of causes. The detection of stenosis on imaging studies is not always easy. It requires multiplanar image reconstruction in sagittal, coronal, and axial planes and three-dimensional image reconstruction. Accurate determination of the length of stenosis is critical for surgical planning and interventional procedures such as stent placement. Stenosis of large airways such as trachea and main bronchi can be asymptomatic and may not lead to other findings on imaging. In contrast, stenosis of lobar and segmental bronchi tends to cause postobstructive pneumonia, collapse, and air trapping (Fig. 4.5).

The causes of stenosis can be divided into causes that are extrinsic and intrinsic to the airway. The common examples of extrinsic cause include a large goiter displacing the trachea and causing stenosis and a hilar lymph nodal mass narrowing the lobar and segmental bronchi (Fig. 4.6).

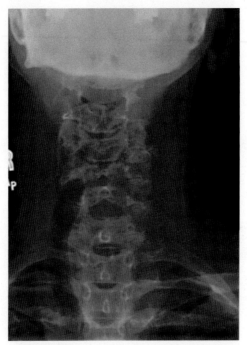

FIGURE 4.6 Rightward displacement of trachea by a large thyroid cancer. Also note the stenosis of subglottic trachea.

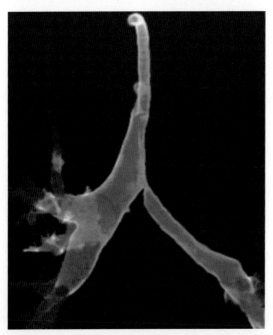

FIGURE 4.7 Three-dimensional external volume rendering image depicting a long segment stricture of distal trachea and left main bronchus, showing the complication of prior granulomatous infection.

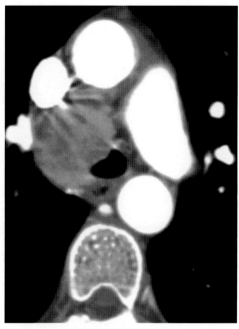

FIGURE 4.8 Squamous cell carcinoma of the trachea causing stenosis of the trachea, note the endoluminal and exophytic mass along the right lateral wall of trachea.

The intrinsic causes can be located in the wall or lumen of the airway. The disease processes that affect the wall can be inflammatory, infectious, or neoplastic. The pattern of wall involvement guides the radiological approach and diagnosis. Long segment stenosis is seen in inflammatory and infectious causes. The narrowed segments are smooth, circumferential, and without intraluminal mass. If there is sparing of the noncartilaginous wall, conditions that only affect the cartilage such as relapsing polychondritis should be suspected. Inflammatory conditions such as Wegener's granulomatosis and infections affect both cartilage and membranous walls of the large airways. Stenosis related to trauma or procedures such as tracheostomy or intubation is site specific and short (Fig. 4.7).

Endoluminal masses that cause stenosis are usually related to neoplastic causes. In children, the commonest tracheal mass is hemangioma. In adults, malignant neoplasms are far more common. The segments involved are usually short, with eccentric luminal narrowing, and endoluminal mass. Invasion into adjacent structures and nodal metastases can be seen. The two most common malignancies of trachea and major bronchi are squamous cell carcinoma and adenoid cystic carcinoma. Squamous carcinoma tends to have exophytic extension outside the trachea and nodal metastases. Adenoid cystic carcinoma tends to spread submucosally and nodal metastasis is relatively less common. The commonest benign tumor is carcinoid and is characterized by intense vascular enhancement in intravenous contrast enhanced imaging studies (Fig. 4.8).

SMALL AIRWAYS DISEASE

Small airways are less than 2 mm in diameter and lack cartilage in their walls. They also include terminal bronchiole, respiratory bronchiole, alveolar ducts, and sacs. Unlike the large airways that are purely responsible for the conduction of air, distal small airways participate in gas exchange. The normal small airways are beyond the resolution of diagnostic imaging, and they cannot be seen on HRCT. The diseases of these airways are inferred by identifying direct and indirect signs on CT. In general, these signs are subtle and require high-resolution CT imaging.

DIRECT SIGNS

Centrilobular opacities: These are subtle ground glass opacities that are seen around the small airways and are mostly the result of inflammatory process around the respiratory bronchiole. These opacities do not make contact with the pleura or fissures (Fig. 4.9).

Tree-in-bud nodules: These are branching V- and Y-shaped nodules that resemble a budding branch of a tree. These are due filling of small bronchioles with inflammatory fluid, debris, and pus (Fig. 4.10).

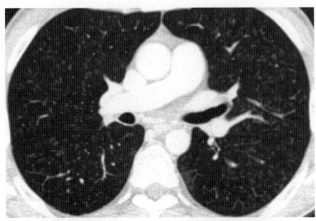

FIGURE 4.9   Subtle diffuse ground glass centrilobular nodules. Note that the nodules do not touch the pleural surface.

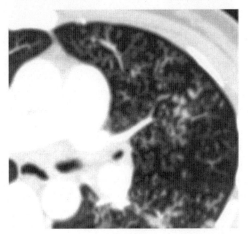

FIGURE 4.10   Branching "V"- and "Y"-shaped nodules in left lung resembling a tree-in-bud.

Bronchiolectasis: This is an uncommon finding where small peripheral bronchioles are seen due to abnormal dilation, from scarring and fibrosis of lung parenchyma.

INDIRECT SIGNS

Mosaic attenuation: Heterogeneous density of lung parenchyma on CT with areas of increased and decreased lucency. In the setting of small airway disease, lucency is due to air trapping. These areas are better demonstrated in expiratory imaging, where lucent areas remain lucent and normal lung increases in density. Sometimes the only evidence of small airway disease can be air-trapping seen only on expiratory imaging (Fig. 4.11).

Bronchiolitis: Bronchiolitis, a collective term for diseases of small airways, is broadly divided into proliferative and obliterative bronchiolitis. Examples of proliferative bronchiolitis include infectious, smoking related, and hypersensitivity bronchiolitis, and these usually manifest with both direct and indirect signs of bronchiolitis. In contrast, the obliterative causes of bronchiolitis cause fibrosis of walls and obliteration of lumens and manifest only as indirect signs of mosaic attenuation and air trapping. The common causes of obliterative bronchiolitis are postinfective obliterative bronchiolitis, chronic rejection after lung transplant, graft versus host reaction after bone marrow transplant, and bronchiolitis obliterans in rheumatoid arthritis (Fig. 4.12).

INFECTION

The morphological pattern of infection in the lungs as seen on imaging, along with clinical history and other lab results, can offer clues in determining the presumptive cause of infection. The pulmonary infections can be broadly classified as lobar pneumonia, bronchopneumonia, interstitial, and hematogenous.

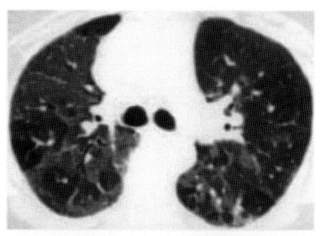

FIGURE 4.11   Mosaic attenuation of lung with areas of increased and decreased lung density.

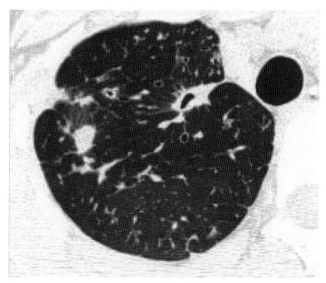

FIGURE 4.12   Cavity in the right upper lobe with tree-in-bud nodules, in a patient with tuberculosis-associated bronchiolitis.

Lobar pneumonia: The infection starts in the distal airspaces such as alveolar sacs and alveoli. The opacities are confluent and contain air bronchograms. The air in the alveoli is replaced by inflammatory fluid and pus. The infectious agent reaches the lung through air droplets in aerosol. The typical examples are *Streptococcus, Klebsiella*, and *Legionella spp*. Radiologically, areas of consolidation (opacity with air bronchograms) involve segments or lobe. In severe disease, this can be bilateral and multifocal. The consolidation can increase the volume of affected lobe and can cause bulging of fissures between the lobes. These can be associated with pleural effusions that are sympathetic or related to empyema (Fig. 4.13).

Bronchopneumonia: The infection starts in the small airways and spreads to adjacent airspaces. The radiological manifestations are bronchial wall thickening, tree-in-bud nodules, and patchy areas of consolidation. The bronchi can be occluded with secretions, and air bronchogram is not always seen. The disease tends to be multifocal and patchy in distribution. The typical examples include viral pneumonia and pneumonia secondary to aspiration (Fig. 4.14).

Interstitial pattern: The predominant pattern is diffuse ground glass opacity with subtle septal thickening and thickening of axial interstitium, the connective tissue that encases the bronchi and pulmonary vessels. This pattern is seen in pneumocystis pneumonia and mycoplasma and viral infections (Fig. 4.15).

Hematogenous infection: The infection reaches the lungs through the blood stream. The source can be an infected right heart valve, a source of sepsis in the body such as an infected indwelling catheter or abscess in the body. Septic emboli in the lungs manifest as multiple predominantly peripheral nodules that tend to cavitate. This may also result in lung infarcts

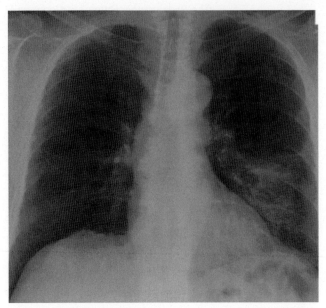

FIGURE 4.13   Community acquired streptococcal pneumonia presenting as lobar consolidation in left lower lobe.

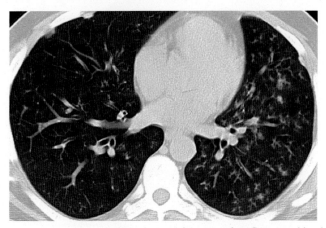

FIGURE 4.14   Multifocal tree-in-bud nodules, bronchial wall thickening, and few areas of confluent opacities, in a patient with respiratory syncytial virus infection.

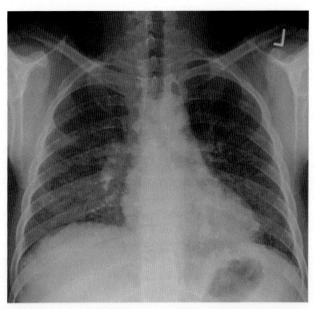

FIGURE 4.15   Perihilar hazy opacities with bronchial cuffing in a patient with pneumocystis pneumonia.

and consolidation. In infections due to tuberculosis and endemic fungi, the nodules are diffuse, innumerable, and less than 3 mm in size. This is referred to as miliary pattern due to its resemblance to millet seeds (Fig. 4.16).

The presence of certain imaging features in the context of infection also helps in determining the causative agent.

Cavitation: The presence of cavity implies necrosis of lung parenchyma, and this is a characteristic feature of post-primary tuberculosis. However, this can be seen in any aggressive necrotizing or angioinvasive infection, such from anaerobic bacteria and in fungal infections from aspergillus and mucormycosis (Fig. 4.17).

Ground glass halo: The "halo" refers to ground glass opacity around a solid nodule. This is considered characteristic of invasive aspergillus infection in neutropenic patients. However, this sign is not specific to aspergillus infection and has been described with many other infections (Fig. 4.18).

Location: The involvement of upper lobes and superior segments is a characteristic feature of postprimary tuberculosis. The involvement of middle lobe and lingula is typical for *Mycobacterium avium* complex infection in immunocompetent middle aged women. The involvement of dependent segments of lungs (posterior segments of upper lobes and superior and posterior basal segments of lower lobes) is a feature of aspiration pneumonia (Fig. 4.19).

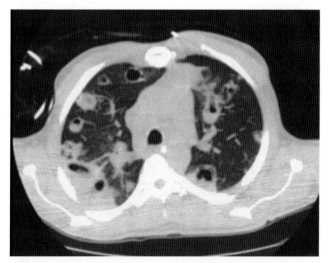

FIGURE 4.16 Multiple peripheral cavitary nodules in a patient with infective tricuspid endocarditis and history of IV drug use.

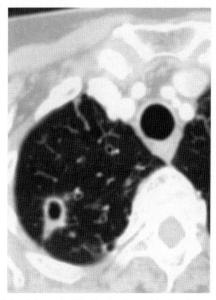

FIGURE 4.17 Cavitary nodule in right upper lobe in a patient with chronic steroid use and *Nocardia* infection.

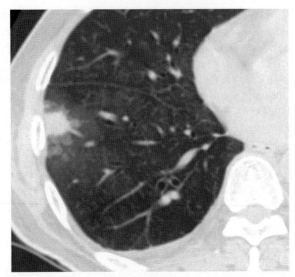

FIGURE 4.18 Peripheral nodule in the right lower lobe with surrounding ground glass opacity (halo) in a neutropenic patient with invasive aspergillus infection.

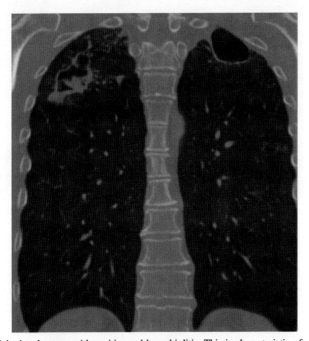

FIGURE 4.19 Upper lobe involvement with cavities and bronchiolitis. This is characteristic of postprimary tuberculosis.

DIFFUSE LUNG DISEASE

Diffuse lung diseases include a diverse group of inflammatory and idiopathic lung conditions that include idiopathic interstitial pneumonias, connective tissue diseases, drug related pneumonitis, vasculitis, sarcoidosis, hypersensitivity pneumonitis, eosinophilic pneumonia, cystic lung diseases, pneumoconiosis, and miscellaneous entities such as amyloidosis. The descriptors used in radiology for these entities carry specific connotations and are useful in narrowing the differential diagnosis along with clinical history and laboratory tests.

The smallest organized anatomical structure that is seen on high-resolution CT is the secondary pulmonary lobule, which is ~2.5 cm. The secondary lobule is hexagonally shaped with the central lobular bronchiole and artery and pulmonary veins in its walls. The lobular bronchus leads to a terminal bronchiole and distal respiratory units comprised of respiratory bronchiole, alveolar ducts, and sacs lined by alveoli (Fig. 4.20).

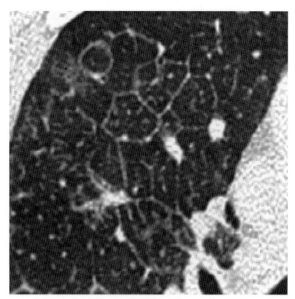

FIGURE 4.20   Hexagon-shaped secondary pulmonary lobules in the right upper lobe.

Septal lines and reticulation: The thickening of the walls of the secondary pulmonary lobule produces septal lines that outline the lobule. The septal lines become apparent when thickened by inflammation or edema of the lung interstitium. Examples include pulmonary edema, infection, infiltrative diseases such as lymphangitic carcinomatosis, and pulmonary alveolar proteinosis. Reticulation refers to fine lines that are seen within the secondary pulmonary lobule and these are commonly seen in both usual and nonspecific interstitial pneumonitis (Fig. 4.21).

Honeycombing: Honeycomb cystic change is a manifestation of irreversible lung damage and fibrosis and is often associated with end stage lung disease. This is a hallmark of usual interstitial pneumonitis, although it has also been described in chronic hypersensitivity pneumonitis, fibrotic interstitial pneumonitis, and healed acute interstitial pneumonitis. The cysts are well-defined spaces, around 1 cm, and arranged in "stacks" at the lung periphery beneath the pleura. These should not be confused with traction bronchiolectasis, where the continuity of the cysts can be traced to tubular bronchioles and bronchi (Fig. 4.22).

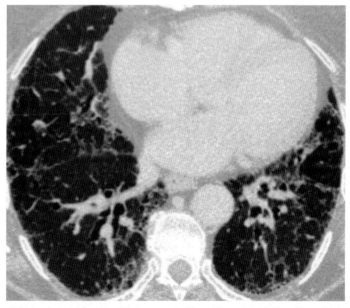

FIGURE 4.21   Septal thickening and fine intralobular reticular lines. Note also subpleural bronchiolectasis in lower lobes.

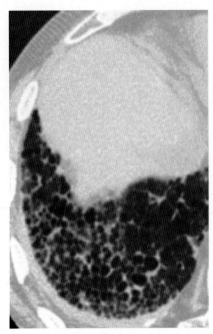

FIGURE 4.22   Stacks of well-defined honeycomb cysts in the subpleural right lower lobe.

Traction bronchiectasis and bronchiolectasis: These terms refer to abnormal dilation of bronchi and bronchioles secondary to parenchymal fibrosis. These findings can be reversible in the early course of the disease and indicate fibrotic process.

Centrilobular nodules: The nodules are commonly seen in diseases associated with inhalation. The nodules are centered on bronchioles, typically around respiratory bronchiole, and are seen in the subacute hypersensitivity pneumonitis and respiratory bronchiolitis. These are of ground glass density, spaced at regular intervals and do not touch the pleural surface.

Perilymphatic nodules: These nodules are distributed along the axial, septal, and subpleural interstitium along the course of lymphatics. These produce thickening of bronchial and pulmonary arteries, nodular septal thickening, and nodules along the pleural surface and fissures. Sarcoidosis, lymphoma, and lymphangitic carcinomatosis are two common conditions associated with perilymphatic nodules (Fig. 4.23).

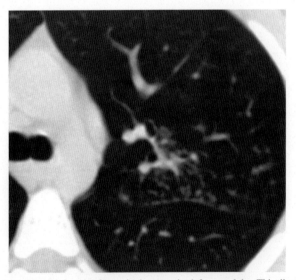

FIGURE 4.23   Small nodules along the bronchi and left major fissure in the posterior left upper lobe. This distribution is characteristic of perilymphatic nodules.

Random nodules: These nodules are in a random pattern that cannot be classified as either centrilobular or having a perilymphatic distribution. They may be numerous or few and are seen in diseases that disseminate through blood such as military infections and sarcomatous metastases (Fig. 4.24).

Calcified nodules: Calcified nodules are seen in amyloidosis, sarcoidosis, and pneumoconiosis such as silicosis and berylliosis (Fig. 4.25).

Lung cysts: A cyst is defined radiographically as an air-filled thin-walled space. Cysts are considered a feature of disease if there are at least six or more. The cysts must be differentiated from centrilobular emphysema, which tends to have an artery in the center of the cyst. The common diseases associated with cysts are lymphangioleiomyomatosis (LAM), Langerhans' cell histiocytosis, Sjogren's disease, desquamative interstitial pneumonitis, and lymphocytic interstitial pneumonia. Other rare congenital entities such as Birt–Hogg–Dube syndrome also have lung cysts (Fig. 4.26).

Peripheral consolidation: Peripheral consolidation is commonly seen in organizing and eosinophilic pneumonia. These are usually fleeting and may not be associated with other features. In chronic setting features of fibrosis can be seen in organizing pneumonia (Fig. 4.27).

Lung volumes: In cystic lung diseases such as LAM and Langerhans' cell histiocytosis, lung volumes are preserved or increased. In fibrotic lung diseases such as usual interstitial pneumonia and fibrotic nonspecific interstitial pneumonia (NSIP), the lung volumes are reduced. However, if there is combined emphysema and fibrotic lung disease, the lung volumes may appear normal (Fig. 4.28).

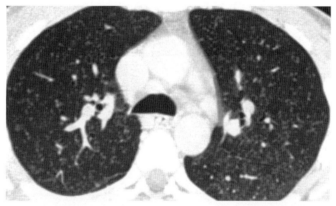

FIGURE 4.24 Innumerable uniform 1- to 3-mm miliary nodules distributed throughout the lung parenchyma. Note the nodules along the fissures and pleural surface.

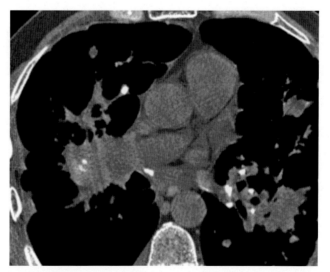

FIGURE 4.25 Calcified mediastinal and hilar lymph nodes and calcified nodules in the background of extensive fibrotic masses in pulmonary beryllium disease.

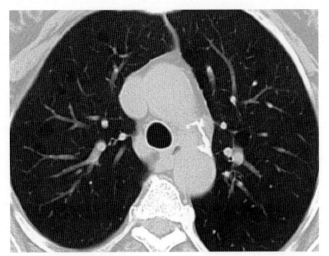

FIGURE 4.26   Note multiple bilateral lung cysts in a middle-aged woman with LAM.

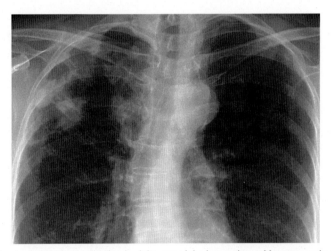

FIGURE 4.27   Peripheral subpleural consolidation in right upper lobe in a patient with cryptogenic organizing pneumonia.

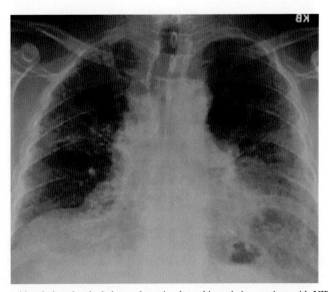

FIGURE 4.28   Low lung volumes with subpleural reticulation and traction bronchiectasis in a patient with UIP in idiopathic pulmonary fibrosis.

Usual interstitial pneumonia (UIP): On imaging, UIP is characterized by subpleural reticulation in lower lobes with evidence of fibrosis in the form of traction bronchiectasis, bronchiolectasis, and honeycombing. If all the features are noted on HRCT, biopsy is not required for confirmation. However, if there are atypical features such as isolated involvement of upper lobes, dominance of ground glass opacity, presence of centrilobular nodules, and absence of honeycombing, biopsy should be performed. UIP can be idiopathic and familial or associated with several diseases such as asbestosis and connective tissue diseases (Fig. 4.29).

Nonspecific interstitial pneumonia: This is characterized by the presence of peribronchial ground glass opacities, traction bronchiectasis subpleural sparing, and absence of honeycombing. In fibrotic type of NSIP, there can be features that may resemble UIP including honeycombing. NSIP is most commonly seen in association with connective tissue diseases and drug toxicity. Idiopathic NSIP is a rare entity (Fig. 4.30).

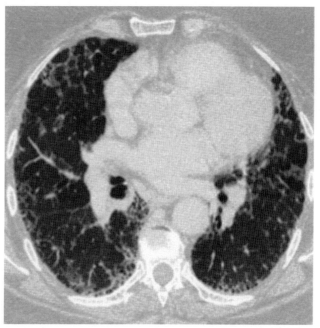

FIGURE 4.29 Reticulation with subpleural honeycombing in the lower lobes in a patient with UIP.

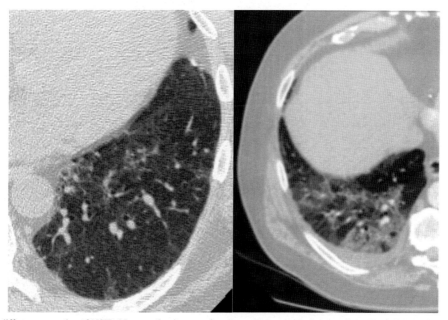

FIGURE 4.30 Two different examples of NSIP. Note predominance of peribronchial ground glass opacity and relative sparing of subpleural lung and absence of honeycombing.

Organizing pneumonia (OP): OP is typically seen as subpleural areas of consolidation and ground glass opacity without honeycombing. The opacities may have a border of increased density called "reverse halo." OP can also present as lung nodules and predominantly with interstitial pattern of reticulation and ground glass opacity with traction bronchiectasis. OP is commonly secondary to lung injury such as drug toxicity, acute interstitial pneumonitis, and connective tissue diseases. Idiopathic OP is called cryptogenic organizing pneumonia (COP) (Fig. 4.31).

Hypersensitivity pneumonitis: The radiological manifestations of acute, subacute, and chronic hypersensitivity pneumonitis are distinct. Acute stage manifests as ground glass opacities and consolidation and without clear history is often misdiagnosed as infection on imaging. Subacute hypersensitivity is characterized by centrilobular nodules, mosaic attenuation of lungs, and patchy air trapping. This constellation of findings is pathognomonic on imaging. Chronic hypersensitivity pneumonitis presents with changes of fibrosis, architecture distortion, traction bronchiectasis, and bronchiolectasis and honeycombing. This can be confused with other causes of pulmonary fibrosis such as UIP, if subtle centrilobular nodules and air trapping are not recognized (Fig. 4.32).

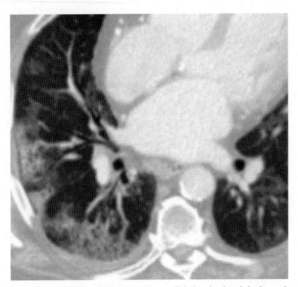

FIGURE 4.31 Subpleural ground glass opacities and consolidation in the right lung, in a patient with COP.

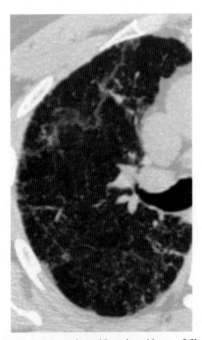

FIGURE 4.32 Diffuse centrilobular nodules with mosaic attenuation with early evidence of fibrosis with reticulation and traction bronchiectasis.

Eosinophilic diseases: Eosinophilic diseases can be "simple" or chronic. Simple eosinophilic pneumonia is usually a response to drug or parasite and is self-limited, with fleeting peripheral ground glass opacities and consolidation in lungs. Chronic eosinophilic pneumonia is idiopathic and presents as peripheral areas of consolidation that resemble opacities seen in OP. These tend to be chronic and can be associated with lymphadenopathy and tend to relapse when steroids are discontinued (Fig. 4.33).

Pulmonary hemorrhage: Pulmonary hemorrhage usually is associated with vasculitis such as granulomatous angiitis (Wegener's granulomatosis) or connective tissue diseases such as systemic lupus erythematosus. However, there are a large number of other causes that can lead to pulmonary hemorrhage such as Goodpasture's syndrome, idiopathic pulmonary hemosiderosis, and capillaritis related to immune complex deposition. In acute hemorrhage, the imaging findings consist of ground glass opacities and areas of consolidation. Following repeated pulmonary hemorrhage there can be changes of fibrosis in the lungs with reticulation, septal thickening, and mild traction bronchiectasis (Figs. 4.34 and 4.35).

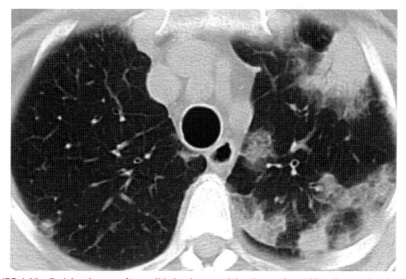

FIGURE 4.33 Peripheral areas of consolidation in upper lobes in a patient with asthma and eosinophilia.

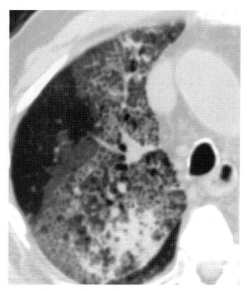

FIGURE 4.34 Acute pulmonary hemorrhage related to vasculitis in a patient with antiphospholipid syndrome.

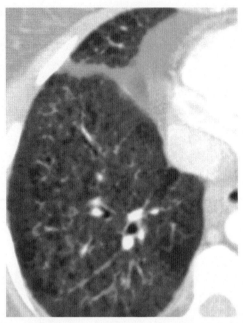

FIGURE 4.35   Changes of reticulation, architecture distortion, and mild traction bronchiectasis in a patient with chronic idiopathic hemosiderosis.

LUNG NEOPLASIA

Lung cancer: Lung cancer is traditionally classified into small and non-small-cell lung cancer. The two main subtypes of non-small-cell carcinoma are adenocarcinoma and squamous carcinoma.

Small cell carcinoma: On imaging, small cell carcinoma is characterized by a small lung nodule with bulky hilar and mediastinal lymphadenopathy and distant metastases. Sometimes the primary in the lung may be difficult to identify. In general, these are aggressive tumors with rapid growth and short tumor doubling time (Fig. 4.36).

Adenocarcinoma: There is correlation between histology and radiological features of adenocarcinoma. Atypical adenomatous hyperplasia (AH), adenocarcinoma in situ (AIS), minimally invasive adenocarcinoma (MIA), and invasive adenocarcinoma can be differentiated based on size and density on CT. Atypical AH shows a ground glass density and typically measures less than 5 mm and remains stable for years. AIS can show up to 3 cm of ground glass density. MIA is of mixed density and has a solid component within the ground glass opacity. Invasive adenocarcinoma tends to be

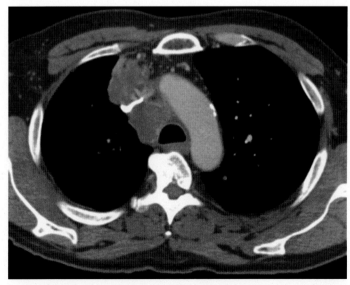

FIGURE 4.36   Bulky right-sided mediastinal lymphadenopathy without a definite lung nodule in small cell lung cancer.

predominantly solid. Adenocarcinomas tend to have spiculated margins and pleural tags. Adenocarcinoma commonly develops in the lung periphery; however, central adenocarcinomas are not uncommon. Occasionally, adenocarcinoma presents as multifocal lung cancer with ground glass opacities, nodules, and consolidations (Figs. 4.37 and 4.38).

Squamous cell carcinoma: Squamous cell carcinoma typically occurs in central location, but up to a third can occur in the periphery. Cavitation is commonly seen in tumors that are greater than 3 cm. The central tumors occlude the bronchi, yielding atelectasis and pneumonia in the distal lung (Fig. 4.39).

Staging of lung cancer. The staging of lung cancer is based on tumor, node, and metastasis. The resected tumor size can be smaller than measured on imaging, due to slight shrinkage of tumor after formalin fixation. The accurate measurement on imaging is also affected if there is adjacent scarring, pneumonia, or fibrosis due to desmoplastic reaction. There are features on imaging that increase the specificity of T staging and improve accuracy. The positive predictive value of pleural

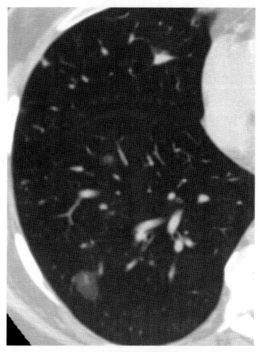

FIGURE 4.37 Atypical AH, minimally invasive adenocarcinoma, and invasive adenocarcinoma.

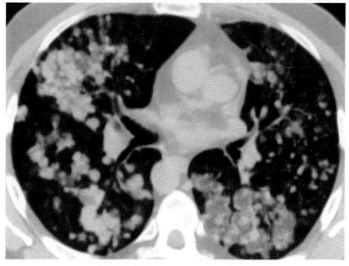

FIGURE 4.38 Multifocal adenocarcinoma.

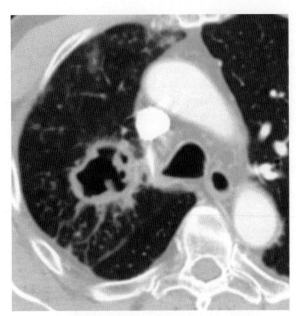

FIGURE 4.39   Central cavitary mass in the right upper lobe in a patient with squamous carcinoma.

and chest wall invasion increases if there is more than 3 cm between pleura and the tumor, the presence of rib or vertebral destruction, or a soft tissue mass in the chest wall muscles. The definite evidence of mediastinal invasion and therefore surgical unresectability is predicted by encasement of mediastinal structures, i.e., the main pulmonary artery, aorta, and arch vessels, esophagus, trachea, and extension into cardiac chambers. However, in many instances the findings of chest wall and mediastinal invasion are not definite and surgical exploration is required to assess resectability (Fig. 4.40).

The lymph nodes are considered enlarged if the short axis measurement is greater than 1 cm. This is a general rule and normal lymph nodes, such as in subcarinal station, may be greater than 1 cm in short axis.

Necrotic lymph nodes that do not demonstrate contrast enhancement are also considered abnormal. However, lymph nodes that are not enlarged can still harbor metastases and lymph nodes can enlarge in conditions other than metastases, such as in reactive hyperplasia due to infection and sarcoid-like reaction to malignancy. The sensitivity and specificity of

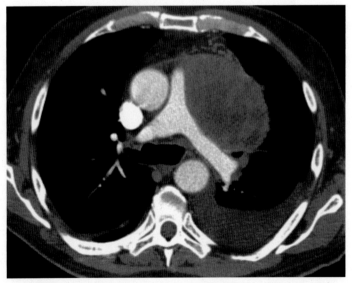

FIGURE 4.40   Mediastinal invasion by large left upper lobe non-small-cell lung cancer.

predicting nodal metastasis on anatomical imaging such as CT are less than 60% and 80%, respectively. The accuracy increases over 80% and 90% if functional imaging such as PET (positron emission tomography) is used (Fig. 4.41).

Whole-body imaging with CT-PET increases the detection of occult metastases and upgrades the staging in up to 5−29%. The most common sites of metastases are the adrenals, bone, brain, liver, and lung.

Pulmonary metastases: Hematogenous metastases to lungs commonly manifest as multiple lung nodules, commonly at lung periphery and in lower lobes. Metastases from primary squamous carcinoma and transitional carcinoma can cavitate and metastatic adenocarcinoma can have solid or ground glass density. Sometimes tumor emboli are spread along the interstitium, wall, and lumen of pulmonary arteries and referred to as tumor emboli. Lymphangitic carcinomatosis can result from metastatic or direct extension of tumor into lymphatics and produces nodular thickening of the axial, septal, and subpleural interstitium (Fig. 4.42).

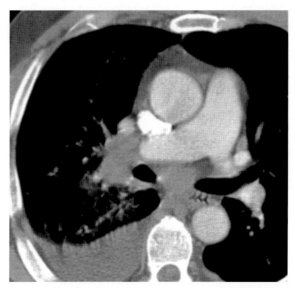

FIGURE 4.41   Enlarged subcarinal and right hilar lymph nodes in a patient with right lower lobe non-small-cell lung cancer.

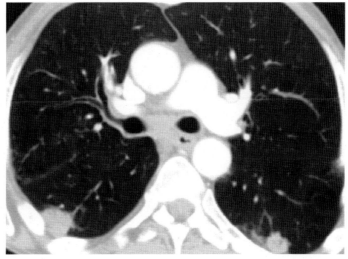

FIGURE 4.42   Bilateral lower lobe peripheral lung metastases from head and neck primary.

FURTHER READING

Kligerman, S., Digumarthy, S., November 2009. Staging of non-small cell lung cancer using integrated PET/CT. AJR Am J Roentgenol 193 (5).

Lichtenberger 3rd, J.P., Digumarthy, S.R., Abbott, G.F., Shepard, J.A., May–June 2014. Sharma A Diffuse pulmonary hemorrhage: clues to the diagnosis. Curr Probl Diagn Radiol 43 (3), 128–139.

Walker, C.M., Abbott, G.F., Greene, R.E., Shepard, J.A., Vummidi, D., Digumarthy, S.R., March 2014. Imaging pulmonary infection: classic signs and patterns. AJR Am J Roentgenol 202 (3), 479–492.

Several papers by the author of the chapter on important topics in radiological diagnosis.

Chapter 5

Diseases of the Airways

Diseases of the airways tend to produce obstruction to airflow (Table 5.1). This can be measured spirometrically by a flow-volume loop (Fig. 5.1; Table 5.2). Significant airways obstruction is evidenced by a decrease in the volume of air forcefully exhaled in 1 second (FEV1) with respect to the forced *vital capacity* (FVC) (Fig. 5.2). An FEV1/FVC ratio of < 0.6 represents a significant obstructive defect. In some disorders, such as bronchiectasis, obstructive airways disease may be accompanied by restriction, due to distal scarring with loss of lung volume.

Changes in respiratory cell number, size, and viscosity of secreted mucus, all play a role in the response to inhaled microbes and other irritants. The proximal airways show a basal layer of reserve cells that are the progenitor stem cells for the cells in the airways (Table 5.3). Mucin-secreting goblet cells normally make up one out of five cells identified in the respiratory epithelium by light microscopy, and an increase in their numbers implies airway irritation. Neuroendocrine cells and surfactant producing *club cells* are also found in the airway epithelium, where they mediate a variety of paracrine activities and reactions to airborne irritants (Fig. 5.3).

Ciliated cells display a stereotypic 9 + 2 arrangement of peripheral and central microtubules, respectively (Figs. 5.4A, B). Cilia contribute to the rotation of organs in embryogenesis, and defects in ciliary function account for congenital Kartagener's syndrome showing *situs inversus* and bronchiectasis. Repeated inflammatory insults to the airways can produce acquired changes in ciliary architecture and function resulting in impaired mucociliary clearance, chronic infection, and bronchiectasis.

The basal cells of the airway both secrete and are anchored to a basement membrane that consists of laminin, type IV collagen, and glycosaminoglycans. This can best be visualized histochemically with Periodic-acid Schiff (PAS) and silver stains. Although thickening of the airway basement membrane is classically associated with asthma (Fig. 5.5), it is common to all chronic airways diseases.

TABLE 5.1 Causes of Pulmonary Obstruction
Bronchospasm
Chronic bronchitis
Emphysema (loss of tethering effect by alveolar elastic fibers)
Bronchiectasis
Chondromalacia
Tumor or foreign body in upper airways
Extrinsic compression of upper airways

TABLE 5.2 Spirometric Analysis
FEV1 = Amount of air forcefully expired in 1 s
FVC = the amount of air forcefully expired until one reaches residual volume
An obstructive defect = ratio of FEV1/FVC < 80% of predicted for height and weight
Restrictive defect = normal to supranormal FEV1/FVC with decreased total lung capacity (TLC) (< 80% of predicted for height and weight)

Understanding Pulmonary Pathology. http://dx.doi.org/10.1016/B978-0-12-801304-5.00005-8

Spirometry at BTPS		ATS ✓		
		Actual	Predicted	% Pred
FEV$_1$	L	3.99	3.82	104
FVC	L	5.16	4.59	112
FEV$_1$ / FVC	%	77	83	93
FEF$_{25-75}$	L/s	3.49	4.11	85
PEFR	L/s	9.20	8.43	109
FIVC	L	4.77	4.59	104
FEF$_{50}$ / FIF$_{50}$		0.71	---	---

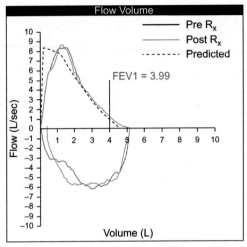

FIGURE 5.1 Normal flow-volume loop.

Spirometry at BTPS		ATS ✓		
		Actual	Predicted	% Pred
FEV$_1$	L	0.75	2.67	28
FVC	L	2.31	3.49	66
FEV$_1$ / FVC	%	32	75	43
FEF$_{25-75}$	L/s	0.26	2.35	11
PEFR	L/s	2.06	6.10	34
FIVC	L	2.30	3.49	66
FEF$_{50}$ / FIF$_{50}$		0.10	---	---

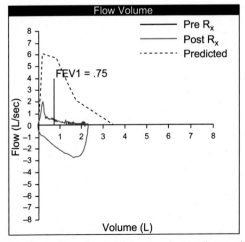

FIGURE 5.2 Flow-volume loop in obstructive airways disease with reduced FEV1/FVC ratio.

TABLE 5.3 Bronchial Epithelium

Ciliated cells

Mucus secreting cells

Multipotential basal cells

Brush cells

Club cells (Clara cells)

Neuroendocrine cells

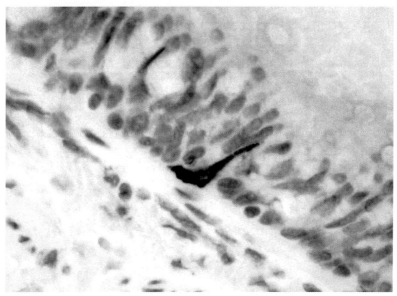

FIGURE 5.3   Neuroendocrine cell in airways (peroxidase).

(A)   **(B)**

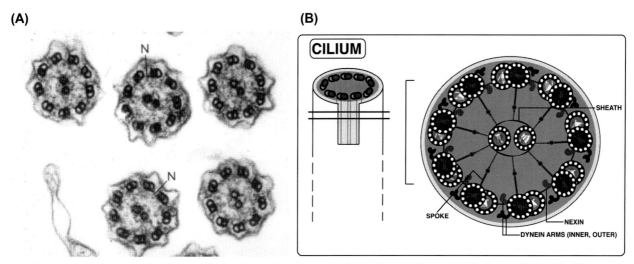

FIGURE 5.4   (A) Ultrastructure of pulmonary cilia showing 9 + 2 arrangements of microfilaments (N = nexin), (B) diagram of cilium showing structural proteins.

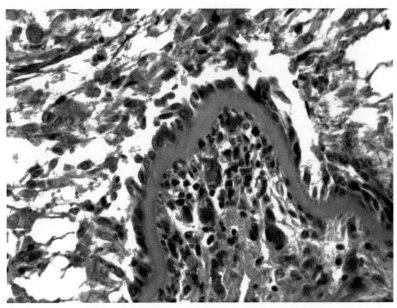

FIGURE 5.5   Thickened airway basement membrane in asthma.

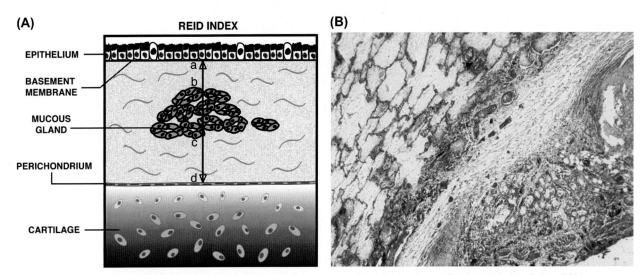

FIGURE 5.6   (A) Diagram of Reid index, (B) increased seromucous glands in the airway of patient with chronic bronchitis.

The walls of the conducting airways are invested with minor salivary glands that include serous and mucin-producing cells that react to chronic irritation. An increase in mucin-secreting cells can lead to mucus inspissation and dilatation of the airway ducts, which may be visualized at bronchoscopy as small "pits" in the bronchial mucosa.

The *Reid index*, i.e., the ratio of height of the seromucous gland layer in a major bronchus to the distance between the epithelial basement membrane and the inner perichondrium of cartilage in a well-oriented section, is normally ~0.3−0.4, and an increased Reid index is the best histologic correlate of the clinical diagnosis of chronic bronchitis (Figs. 5.6A, B). The airway walls in chronic bronchitis are grossly thickened (Fig. 5.7A) and show increased mucus secretion (Fig. 5.7B). Late in disease, they may become scarred. In chronic airways infection and in autoimmune disorders, e.g., Sjögren's syndrome, the seromucous glands are chronically inflamed and eventually undergo atrophy and are replaced by scar tissue leading to bronchiectasis. This can be seen at bronchoscopy as thinning of the normal mucosa with prominence of the cartilaginous plates or by the presence of floppy malacic airway segments.

The airway wall is invested with smooth muscle that mediates bronchomotor tone. Airway muscle tends to contract under vagal parasympathetic cholinergic stimulation and to dilate in response to sympathetic adrenergic stimulation. In

(A)

(B)

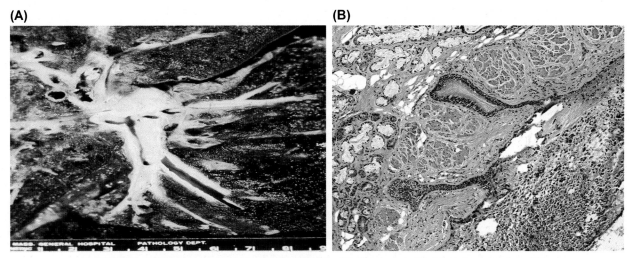

FIGURE 5.7   (A) Thickened airways in chronic bronchitis, (B) dilated mucin-filled ducts in chronic bronchitis.

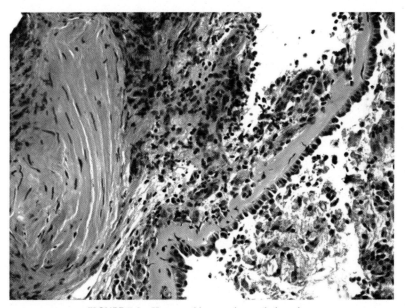

FIGURE 5.8   Hypertrophic smooth muscle in asthma.

asthma, and to a lesser extent in chronic bronchitis, chronically increased bronchomotor tone results in smooth muscle hypertrophy and bronchospasm (Fig. 5.8). Narrowing of the airways by bronchoconstriction and increased secretions produce turbulent airflow detected by auscultation as high pitched wheezes. Rhonchorous sounds on auscultation primarily reflect secretions in the airways, which may clear with cough, and these sounds are cardinal auscultatory features of obstructive airways disease.

ASTHMA

Asthma is a *syndrome* defined physiologically as reversible airways obstruction (Fig. 5.9) (Table 5.4). Patients who present for medical attention after symptoms have abated may show normal airflow rates by pulmonary function tests (PFTs). In such cases, methacholine bronchoprovocation can be used to induce bronchospasm in patients who have underlying airway irritability. However, viral and bacterial infections of the airways also increase bronchomotor irritability, and wheezing and bronchoconstriction may persist for weeks before spontaneously resolving, so that a diagnosis of asthma should be made with caution in the setting of acute airway infection. There are numerous factors that can cause and exacerbate asthma and these must be considered whenever a new patient presents with wheezing (Table 5.5).

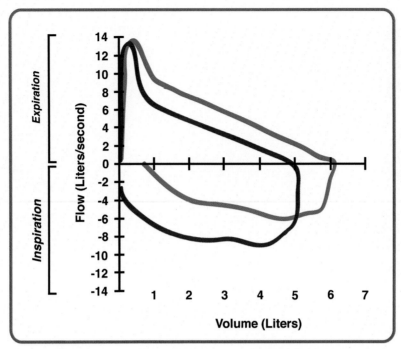

FIGURE 5.9   Flow-volume loop before (*red*) and after (*blue*) bronchodilator use.

TABLE 5.4 Features of Asthma

Reversible airways obstruction (to be differentiated from irreversible causes of COPD)

Increased viscid mucus secretion

Increased bronchomotor tone and airway muscular hypertrophy

Thickening of basement membrane

Mild to moderate lymphocytic inflammation

Airway eosinophilia

Increased mast cells

Cases of asthma can be divided into those associated with atopic disease and those that are not. The former may be referred to as extrinsic asthma, as symptoms are caused by the inhalation of environmental antigens. Atopic asthma occurs commonly in children but may develop at any time in life. In childhood it is often associated with seasonal allergies, eczema, blood eosinophilia, and elevated serum IgE levels. The role of allergy testing is controversial but can be helpful if an allergic factor is likely but not apparent.

Some patients with asthma, particularly adults, show no evidence of underlying allergy. They are referred to as having "intrinsic" asthma. However, it is important to consider other etiologies including viral infection, chronic gastroesophageal reflux, non-steroidal anti-inflammatory drugs (NSAID) use, immune responses to food and drug dyes, exercise, and cold air. In early congestive heart failure edema fluid accumulates in the airway wall and can mimic asthma.

The pathology of asthma includes several characteristic changes. The epithelial lining cells show mucinous metaplasia and the epithelial basement membrane is thickened. The wall of the asthmatic airway is edematous and displays mast cells and eosinophils in various stages of degranulation (Fig. 5.10). There may be foci of eosinophilic pneumonia. The presence of Charcot–Leyden crystals in tissue or expectorated mucus reflects eosinophil degranulation (Fig. 5.11). Curschmann's spirals are mucus plugs expectorated from small airways in asthma and chronic bronchitis (Fig. 5.11).

TABLE 5.5 Clinical Triggers of Asthma

Atopic allergy
Viral infection
Exercise
Strong emotions
Sensitivity to cold air and humidity (vasomotor disease)
Hyphate fungal colonization (bronchopulmonary aspergillosis)
Haptenated proteins
Drug induced, e.g., aspirin and non-steroidal anti-inflammatory drugs (NSAID)
Churg–Strauss granulomatosis (ANCA or drug related)
Subset of patients with "COPD"

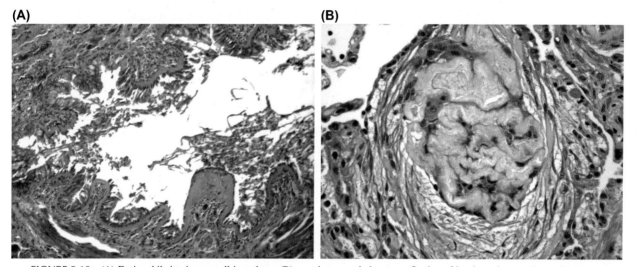

FIGURE 5.10   (A) Eosinophils in airway wall in asthma, (B) curschmann spirals are a reflection of inspissated mucus in small airways.

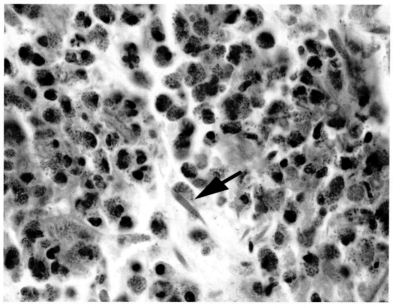

FIGURE 5.11   Charcot–Leyden crystals (*arrow*).

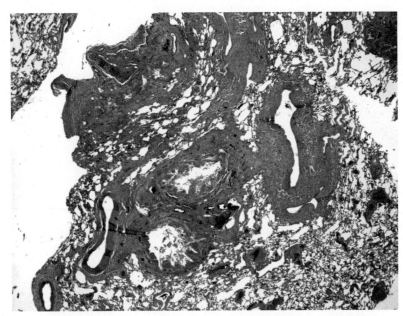

FIGURE 5.12   Airway thickening and mucus plugging in status asthmaticus.

Although it is currently recognized that Th2 lymphocytes play an important role in asthma, lymphocytic inflammation is rarely a prominent feature and may be sparse. In chronic asthma, the bronchial wall can show scarring with bronchiectasis. This explains why some asthmatics develop fixed obstructive defects that no longer respond to bronchodilators. The lumen of the asthmatic airway exhibits mucus plugging and this is an underrecognized mechanical cause of airways obstruction and death in patients with status asthmaticus (Fig. 5.12).

In addition to showing extensive mucus plugging of airways, the lungs of patients dying of status asthmaticus show alternating areas of lung hyperexpansion and atelectasis due to differential regional ventilation and airway time constants. This leads to an increase in V/Q mismatch, shunt, and dead space. Asthmatic patients also suffer from an increase in the work of breathing due to high airways resistance. Terminal failure in respiratory drive due to central and peripheral muscular fatigue leads to elevated pCO_2, respiratory acidosis, cerebral vasodilation, and intracranial edema if mechanical ventilation is not instituted.

Hydration, short- and long-acting β2 agonists, anticholinergics, lipoxygenase inhibitors, and corticosteroids are the mainstays of asthma treatment, and in selected cases monoclonal anti-IgE therapy may have a role. Drugs that dry airway secretions, including antihistamines, should be avoided in patients with moderately severe disease and during asthma exacerbations.

Although the airways in asthma are rarely biopsied for diagnosis, evidence of asthma is frequently seen in resections of lung tissue for other causes. In such cases, the pathologist should make the diagnosis of "asthmatic bronchitis," and suggest clinical correlation. This is clinically important, as some patients with COPD are not recognized as having a potentially reversible component to their disease and may not be receiving optimal treatment with bronchodilators or corticosteroids.

Specific etiologies of asthma can sometimes be established by pathological examination. Allergic bronchopulmonary aspergillosis is associated with asthma that is clinically difficult to control, tissue and peripheral blood eosinophilia, proximal cystic bronchiectasis with mucoid impaction of the airways, and elevated IgE levels directed against *Aspergillus spp*. The migration of nematode larvae through the lung is a common cause of "asthma" and pulmonary eosinophilia in third world countries (see Chapter 9).

Drug-related atopic responses also cause asthma. Aspirin is associated with asthma in patients with concomitant allergic nasal polyps. The Churg—Strauss syndrome (CSS) includes asthma, eosinophilia, and multisystem involvement by granulomatous inflammation with eosinophil vasculitis, confirmed by lung biopsy or by biopsy of another involved organ (Fig. 5.13). About 50% of patients with CSS have an associated perinuclear antineutrophilic cytoplasmic antibody (pANCA) with antimyeloperoxidase specificity. Certain drugs used in the treatment of asthma, specifically the slow-reacting substance of anaphylaxis (SRS-A) antagonist montelukast (Singulair), can yield an iatrogenic CSS, and this diagnosis may be difficult to discern from the underlying asthma if not considered.

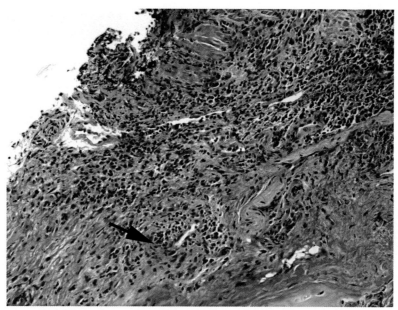

FIGURE 5.13 Eosinophilic angiitis (*arrow*) in Churg—Strauss granulomatosis.

TABLE 5.6 Chronic Obstructive Pulmonary Disease
Chronic bronchitis
Emphysema
Bronchiectasis

CHRONIC OBSTRUCTIVE PULMONARY DISEASE

Chronic obstructive pulmonary disease (COPD) is a rubric applied to a set of chronic airways diseases without clear evidence of reversible airways obstruction (Table 5.6). Most cases are caused by cigarette smoking. COPD includes chronic bronchitis, emphysema, and bronchiectasis, and asthma should be considered as distinct from this rubric unless it has progressed to irreversible disease.

The primary utility of the term is that most smokers show overlapping clinical and anatomic features of these disorders. Treatment includes smoking cessation, β-adrenergic agonists, acetylcholine antagonists, corticosteroids, and antibiotics.

Pulmonary function testing is helpful in discerning how COPD can be parsed into its components. Evidence of airways obstruction as evidenced by a decrease in the FEV1/FVC ratio < 0.6 is a *sine qua non*. Air trapping as evidenced by an increase in the residual volume (RV)/total lung capacity (TLC) ratio is generally also present and airways resistance may be increased. Emphysema and to a lesser degree bronchiectasis can produce a decrease in the diffusing capacity (DLCO) due to the loss of microvasculature (Fig. 5.14A). Some patients meet the radiographic criteria for emphysema without evidence of airways obstruction, but in such cases the DLCO is rarely normal (Fig. 5.14B).

CHRONIC BRONCHITIS

Chronic bronchitis is a *clinical syndrome* defined by the chronic presence of chronic cough and sputum production, not by histologic evidence of inflammation. The body habitus of these patients is often different than what is seen in emphysema. Chronic bronchitics tends to be mesomorphs. In the late stages of the disease, their central drive for ventilation is blunted leading to an inadequate response to hypercarbia and respiratory acidosis. Under stable chronic conditions, this is compensated by renal retention of HCO_3^- with maintenance of a near normal plasma pH.

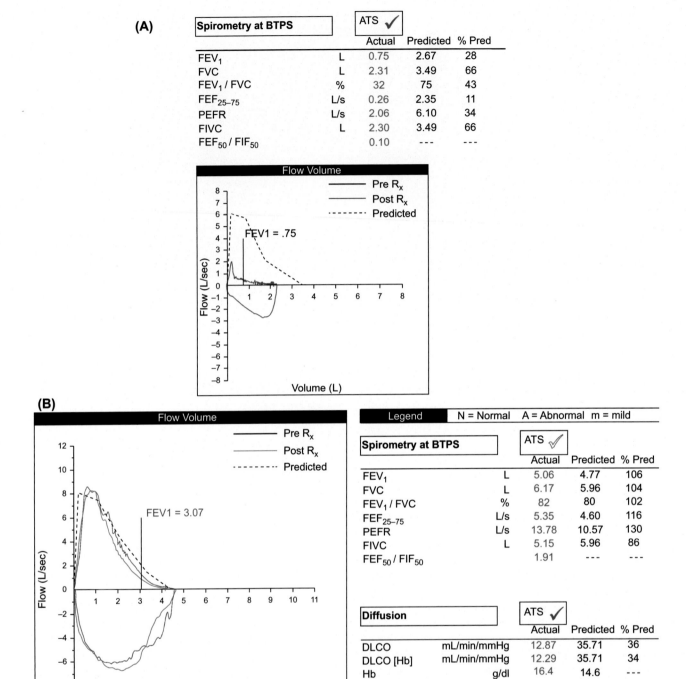

FIGURE 5.14 (A) Flow-volume loop in severe emphysema shows reduced FEV1/FVC with prolonged expiratory phase and bowing of the inspiratory loop, (B) patient with severe emphysema radiographically with normal flow rates but reduced diffusing capacity.

These patients are prone to develop hypoxic pulmonary hypertension with right ventricular failure (cor pulmonale), and cyanosis, yielding the "blue-bloater" phenotype as opposed to the emphysematous asthenic "pink puffer." Polycythemia with increased hematocrit ($>52\%$ in men and $>48\%$ in women) can develop as a compensatory change to chronic hypoxemia, increasing the red cell mass and oxygen carrying capacity of the circulating blood but also leading to intravascular sludging and cerebrovascular ischemia. It is noteworthy that with better management the blue bloaters with respiratory failure and cor pulmonale are far less frequently encountered than in the past.

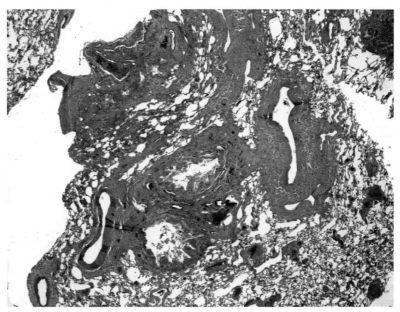

FIGURE 5.15   Airway and peribronchiolar fibrosis in a smoker.

The most consistent pathological feature of chronic bronchitis is the presence of mucus metaplasia and an increased *Reid index*. The presence of eosinophils suggests a concomitant atopic element that may be responsive to bronchodilators and corticosteroids. Basement membrane thickening is virtually always present.

Smoking is the most common cause of chronic bronchitis, but virtually any persistent cause of airway irritation can yield comparable changes. Fibrosis is commonly seen in the respiratory bronchioles and may extend into the peribronchiolar alveoli (Fig. 5.15). This complicates the diagnosis of diffuse interstitial lung fibrosis by transbronchial biopsies in smokers. However, the introduction of large sampling cryobiopsies may lead to a reevaluation of the value of TBB in this setting.

EMPHYSEMA

Emphysema is in truth an interstitial lung disease that results from the patchy loss of the alveolar elastic tissue in the gas-exchanging structures of the proximal secondary lobule (Fig. 5.16). It predominates in the upper lobes of the lung (Fig. 5.17). It is caused by the proinflammatory effects of cigarette smoke, but other toxic fumes including crack cocaine, cadmium, tungsten, as well as infectious agents, e.g., HIV-1 produce comparable changes (Table 5.7).

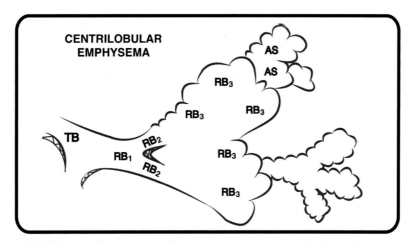

FIGURE 5.16   Changes in the pulmonary acinus in centriacinar (lobular) emphysema.

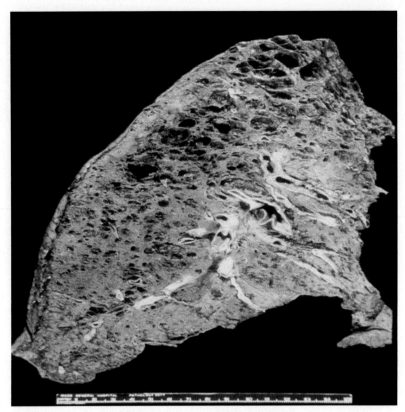

FIGURE 5.17   Centrilobular emphysema.

TABLE 5.7 Emphysema

Types	Known Causes of Emphysema
Centrilobular	Cigarette smoke
Panlobular	Air pollution
Paraseptal	Crack cocaine
Traction	HIV-1
Senile	Tungsten
	α1-antitrypsin deficiency
	Chronic air trapping
	Collagen disorders
	Focal scarring
	Aging

Unlike chronic bronchitis, emphysema is defined and diagnosed *anatomically*. A small percentage of patients with emphysema on computed tomography (CT) scans are asymptomatic and lack pulmonary functional abnormalities. As noted in Chapter 3, airways obstruction results from the loss of the tethering effect of the small airways by the loss of elastic fibers in the alveoli. This leads to the narrowing of the adjacent small airways during expiration, and to increased airways resistance. The alveolar wall in emphysema is rarely histologically inflamed, but there is a measurable increased transit of neutrophils through the alveolar wall that can be detected in the bronchoalveolar lavage fluid.

As the diagnosis of emphysema is anatomical, it can be established radiographically by the presence of dilated air spaces on CT scans. Mild emphysema often goes undetected by radiographic imaging. By the time that emphysematous

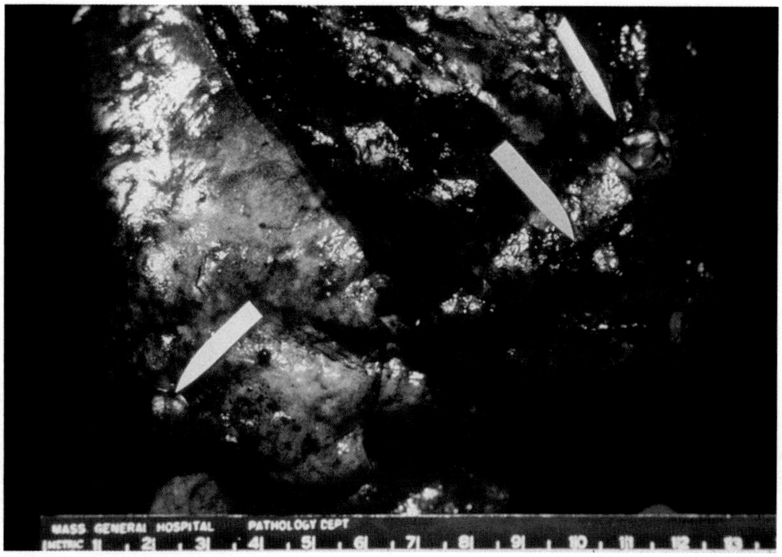

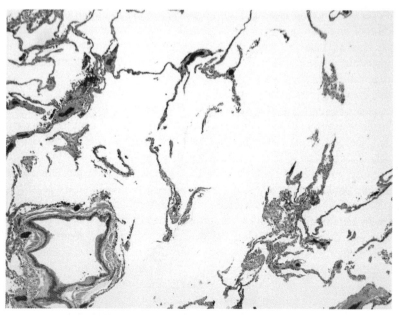

FIGURE 5.18   Hyperinflation of lung with bullous emphysema and subpleural blebs (*arrowheads*).

changes are seen in conventional chest radiographs, patients have extensive bullous disease, i.e., emphysematous spaces ≥ 1 cm, and will have lost almost half of their lung parenchyma.

Lung hyperinflation, an increased antero-posterior (A-P) diameter, diaphragmatic flattening, and an enlarged retro-sternal airspace are radiographic features of emphysema, although they do not necessarily distinguish it from other causes of hyperinflation, e.g., asthma. An important radiographic feature of emphysema is the loss of pulmonary vasculature in the peripheral lung leading to hyperlucency or mosaicism seen on CT scans.

The pathologic diagnosis of emphysema can be made by gross observance of the lung at autopsy. Air trapping is evidenced by failure of the lung to collapse at atmospheric pressure (Fig. 5.18). The cut surface of the lung shows microcystic changes and elevation of the small airways and vessels above the cut surface due to loss of elastic tethering. The histological diagnosis is established by the finding of dilated air spaces and the presence of "floating fragments" of alveolar septa, a change attributable to a disrupted three-dimensional alveolar network viewed in two dimensions (Fig. 5.19). In the absence of this finding, it is impossible to distinguish emphysema from age-related senile distensive changes or hyperinflation due to air trapping from other causes.

FIGURE 5.19   Floating fragments of lung in emphysema.

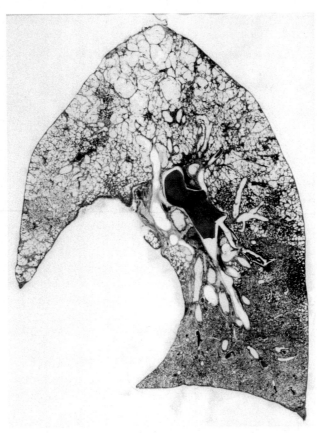

FIGURE 5.20 Upper lobe predominance and anthracotic pigment in thin cross section of emphysematous lung.

Most cases of emphysema caused by cigarette smoking show centrilobular destruction of the acinus including the respiratory bronchioles and alveolar ducts, with sparing of distal alveoli. Even in severe bullous disease, one can still discern intact alveoli surrounding the dilated airspace (Fig. 5.20). When bullae becomes sufficiently large, they can compress the surrounding lung leading to severe V/Q abnormalities (Fig. 5.21). Bullae increases the RV, functional residual capacity (FRC), and the work of breathing, leading to severe dyspnea. Bullectomy in selected patients with limited upper lobe disease and relatively preserved exercise function may benefit from this procedure. A recent nonoperative therapeutic approach is the bronchoscopic insertion of one-way air valves into a feeding airway, but complication rates are currently high.

In some cases, diffuse emphysematous change can produce pulmonary restriction on PFTs and show changes on radiographs that mimic interstitial disease. Biopsies in these cases show diffuse alternating areas of hyperinflation and atelectasis reflecting different time constants of alveolar filling (Fig. 5.22).

Panlobular emphysema is usually associated with alpha-1 antitrypsin deficiency (Fig. 5.23). The disease, which shows recessive inheritance, varies in severity depending on whether patients are heterozygotes (MZ) or homozygous (ZZ) for the gene. Severe deficiency may present as hepatitis in childhood due to engorgement of abnormal protein in hepatocytes. When the lung is affected, the emphysematous changes tend to be distributed mostly in the lower lobes and the entire acinus is involved, which distinguishes it from centrilobular emphysema. Disease tends to develop more rapidly and severely in patients who smoke, so abstinence is a critical factor in treatment and counseling. Replacement of the deficient protein may slow progression, but lung transplantation may be necessary, although the disease has recurred in allografts. It noteworthy that a small percentage of cases of severe emphysema occur in never smokers without a history of inhaled toxins, or recognized genetic abnormalities, and are currently idiopathic.

A variety of anatomic variants of emphysema have been recognized. One relatively common form is paraseptal emphysema (Fig. 5.24). Its relationship to smoking is less robust than that of centrilobular disease. One observes dilated airspaces lined up along the visceral pleural, or along collagenous fissures. Rupture of these emphysematous spaces into the pleural space is a known cause of pneumothorax. When an emphysematous space ruptures into a fissure, air may dissect within the pleura to form a *bleb*, i.e., *an intrapleural collection of air*, which should be distinguished from bulla. Both bullae and blebs are often seen in patients with pneumothoraces (Fig. 5.18, arrow).

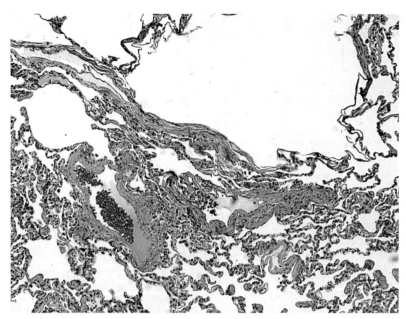

FIGURE 5.21   Areas of hyperinflation and atelectasis in bullous emphysema.

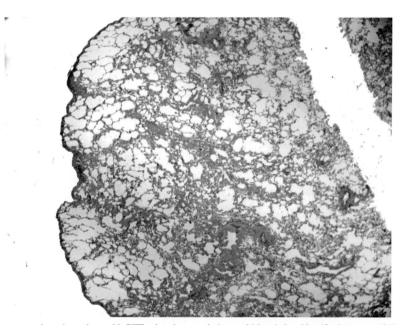

FIGURE 5.22   Emphysematous lung in patient with PFTs showing restriction and biopsied to identify the cause of "interstitial lung disease."

Bullae can become infected and mimic lung abscess with air fluid levels seen radiographically (Fig. 5.25). In patients receiving anticoagulation, bleeding into a bulla can appear as a fluid-filled cavity. Bullous emphysema can also distort the radiographic appearance of common lung pathologies, e.g., cardiogenic pulmonary edema. The radiographic finding of cystic changes in patients with lepidic (bronchioloalveolar) adenocarcinoma in my experience is attributable to tumor growth in an area of bullous change.

Traction emphysema develops adjacent to a preexisting scar and represents chronic hyperinflation with air trapping due to an anatomically distorted airway. It is most commonly seen in the upper lobes of the lung as *fibrobullous* disease (Fig. 5.25B). At times, isolated cystic bullae develop within lung parenchyma. Some are idiopathic, although there is an increasing awareness that a variety of autoimmune lung diseases can cause cystic changes. Other causes of lung cysts, e.g., lymphangioleiomyomatosis and Birt−Hogg−Dube, are due to genetic abnormalities of the tuberous sclerosis and of

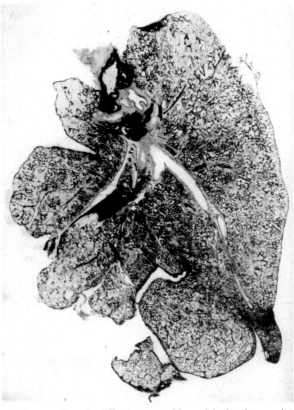

FIGURE 5.23   Panlobular emphysema, Note the diffuse nature and lower lobe involvement in $\alpha-1$ antitrypsin deficiency.

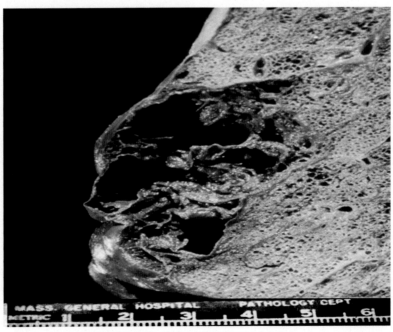

FIGURE 5.24   Paraseptal bullous emphysema.

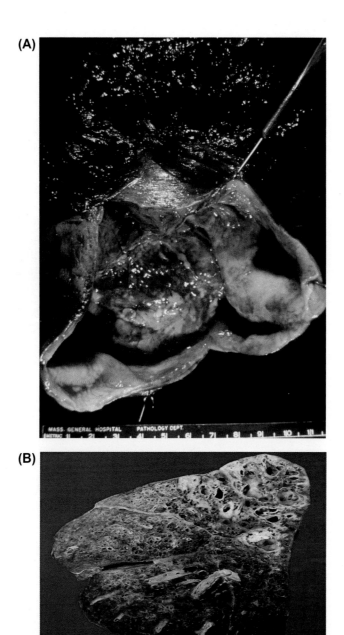

FIGURE 5.25   (A) Infected bullae, (B) fibrobullous disease in upper lobes.

the folliculin genes, respectively. Langerhans' cell histiocytosis is another common cause of cystic and nodular lung disease.

EXACERBATIONS OF COPD

Exacerbations account for a large percentage of hospitalizations in COPD. The cause of most exacerbations is uncertain and they are likely multifactorial reflecting ambient air pollution, viral and other atypical infections, or superimposition of bronchospastic changes. The pathology of COPD exacerbations is currently uncertain as airway and lung biopsies are rarely pursued. Most cases respond to a combination of antibiotics and corticosteroids, and recent research has been focused on identifying whether there is a phenotype that is prone to frequent exacerbations. Some studies suggest that atopy with blood and airway eosinophilia may play a role, at least in some patients.

BRONCHIECTASIS

Bronchiectasis can be conceived as the end-stage of large airway disease. The bronchiectatic airway shows a loss of cartilaginous support, chronic inflammation, fibrosis, and atrophy of the seromucous glands (Fig. 5.26). The disorder may be divided clinically into two forms: dry bronchiectasis, in which the airways are dilated, scarred, but minimally inflamed; and wet bronchiectasis where they are chronically inflamed and infected (Table 5.8).

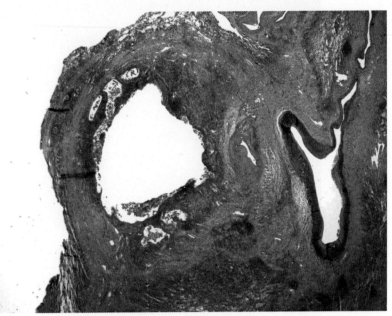

FIGURE 5.26   Bronchiectatic airway.

TABLE 5.8 Bronchiectasis

Types	Cystic
	Varicose
	Cylindrical
Causes	Postinfectious
	Postobstructive change
	Immunoglobulin deficiency
	Cystic fibrosis
	α1 antitrypsin deficiency
	Mannose binding protein deficiency
	Yellow nail syndrome
	Eating disorder
	Malnutrition
	Traction bronchiectasis

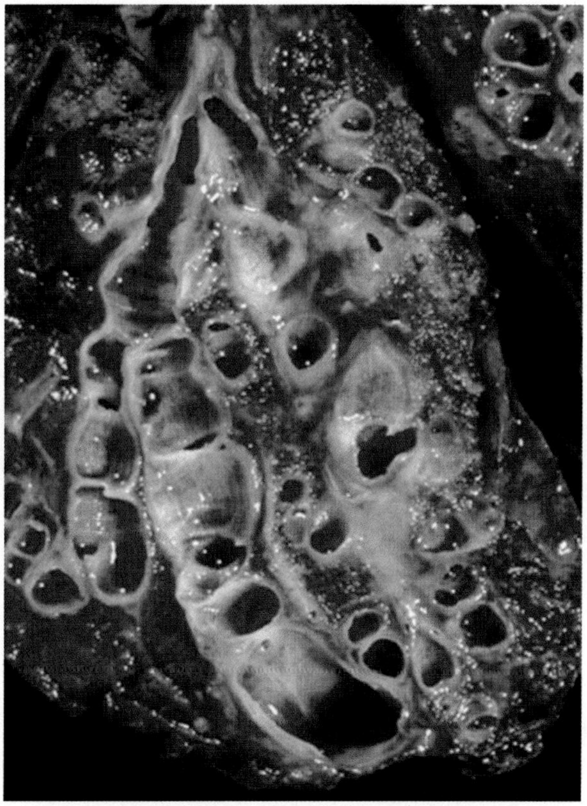

FIGURE 5.27   Saccular bronchiectasis.

Wet bronchiectasis is distinguished by cough and production of mucopurulent sputum that tends to sediment out into layers of pus, mucus, and serous secretions. Bronchiectasis has multiple etiologies. Most cases of wet bronchiectasis are attributable to prior and persistent airways infection. Prior to the era of widespread childhood vaccination, *Bordetella pertussis* and measles were common causes. Currently, most cases of bronchiectasis are the sequelae of previous bacterial pneumonias, long-standing asthma, aspiration, chronic bronchitis, and emphysema. Acquired hypogammaglobulinemia (common variable immunodeficiency disorder or CVID), collagen-vascular disease, inflammatory bowel disease, alpha-1 antitrypsin deficiency, primary ciliary dyskinesia, severe eating disorders, scurvy, and the yellow nail syndrome are other causes of bronchiectasis.

Reid and others noted that the severity of bronchiectasis tends to vary with the level of airway involvement. The more proximal the bronchiectatic changes, the more likely they are to be cystic and *saccular* (Fig. 5.27) and to yield atrophic scarring of the adjacent and distal lung (Fig. 5.28). Interestingly, the small airways in the distally scarred lung tend to be narrowed and entrapped by eccentric fibrosis (Fig. 5.29). The scarring of gas-exchanging parenchyma accounts for the restrictive changes that are seen on pulmonary function testing in these patients. Intermediate forms are termed *varicose* bronchiectasis, and the most common and least severe form with respect to loss of lung function is *cylindrical* bronchiectasis.

When cylindrical bronchiectasis is detected radiographically in the presence of an acute pneumonia it may prove to be reversible and no intervention is warranted until there is evidence of its persistence months following the eradication of the acute infection.

CYSTIC FIBROSIS

Cystic fibrosis is caused by mutations in the cystic fibrosis transmembrane conductance regulator (CFTR) gene, most commonly due to the point mutation, $\Delta 508$ (Fig. 5.30). This leads to a defective chloride channel and increased mucus viscosity, with mucus plugging of airways and other tissues, including the pancreas, sinuses, and vas deferens in males. The pulmonary airways become plugged and dilated by mucin most prominently in the upper lung zones, and subsequently superinfected. Patients suspected of having the disorder can be diagnosed by an abnormal sweat test and/or by genetic sequencing.

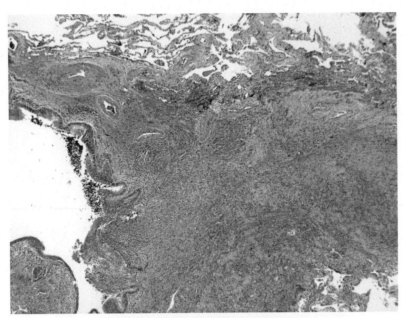

FIGURE 5.28 Peribronchiolar fibrosis.

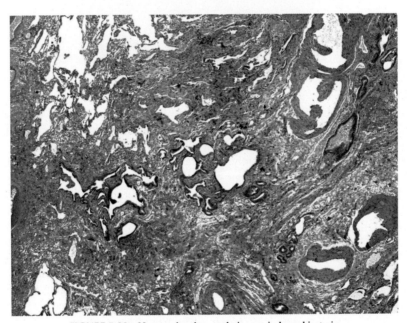

FIGURE 5.29 Narrowed and scarred airways in bronchiectasis.

The CFTR gene is heterozygous at a high frequency in Caucasians (approximately 1 in 25). Improved pulmonary toilet and targeted antibiotic use has resulted in a substantial improvement in longevity, with many patients living into their fourth, fifth, and even sixth decades, so that currently CF is both a pediatric and adult pulmonary problem. It is noteworthy that addiction to narcotics is a common clinical problem in these patients for reasons that may not be solely explained by psychological factors.

In the "wet bronchiectasis" that is virtually always present in CF, one sees acute and chronic inflammation of the airway wall with acute bronchitis/bronchiolitis and bronchopneumonia (Fig. 5.31). Tissue gram stains may demonstrate Gram-positive and Gram-negative organisms, compartmentalized within areas of mucus inspissation, where they apparently grow as a biofilm that is difficult to eradicate (Fig. 5.32). Early in the disease process, infection tends to be with organisms such as *Hemophilus influenzae*, but with time methicillin-sensitive and resistant *Staphylococcal spp.* as well as Gram-

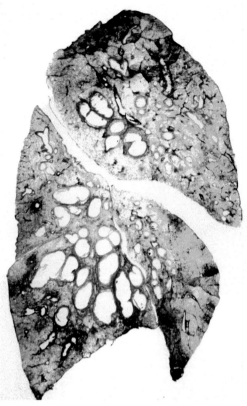

FIGURE 5.30   Thin cross section of lung in cystic fibrosis. Note the loss of lung parenchyma.

negative bacilli colonize and infect the airways. Indeed, the finding of these organisms in patients with chronic airways disease almost certainly indicates the presence of bronchiectasis, even if this is not radiographically obvious.

Mucoid strains of *Pseudomonas* tend to occur later in disease and they can be difficult to treat due to high resistance patterns. In hospital "clean-outs" with bronchodilators and antibiotics based on bacterial susceptibilities have greatly improved the prognosis in these patients. However, the microbial flora is never fully eradicated due to the continued severe anatomic distortion, defective clearance mechanisms, and the development of biofilms. Genetic studies of the microbiome in cystic fibrosis may yield increased insight into the patterns of infection.

In addition to the inflammation of the airways, there is an increase in angioneogenesis from the systemic circulation that supports the remodeling of the ectatic variceal bronchial arteries arise from branches of the aorta and intercostal arteries and can be seen by CT with contrast or by angiography. Erosion of an inflamed airway into a vessel is a common cause of hemoptysis in bronchiectasis. As the bronchial circulation is at systemic blood pressure, bleeding can be severe and lead to death by aspiration into the lungs. For this reason, it is also best to avoid pursuing bronchial or transbronchial biopsies in bronchiectatic patients, as the bronchial arteries course directly beneath an atrophic bronchial wall and can easily be inadvertently lacerated leading to life-threatening pulmonary hemorrhage (Fig. 5.33).

Treatment of bronchiectasis generally requires appropriate antibiotic coverage, postural drainage, the addition of a flutter valve and hypertonic saline to help mobilize secretions.

"Dry" bronchiectasis occurs as a result of fibrosis-induced traction of the airways, which are not chronically infected. This change may be seen posttuberculous infection, in late stages of parenchymal sarcoid, other chronic fibrosing disorders, or as a complication of chronic bronchitis. At times, hemoptysis may occur in dry bronchiectasis.

Nontuberculous mycobacteria tend to colonize bronchiectatic airways but also promote airway inflammation. This is discussed further in the chapter on infection.

Bronchiectasis is generally present in the *right middle lobe syndrome* (or lingular segment of the left upper lobe). This is a poorly defined entity in which there is chronic or intermittent acute partial or complete collapse of the right middle lobe or the left lingular segment. It is caused by several factors including (1) the acute angle take-off of the feeding middle lobar bronchus, (2) a malacic bronchiectatic segment, (3) regional reactive hilar adenopathy that compresses the supplying airway, and (4) poor collateral ventilation due to variations in lobar septation which can limit collateral ventilation from other lung lobes. In such cases the right middle lobe shows a variety of changes. There is invariably bronchiolectasis and

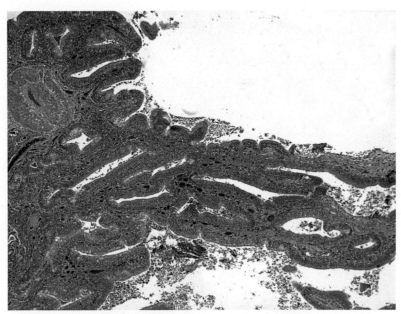

FIGURE 5.31   Acute inflammation in polypoid respiratory epithelium in cystic fibrosis.

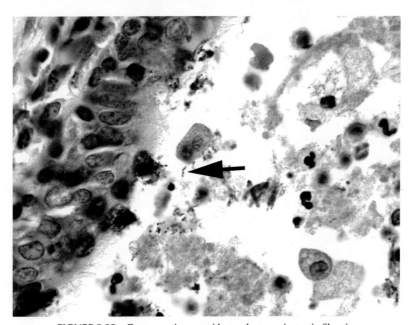

FIGURE 5.32   Gram-negative mucoid *pseudomonas* in cystic fibrosis.

chronic inflammation (Fig. 5.34). There may be changes of chronic organizing pneumonia and lipoid pneumonia, reflecting chronic small airways obstruction. Some cases show atypical mycobacteria infection complicating chronic bronchiectasis.

Proximal obstruction by an endobronchial tumor or foreign body is a cause of postobstructive bronchiectasis. It produces either mucus inspissation with dilated but noninflamed airways, or postobstructive lipoid pneumonia due to the accumulation of lipid in alveolar macrophages presumably due to decreased clearance of lipoproteins secreted into the airways. Most cases of postobstructive pneumonia are sterile by routine cultures but bacterial infections can develop.

Idiopathic forms of *endogenous lipoid pneumonia* occur rarely and are characterized by the presence of lipid-engorged macrophages, cholesterol breakdown products, and giant cell reactions to cholesterol (Fig. 5.35). The lung grossly takes on a yellow color, so-called "golden pneumonia" and the diagnosis can be suggested at the time of surgical biopsy or resection. Some of these cases have been associated with elevated levels of IgG4 production, when there is associated prominent plasma cell infiltration. Immunostaining for IgG4 establishes the diagnosis.

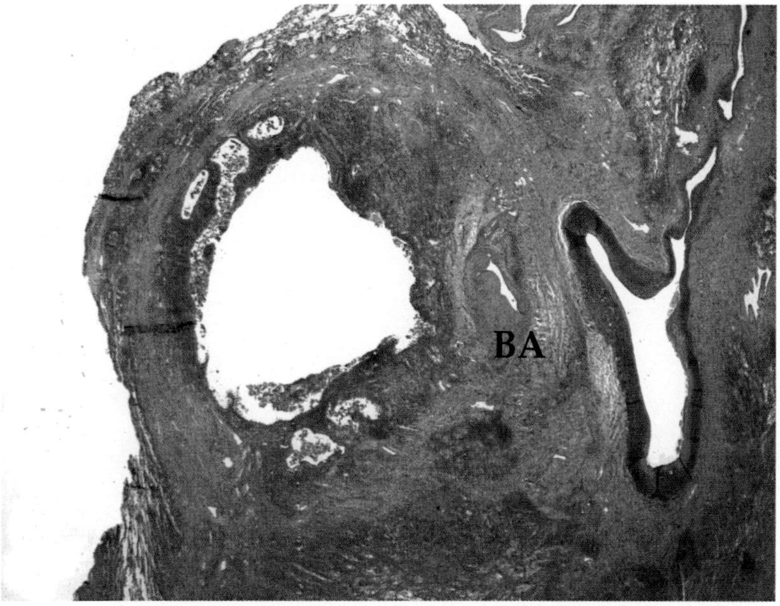

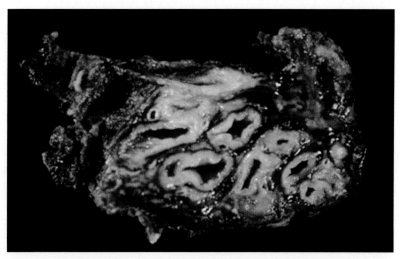

FIGURE 5.33   Bronchial artery (BA) hypertrophy in bronchiectasis.

FIGURE 5.34   Atelectatic lung in the right middle lobe syndrome.

THE SMALL AIRWAYS

Bronchiolitis

Infections, aspiration, and toxic exposures can produce inflammation of the membranous bronchioles and the pathologic findings are generally nonspecific. Patients complain of cough, fever, and dyspnea. The characteristic radiographic appearance is that of a "tree-in-bud," which is a nonspecific indicator of bronchiolar disease. In most cases the airway wall shows chronic inflammation but there may be acute neutrophilic inflammation with mucus in the airway lumens (Fig. 5.36).

Diffuse pan-bronchiolitis is a distinctive lesion seen mostly in patients of Japanese lineage who have an increased incidence of HLA-A11 and HLA-B54 antigens. Patients present with cough, wheezing, and dyspnea. Close relatives may present with chronic sinusitis. Chest imaging shows diffuse 1−3 cm centrilobular nodules. Histologically, the membranous bronchioles show marked lymphoplasmacytic inflammation, and there may be surrounding areas of organizing pneumonia. The bronchiolar wall and the peribronchial lung interstitium are infiltrated by foamy macrophages (Fig. 5.37). Acute

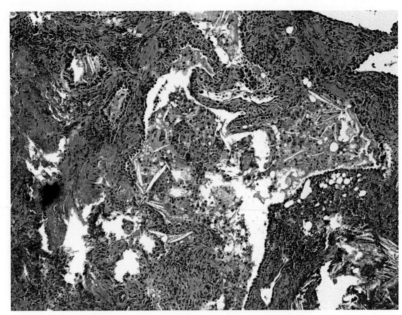

FIGURE 5.35 Postobstructive "endogenous" lipoid pneumonia.

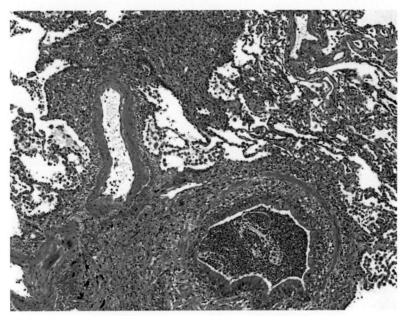

FIGURE 5.36 Acute and chronic bronchiolitis.

inflammation in the airway lumen is often present. In chronic cases, proximal bronchiectasis may develop and this is generically the case for all disorders that produce bronchiolar obliteration. The macrolide antibiotics appear to benefit patients with pan bronchiolitis possibly via their anti-inflammatory rather than antimicrobial effects.

OBLITERATIVE BRONCHIOLITIS

Obliteration of the small airways is caused by a variety of disorders, including infections, inhaled toxins, diacetyl used in the production of buttered popcorn, drugs, collagen vascular disease, and immune mediated processes such as graft versus host disease and chronic pulmonary allograft rejection. Patients complain of dyspnea on exertion with minimal findings on conventional chest radiographs. Air trapping may be seen in the exhalation scans by high resolution computerized

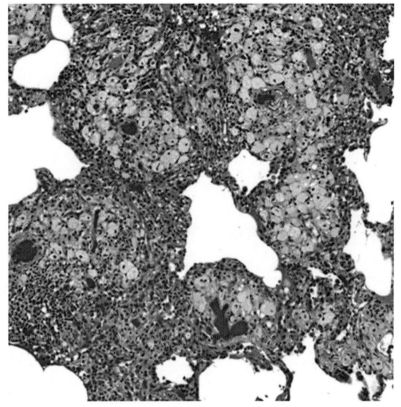

FIGURE 5.37   Diffuse panbronchiolitis showing interstitial foamy macrophages.

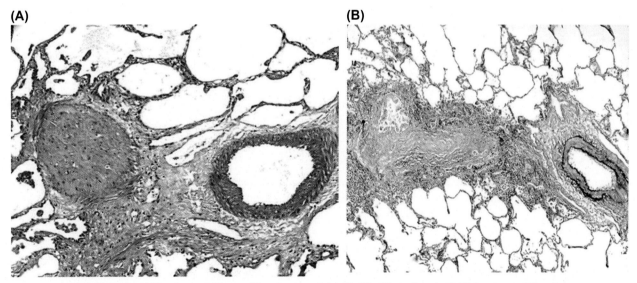

FIGURE 5.38   (A) Intralumenal fibrous obliteration in patient with OB, (B) elastic stain highlights airway obliteration.

tomography (HRCT) scans and PFTs show increased residual volume with a total lung capacity that is either normal or reduced. The diagnosis is suggested by the clinical picture but definitively requires lung biopsy confirmation.

Obliterated airways show either intraluminal tufts of fibrous tissue (Fig. 5.38) or are narrowed by fibrosis in the airway wall encroaching on its lumen, i.e., *constrictive bronchiolitis*. The presence or absence of inflammation predicts the response to anti-inflammatory interventions (Fig. 5.39). Treatment depends on severity and generally includes macrolide

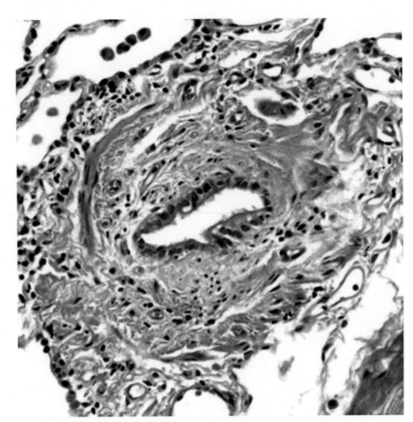

FIGURE 5.39 Constrictive bronchiolitis with concentric scarring narrowing the airway lumen.

antibiotics and potentially corticosteroids; however, lung transplantation is indicated in younger and more severely ill patients.

The pathologic diagnosis of obliterative bronchiolitis (OB) can be hampered by a paucity of findings in hematoxylin and eosin stained sections. The pathologist must evaluate the lung for small scars and determine whether there is evidence of small airways drop out. The latter is suggested by the presence of pulmonary muscular arteries that are not accompanied by an associated small airway. Where an airway is expected there may be nothing more than a small scar (Fig. 5.40). The obliteration of small airways is confirmed by the elastic stain, which reveals remnants of the elastic network of the airway (Fig. 5.38B) within the scar. One also may discern areas of hyperinflation alternating with foci of atelectasis in the surrounding lung. The larger caliber airways can develop bronchiectasis.

BRONCHIOLITIS OBLITERANS ORGANIZING PNEUMONIA

Bronchiolitis obliterans organizing pneumonia (BOOP) or cryptogenic organizing pneumonia (COP) is generally classified with the interstitial lung diseases, but as its name suggests, the vast majority of cases show primary involvement of the terminal airways. The disorder is a pathological syndrome and except when there are pathognomonic features of infection, clinico-pathological correlation is required to establish one of the many etiologies of BOOP before terming it COP.

A finding of organizing pneumonia in the lung is highly nonspecific, as it is a common reaction pattern to a variety of injuries. Areas adjacent to malignancies or infection may show prominent organizing pneumonia, so it is important to correlate the clinical and radiographic picture with the biopsy findings before accepting organizing pneumonia (OP) as a final definitive diagnosis. Clinically patients present nonspecifically with shortness of breath and/or cough, and radiographically show areas of consolidations often in the upper lung zones that may wax and wane, associated with branching airways.

The diagnosis is most reliably made by VATS biopsy, although transbronchial biopsy (TBB) can also be effective if sampling is ample. Bronchoalveolar lavage can show a modest increase in eosinophils but is otherwise nondiagnostic in the absence of infection. The biopsies show a branching pattern of myxoid fibrosis (*Masson bodies*) obliterating the lumens of terminal airways (Fig. 5.41). This is associated with interstitial inflammation that includes lymphocytes, plasma cells, and

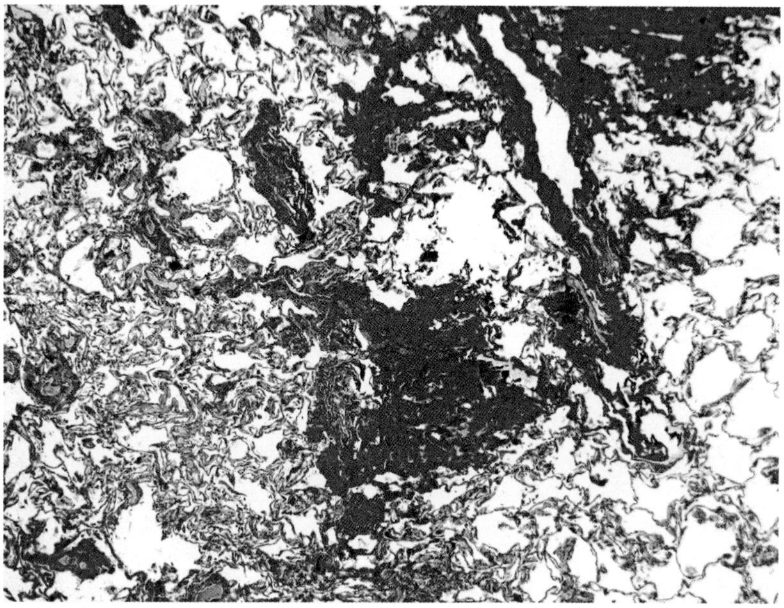

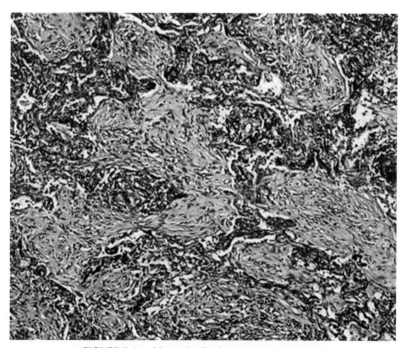

FIGURE 5.40   Focal peri-arterial scar is all that remains of an obliterated airway.

FIGURE 5.41   Masson bodies in organizing pneumonia.

macrophages. In some cases, the interstitial component may be disproportionately prominent for the degree of OP, and this suggests the possibility of an associated collagen vascular disease. Increased eosinophils are often present (Fig. 5.42) and a spectral change exists between OP with eosinophils and chronic eosinophilic pneumonia. In general, the presence of large numbers of eosinophils in the biopsy suggests a corticosteroid responsive lesion.

A variant of OP exists that shows exclusively interstitial fibroblastic foci rather than intraluminal tufts of fibrosis. This was termed *cryptogenic fibrosing alveolitis* in the older British literature. Whether this entity is mechanistically a variant of OP is questionable but currently it has no distinct diagnostic category. However, when observed, a careful search should be made both radiographically and by examining the biopsy for evidence of preexisting old areas of interstitial lung disease, as accelerated interstitial lung disease can show an OP pattern with numerous fibroblastic foci.

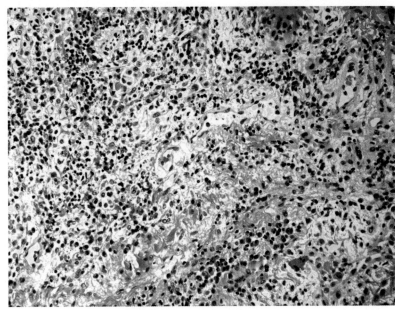

FIGURE 5.42   Eosinophils in OP.

When OP is diffuse and the patient develops respiratory failure, it is most often due to a failure to recognize the presence of diffuse alveolar damage (DAD). Important differential points include the presence of hyaline membranes that may be difficult to discern in the late fibrosing phase of DAD (Fig. 5.43). In such cases, hyaline membranes are often located at the periphery of the areas of organizing fibrosis.

The importance of distinguishing OP from DAD is their potential response to corticosteroids, which is good in OP but not in DAD. A prognostic factor appears to be the degree of fibrin deposition in the biopsy of patients with OP. Response to corticosteroids in OP is generally prompt. However, the duration of response is inversely proportional to the degree of fibrin deposition in the biopsy, as well as to the number of lobes of lung radiographically involved by the process. Early responses to corticosteroids in such settings may be complicated by the relapsing disease if the corticosteroids are tapered rapidly.

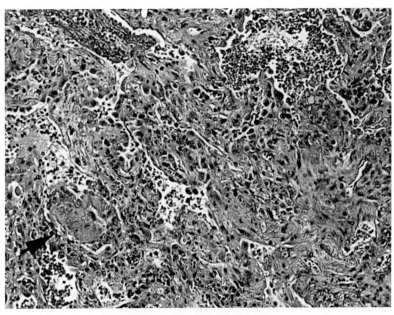

FIGURE 5.43   OP with hyaline membranes.

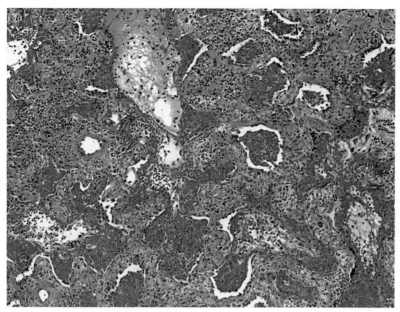

FIGURE 5.44   Acute fibrinous and organizing pneumonia.

The extreme variant is *acute fibrinous and organizing pneumonia* (AFOP) (Fig. 5.44). Although originally described as a form of DAD with a poor prognosis, when alveolar filling by fibrin is present in patients who are not severely hypoxemic and show limited radiographic disease (less than three lobes involved), these patients almost always respond well to corticosteroids, but tend to relapse if corticosteroids are rapidly tapered. On the other hand, patients that present in respiratory failure with AFOP almost always have hyaline membranes in their biopsies indicating DAD. The prognosis of DAD with a prominent (> 25%) OP component is better than for DAD alone, but it remains guarded.

There are multiple causes of OP. Infection should be excluded by probing for viruses, bacteria, *mycoplasma*, chlamydial disease, and Q-fever (Table 5.9). Patients should be screened by history for gastroesophageal reflux disease (GERD) and the biopsy examined for the presence of multinucleated giant cells that can develop with aspiration (Fig. 5.45). Some inconclusive studies suggest that the presence of lipid-laden macrophages may be a marker of aspiration. Dependent lung zone predominance and evidence of a patulous esophagus or hiatal hernia, even in the absence of symptoms of reflux, may be clues to GERD as the underlying cause of OP. pH testing and modified barium swallows are of limited diagnostic benefit, as GERD is a disorder that waxes and wanes with the tone of the lower esophageal sphincter (LES), which is dynamically dependent on diet and stress.

Evidence of old areas of chronic inflammation with scarring around small airways or the presence of *Lambertosis*, i.e., bronchiolar cell metaplasia that extends along adjacent alveolar septa (Fig. 5.46), are clues to the diagnosis of chronic acid aspiration. However, *Lambertosis* is also seen in smokers, hypersensitivity pneumonitis, and a variant of interstitial lung disease termed *airway-centered interstitial fibrosis*. This benign metaplasia should not be mistaken for adenocarcinoma, as the cells are invariably benign appearing and ciliated.

A long list of drugs is capable of producing OP with or without eosinophilia or alveolar type II cell dysplasia. OP can also be seen as part of the spectral changes of hypersensitivity pneumonitis, autoimmune disease, uremia, inflammatory bowel disease, celiac disease, chronic pulmonary allograft rejection, and graft versus host disease. In most cases, the clinical features will be obvious, but at times they may be subtle and escape attention unless aggressively pursued.

FOLLICULAR BRONCHIOLITIS

Lymphoid follicles with prominent germinal centers can develop in the walls of the small airways causing bronchiolar narrowing, distal air trapping, and post obstructive cyst-like changes (Fig. 5.47). Follicular bronchiolitis appears to represent a hyperplastic response of the bronchial-associated lymphoid tissue. It may be seen in autoimmune disease, particularly Sjögren's syndrome, rheumatoid disease, and systemic lupus erythematosus, in the children with HIV-1 infection, and in patients with common variable immunodeficiency, in which immunoglobulin production is abnormally

TABLE 5.9 Organizing Pneumonia

Cryptogenic organizing pneumonia	Idiopathic
Bronchiolitis obliterans organizing pneumonia	Postinfectious
	Autoimmune disease
	Drug toxicity
	Aspiration
	Chronic lung allograft rejection
	Postbone marrow transplant
Acute fibrinous organizing pneumonia	Idiopathic
	Postinfectious
	Autoimmune disease
	Acute hypersensitivity pneumonitis
	Drug toxicity
	Aspiration
	Chronic lung allograft rejection
	Postbone marrow transplant
	Acute respiratory distress syndrome (ARDS)

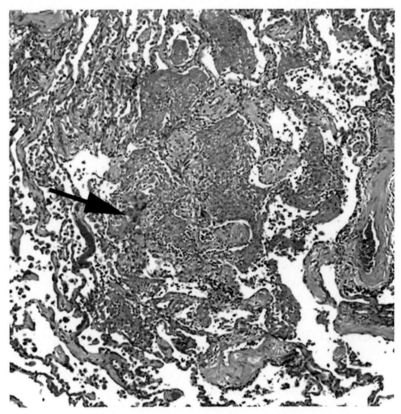

FIGURE 5.45   Foreign body giant cells (*arrow*) in aspiration pneumonia.

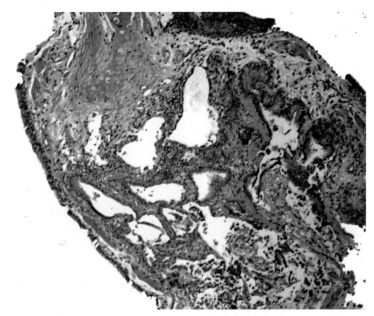

FIGURE 5.46   Bronchioloalveolar metaplasia (Lambertosis).

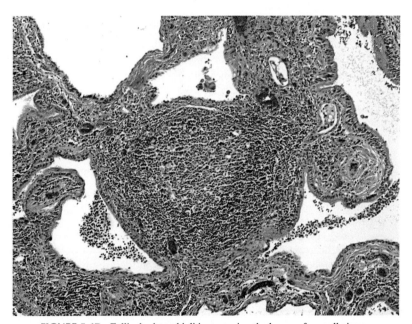

FIGURE 5.47   Follicular bronchiolitis narrowing the lumen of a small airway.

decreased. Chronic systemic infections that cause generalized lymphoid hyperplasia, e.g., circulating immune complexes and hepatitis C infection, can also produce follicular bronchitis.

Nodules of benign lymphoid tissue generally 1–2 cm in size are termed *nodular lymphoid hyperplasia* (NLH) and must be distinguished from B-cell lymphomas by the former's polytypic light chain expression (Fig. 5.48). At times, the nodules may be seen with organizing pneumonia and fibrosis, and these lesions were previously referred to as *pseudolymphomas*.

When the interstitial compartment is infiltrated diffusely, the disorder is termed lymphoid interstitial pneumonitis (LIP) and it shows a high degree of association with dysgammaglobulinemia and Sjögren's syndrome. Small cysts may be seen on CT scan in LIP but are less often appreciated under the microscope. Some of these cysts reflect hyperinflation behind a "ball-valve" effect from follicular bronchiolitis.

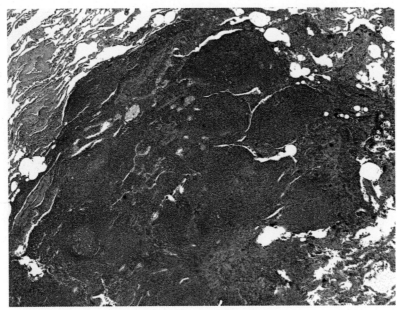

FIGURE 5.48   Nodular lymphoid hyperplasia.

One must not be overly complacent when faced with patients who show lymphoproliferative changes in the lung. It may be wise to examine these lesions by tissue PCR for evidence of light chain monoclonality that may go undetected by immunohistology or in situ hybridization. The presence of light chain deposition or congophilic AL amyloid in a lymphoid lesion is a clue that a monoclonal proliferation is likely present. Careful clinical and radiographic follow-up is indicated.

An example of this dilemma is seen in patients with CVID and lung disease. Although these patients have an increased risk of lymphoid malignancy, they may show waxing and waning infiltrates and nodular lesions attributable to BOOP with prominent lymphoid and granulomatous infiltrates. Concerns can lead to repetitive lung biopsies that rarely show a malignant transformation. As these patients may be receiving therapeutic intravenous immunoglobulin, peripheral reactive lymphoid hyperplasia and hepatosplenomegaly can develop with time, further complicating the exclusion of lymphoid malignancy.

DISEASES OF THE UPPER AIRWAYS

Most benign diseases of the upper airways present as obstruction to airflow with reduced FEV1 and a plateau in the expiratory curve (Fig. 5.49). This can be acute and life threatening, as in the aspiration of a foreign body, or gradual and progressive, as in slowly progressive posttraumatic tracheal stenosis. Infectious and malignant disorders of the trachea will be discussed elsewhere.

TRACHEAL STENOSIS

Most cases of tracheal stenosis are caused by trauma as a consequence of prolonged intubation with a high-pressure endotracheal cuff. This leads to ischemia of the airway wall and to ulceration with damage to the cartilaginous rings. The tracheal lumen is narrowed by fibrosis and often lined by metaplastic squamous epithelium. There is mucosal granulation tissue, i.e., a proliferation of small blood vessels into a matrix of chronically inflamed fibrosis (Fig. 5.50A). Recognition of the role of prolonged intubation and excessive cuff inflation has reduced the incidence of this disorder.

Idiopathic tracheal stenosis is an uncommon disorder seen in young women. Its pathology differs from that of posttraumatic stenosis as judged by the preservation of near normal cartilaginous rings (Fig. 5.50B), unless there have been repeated endoscopic dilatory procedures, at which point distinguishing the pathology of idiopathic from posttraumatic tracheal stenosis may be impossible except by history. The matrix cells in idiopathic disease express receptors for both estrogen and progesterone, and this together with the female prevalence and absence of a trauma history suggests a genetic or endocrinological component in the disorder.

FIGURE 5.49   Plateau in flow-volume loop in tracheal stenosis.

(A)   **(B)**

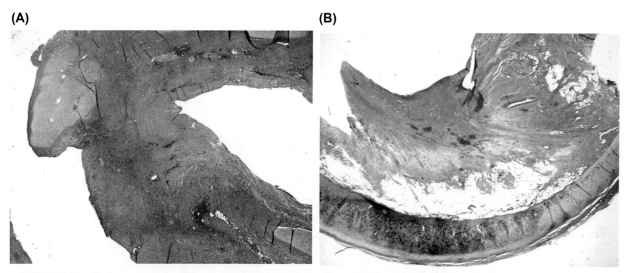

FIGURE 5.50   (A) Posttraumatic tracheal stenosis showing reactive cartilage, (B) idiopathic tracheal stenosis with intact cartilaginous plate.

OTHER TRACHEOBRONCHIAL DISORDERS

Amyloidosis

Tracheobronchial amyloidosis is an unusual cause of airway disease. The trachea may be focally, or more commonly, diffusely involved. Patients present with dyspnea or stridor and PFT evidence of upper airway obstruction. Most cases are due to the deposition of amyloid light chain (AL) amyloid, and plasma cell dyscrasia should be excluded (Fig. 5.51).

Granulomatous polyangiitis (GPA), or what was referred to as Wegener's granulomatosis, can involve the upper airways but is a difficult diagnosis to establish pathologically (Fig. 5.52). When the entire syndrome of pan-sinusitis, pulmonary disease, and acute renal failure is present, the serum ANCA is almost always positive, but serologic positivity may be absent in almost a quarter of cases of disease limited to the airways and lung. If biopsy of the airways is pursued, the findings are frequently those of nonspecific ulceration; but the presence of focal granulomatous inflammation suggests the diagnosis. Some patients have a superficial slowly spreading form of the disease that primarily involves the airway

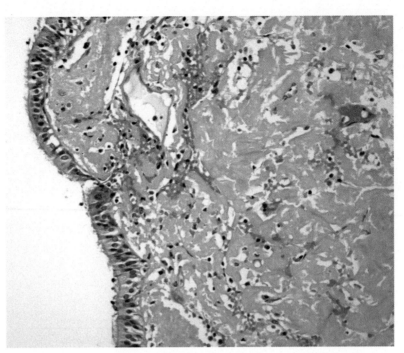

FIGURE 5.51 Tracheal amyloidosis showing amorphous eosinophilic deposits.

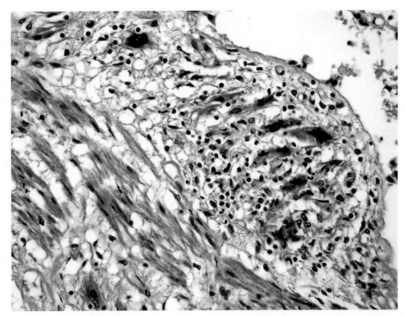

FIGURE 5.52 Airway granulomatous inflammation in granulomatous polyangiitis (Wegener's).

mucosa and may have a protracted clinical course. If a biopsy is required to establish the diagnosis, and both the lung and kidney are involved, the latter is the preferred site for biopsy due to the near universal presence of focal necrotizing glomerulitis.

It is important for pulmonologists to recognize that the term *granulomatous* inflammation is used in several ways by pathologists. The strict definition of a *granuloma* connotes a nodule of epithelioid macrophages, as is seen in sarcoidosis, or in mycobacterial and fungal infections. However, pathologists also use the term *granulomatous inflammation* to refer to

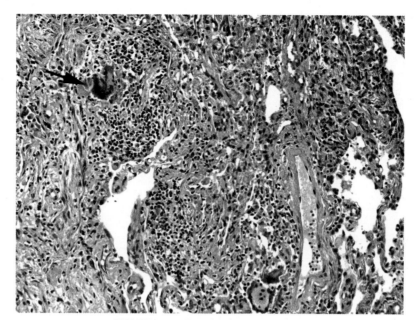

FIGURE 5.53   Granulomatous inflammation.

a polymorphic diffuse cellular infiltrate that is rich in lymphocytes, macrophages, plasma cells, and multinucleated giant cells (Fig. 5.53). This is the type of inflammation characteristically seen in GPA in which true granulomas are uncommon. Granulomatous inflammation can be observed in infections and other diseases. A detailed description of the findings in GPA/Wegener's is given in the section on vascular diseases.

RELAPSING POLYCHONDRITIS

Relapsing polychondritis can affect the larynx and tracheobronchial tree. The disease may be limited or systemic with uveitis and involvement of the cartilage of the ear and other sites. The pathology shows acute and chronic inflammation that targets the respiratory cartilage leading to chondromalacia and collapse of the airways particularly during exhalation.

TRACHEOPATHIA OSTEOPLASTICA

This is an unusual disorder that is usually identified radiographically and reflects metaplastic bone formation in the airways. There is an association with tracheobronchial amyloidosis. Patients may be symptomatic with airways obstruction or asymptomatic.

Tracheobronchomegaly, i.e., Mounier–Kuhn disease and saber-sheath trachea due to emphysema and loss of collagenous structural support of the airways, is detected radiographically and generally does not cause symptoms.

TRACHEOBRONCHOMALACIA

This is a generic term that refers to collapse and obstruction of the airways during expiration. The disease primarily involves the cartilaginous plates of the trachea and bronchi. It may be seen as part of the spectrum of COPD if bronchiectasis produces segmental or diffuse collapse, or with chondromalacia from infectious or other causes. The appearance on pulmonary function test is a diminished peak expiratory flow rate with flattening of the expiratory flow-volume loop. Treatment includes either airway stenting or surgical repair. However, caution is required before making this diagnosis as near total collapse of the membranous wall is common in forced expiration, and interventions should be directed towards medical treatment when there is potentially reversible disease, e.g., asthma or COPD, before resorting to instrumental or surgical intervention.

FURTHER READING

Celli, B.R., MacNee, W., ATS/ERS Task Force, 2004. Standards for the diagnosis and treatment of patients with COPD: a summary of the ATS/ERS position paper. Eur Respir J 23 (6), 932–946.

A recent review on the diagnosis and treatment of COPD.

Thurlbeck, W.M., Wright, J.L. (Eds.), 1999. Thurlbeck's Chronic Airflow Obstruction. B.C. Decker, London.

Perhaps the most thorough text on the pathology of airway diseases.

Strausbaugh, S.D., Davis, P.B., 2007. Cystic fibrosis: A review of epidemiology and pathobiology. Clin Chest Med 28 (2), 279–288.

A review of the genetics and pathobiology of this disease.

Chapter 6

Interstitial Lung Disease

Diseases of the pulmonary interstitium constitute a challenging aspect of both pulmonary medical and pathology practice. The term interstitial lung disease (ILD) properly refers to disorders of the gas-exchanging alveolar walls. The normal excursion of the lung is diminished due to the infiltration of inflammatory cells and deposition of collagen and elastic fibers in a stiffened alveolar wall (Table 6.1). The total lung capacity and other lung volumes, including the residual volume, are reduced (Table 6.2), and the flow-volume loop is restricted (Fig. 6.1A) with closely matched decrements in both the FEV1 and FVC, and an increased or normal FEV1/FVC ratio. The microvasculature is lost in the thickened alveolar walls with a corresponding decrease in the DLCO that does not correct for the loss of lung volume (Table 6.3).

ILD can be seen in chronic infection, autoimmune disorders, hypersensitivity reactions, drug-induced injury, and with genetic abnormalities. In many cases, no specific etiology is identified. In some cases, the histopathological appearance may be diagnostic and an etiology can be suggested, but this is often not the case.

The diagnosis of ILD rests on the correlation of radiographic, clinical, serological, physiological, and pathological abnormalities (Table 6.4). High-resolution computed tomography (CT) scanning has added sensitivity and specificity to the noninvasive diagnosis of ILD. However, pathologists must resist relying too heavily on ancillary approaches, and limit their initial efforts to establishing a pathological diagnosis independently. In recent years, this has become an issue with respect to the diagnosis of usual interstitial pneumonitis/idiopathic pulmonary fibrosis (UIP/IPF), as will be discussed.

Clinicians and pathologists should recognize that despite improvements in diagnostic categorization, the diagnosis of the interstitial disorders in roughly a quarter of cases lacks precision, and more work needs to be done to define the specific pathological criteria for diagnosis. Ultimately, molecular genetic or proteomic profiling may also be required, but neither approach is currently sensitive or specific for the classification of ILD.

The pulmonologist should establish a working knowledge of the clinical, radiographic, and pathological features of ILD. Clinicians should routinely question the pathologist as to how *confident* he or she is concerning the diagnosis, before making a definitive treatment decision. In this regard, the pathologist should be able to explain the criteria upon which the diagnosis is based and exhibit consistency in this regard.

The corollary is that when an experienced pathologist makes a diagnosis, e.g., of UIP with a high degree of confidence, and if the clinical and radiographic findings do not strongly conflict with that diagnosis, substantial emphasis should be placed on the pathologist's interpretation, with radiographic opinions given less weight. This is because even sensitive imaging techniques do not achieve the degree of resolution of the microscopic exam. However, in all cases of ILD, other factors including serological testing and family history should be explored.

USUAL INTERSTITIAL PNEUMONITIS

UIP is a histopathological syndrome with several etiologies. It is a moniker originally adopted to describe the "usual" form of interstitial fibrosis. UIP represents the pathological *sine qua non* for what pulmonologists refer to as IPF, which has an ominous prognosis. However, a diagnosis of UIP does not equate with IPF, and other etiologies must be excluded.

TABLE 6.1 Causes of Restrictive Defect

Pulmonary edema

Pulmonary fibrosis

Muscular weakness

Extrinsic compression by effusion or fibrothorax

Pulmonary resection

Understanding Pulmonary Pathology. http://dx.doi.org/10.1016/B978-0-12-801304-5.00006-X

TABLE 6.2 Pulmonary Volumes and Capacities

Volumes

Tidal volume (TV) = amount of air exchanged with each normal breath ∼500 mL

Expiratory reserve volume = amount of air that can be forcefully exhaled from end of a tidal breath

Residual volume (RV) = amount of air retained in the lung at end of forced expiration. It represents the termination of the expiratory phase of the flow volume loop

Total lung capacity (TLC) = amount of air in lung after maximal inspiration

Functional residual capacity = volume of air after end tidal volume

Forced vital capacity = volume of air in a forced exhalation from TLC

(A)

Oximetry	O$_2$ Delivery	O$_2$ Sat	Pulse
Normal room air values	L/Min	95 to 98 %	
Rest (Room Air)	None	94.0	88
Rest			
Exercise		88.0	110

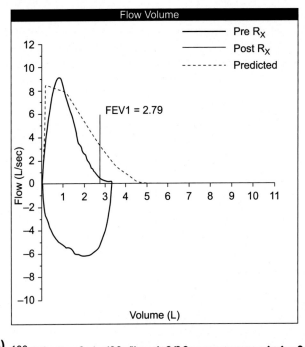

(B)

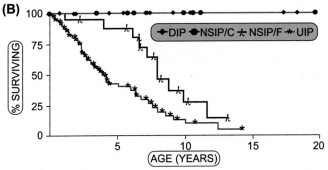

FIGURE 6.1   (A) Flow-volume loop in restrictive interstitial disease. Note the table showing the fall in O$_2$ saturation with exercise and (B) Prognosis in usual interstitial pneumonitis and subtypes of nonspecific interstitial pneumonitis.

TABLE 6.3 Causes of Decreased Diffusing Capacity

Interstitial thickening
Interstitial inflammation
Loss of microvascular bed
Pulmonary hypertension
Pulmonary edema

TABLE 6.4 Classification of Chronic Interstitial Diseases

Pathology	Etiology
Usual interstitial pneumonitis	Idiopathic pulmonary fibrosis
	Autoimmune disease (RA, scleroderma, SLE, mixed connective tissue, myositis syndrome)
	Drug
	Hermansky—Pudlak disease
	Telomerase deficiency
	Chronic hypersensitivity pneumonitis
Nonspecific interstitial pneumonitis	Autoimmune diseases
	Hypersensitivity pneumonitis
	Drug reactions
	ANCA associated
	Post acute respiratory distress syndrome
	Idiopathic
Lymphoid interstitial pneumonitis	Autoimmune disorders (Sjogren's, etc.)
	Dysgammaglobulinemia
	HIV-1 infection
Acute interstitial pneumonitis	Idiopathic
Pleuropulmonary fibroelastosis	Idiopathic

SLE, systemic lupus erythematosis.

Most patients with UIP/IPF will die within 5 years of presentation from complications of end-stage lung disease, superimposed infection, or accelerated disease with elements of diffuse alveolar damage and organizing pneumonia (Fig. 6.1B). An accelerated phase accounts for approximately half of the mortality seen in IPF.

Patients with UIP/IPF are usually in their sixth or seventh decade of life with male predominance. Many patients with UIP/IPF are former smokers and the pathogenic role of smoking remains to be elucidated. It is wise to consider age before rendering a pathological diagnosis of IPF (Table 6.5). If a patient is young, other etiologies should be excluded, including collagen vascular disease, chronic hypersensitivity pneumonitis, Langerhans' cell histiocytosis, old sarcoidosis, pneumoconiosis, telomerase deficiency, tyrosinase deficiency (Hermansky—Pudlak disease), and rare familial forms (Table 6.6). In older patients, chronic aspiration into the lower lobes can mimic UIP.

Patients with UIP/IPF present with the gradual onset of cough and dyspnea on exertion. Pulmonary function tests show a restrictive abnormality, but at times, lung volumes may be preserved and only a decreased diffusing capacity (DLCO) is present. On average, the disease requires more than a year before it is eventually diagnosed by a pulmonologist.

The surface of the lung in UIP shows a cobblestone pattern that has euphemistically been termed pulmonary cirrhosis, as it mimics the changes seen grossly in chronic liver disease (Fig. 6.2). The pathology of UIP (Table 6.5) is patchy but

TABLE 6.5 Features of Usual Interstitial Pneumonitis

Advanced age

Basilar distribution

Subpleural distribution

Active interstitial fibrosis (fibroblastic foci)

Chronic scarring

Smooth muscle hyperplasia

Fatty metaplasia beneath visceral pleura

End-stage lung (cystic bronchiolectasis with intralobular fibrosis)

Traction bronchiectasis with anatomic distortion

Variable interstitial chronic inflammation

TABLE 6.6 Morphologic Mimics of Usual Interstitial Pneumonitis/Idiopathic Pulmonary Fibrosis

Chronic aspiration with bronchiectasis and scarring

Chronic hypersensitivity pneumonitis

Pneumoconioses (asbestosis)

Fibrotic sarcoidosis

Chronic Langerhans' cell histiocytosis

Radiation fibrosis

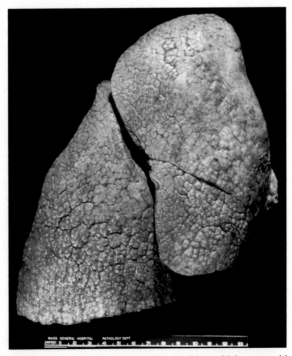

FIGURE 6.2 "Pulmonary cirrhosis" in usual interstitial pneumonitis.

diffuse. This means that microscopically the pathologist can find areas of near normal lung intervening between areas of fibrosis. The disease is peripheral and is distributed immediately subjacent to pleural or septal surfaces where alveolar acini normally terminate (Fig. 6.3A—C). The distribution of disease is predominantly in the basilar portions of the lung and most severe adjacent to the diaphragmatic surfaces. This is an important feature on HRCT where evidence of honeycomb cystic lung is almost always present at the bases. However, rare cases meet the criteria for UIP/IPF with changes that are dominant in the upper lung zones. Many of these may be unrecognized cases of chronic hypersensitivity pneumonitis. There is an unusual predilection for the subpleural lung to show fatty metaplasia in UIP (Fig. 6.4A) and for the peri-bronchial lung to show smooth muscle proliferation (Fig. 6.4B). These changes are generally not often seen in other interstitial lung disorders. Finally, the disease is temporally heterogenous, which means that one can find both areas of established scarring with architectural distortion together active areas of interstitial *fibroblastic foci* (Fig. 6.5).

When radiologists examine HRCT scans for evidence of UIP, they apply many of the same criteria as the pathologist. They are looking for patchy subpleural areas of scarring with architectural distortion, as well as for evidence of "end-stage" honeycomb lung. In my experience, the criterion used by many radiologists to diagnose honeycomb lung ($\geq$3 cystic spaces stacked together) is overly conservative and tends to underdiagnose a substantial number of cases. Fibroblastic foci are too small to be localized by radiographic techniques.

End-stage honeycomb lung is a descriptive term and its changes can be seen in a variety of disorders but are virtually always required before a diagnosis of UIP/IPF can be established. Honeycomb lung shows stacking of subpleural cysts that resemble a beehive honeycomb with architectural distortion as evidenced by traction bronchiectasis. Unfortunately, there is currently no consensus among pathologists as to what the earliest features of end-stage "honeycomb lung" are.

A consideration of pulmonary embryogenesis may be helpful in this regard. Embryologically, the gas exchanging alveolated acinus is the last element to differentiate in normal lung development. This process is reversed in UIP via what can be conceived of as a fibrotic "dying back" of the alveolated acinus to recapitulate earlier simplified lung structure.

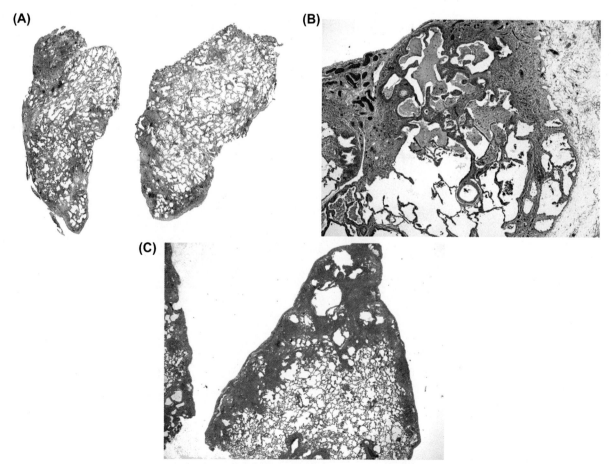

FIGURE 6.3   (A) Patchy subpleural fibrosis and cysts in sectioned lung of usual interstitial pneumonitis (UIP), (B) cystic bronchiolectasis and intra-lobular scarring (Honeycomb lung), and (C) young man with UIP pattern of lung injury attributable to myositis syndrome.

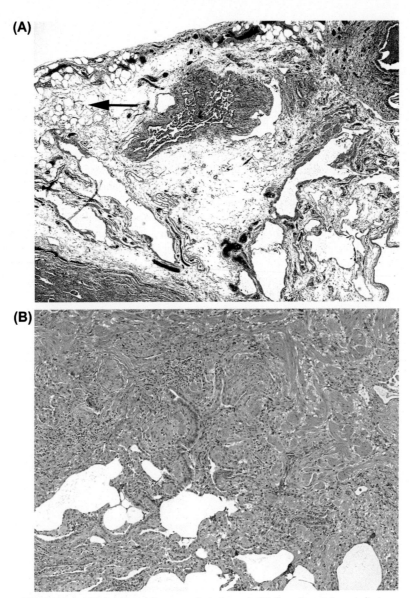

FIGURE 6.4 (A) Subpleural fatty metaplasia (*arrow*). (B) Smooth muscle metaplasia.

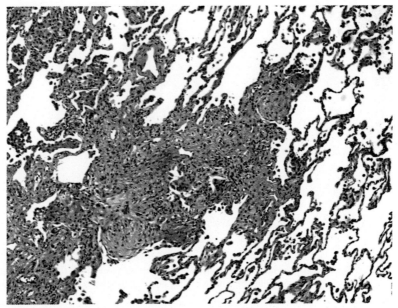

FIGURE 6.5 Multiple fibroblastic foci in usual interstitial pneumonitis with rapid functional decline.

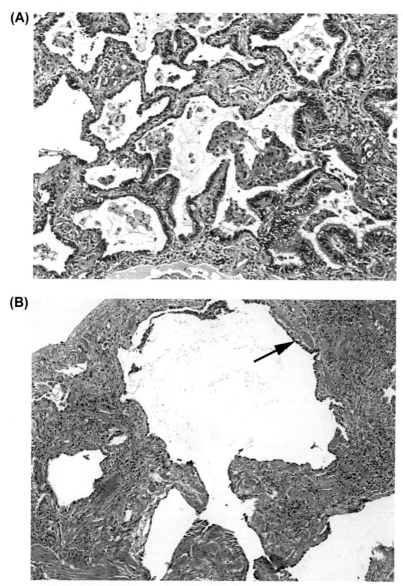

FIGURE 6.6 (A) Honeycomb lung in usual interstitial pneumonitis lined by bronchiolar epithelium and (B) cystic space lined by low cuboidal epithelium on patient with autoimmune fibrotic nonspecific interstitial pneumonitis.

"Honeycomb lung" is a terminal bronchiole that has become dilated and distorted by surrounding fibrosis, i.e., terminal bronchiolectasis. The absence of gas exchanging alveoli with bronchiolectasis and intralobular fibrosis is the best microscopic definition of "honeycomb lung."

As acini are increasingly lost, the cystic bronchioles "stack" up to give the gross appearance of a honeycomb that can be detected radiographically. These cystic spaces should be lined by respiratory epithelium although foci of cuboidal epithelium may coexist. However, when cysts are lined only by cuboidal epithelium without bronchiolar differentiation, the diagnosis of UIP should be made with caution, as there is frequently another etiology, such as autoimmune disease (Fig. 6.6A and B).

When lungs in UIP are extensively sampled, one detects, in approximately 25% of cases, changes that do not meet all of the diagnostic criteria set forth above, and instead shows areas that are best characterized as "nonspecific interstitial pneumonitis (NSIP), fibrotic type." However, if the characteristic features of UIP are found in any of the lung biopsies, the proper diagnosis is UIP. The fact is that NSIP, fibrotic type, even when it is the sole feature, has a poor prognosis and most patients are also dead within 10 years, although the short-term prognosis is better than for UIP.

The pathogenesis of UIP/IPF is unknown. Researchers have not yet discovered an animal model that mimics this disease. Bleomycin-induced fibrosis, perhaps the most frequently adopted animal model, recapitulates interstitial fibrosis, but does not show the characteristic distribution of the disease or temporal heterogeneity, and may ultimately be of limited value in the study of this disorder.

Research indicates that at least some patients with IPF/UIP have genetic changes in the MUC3 gene that encodes for a mucin product. It has been suggested that differences in airway pressures in the distal portion of the lung with tethering may account for the unusual distribution of this disorder. There are currently no clear molecular fingerprints for UIP that are potentially helpful in diagnosis.

Esophageal reflux and acid aspiration have been implicated in the progression of this disorder and all patients with ILD should receive proton-pump inhibitors, unless there is a contraindication to their use. The lung in UIP shows great variations with respect to interstitial inflammation, although up until now, there has been no evidence that anti-inflammatory regimens are helpful in most cases. In addition, it must be recognized that other diseases can at times histologically mimic UIP/IPF and clinical correlation may be required for accurate assessment.

MANAGEMENT

Despite the lack of evidence to support a role for corticosteroids in clinical trials, virtually all pulmonologists have anecdotal cases of individual who showed radiographic and pulmonary function improvement at least for a period of time, and it is wrong to adopt a nihilistic attitude toward corticosteroid therapy without a brief trial, especially if a biopsy shows inflammation and minimal end-stage disease. Two new drugs, pirfenidone, a TGF-beta antagonist, and nintedanib, and tyrosine kinase inhibitor, have recently been approved for the treatment of IPF, as they appear to slow the decrement in pulmonary function, although there is no mortality benefit. Whether these drugs have specific efficacy for IPF or may be therapeutic for a variety of fibrotic forms of interstitial disease is currently being investigated.

Attention to optimal pulmonary toilet is required. When cases of UIP are examined pathologically many show evidence of acute mucopurulent inflammation in terminal cystic spaces. Whether this represents low-grade infection or noninfectious inflammation is uncertain, but it does suggest that increased attention to airways inflammation, mucociliary clearance, and infection could potentially slow the progression of disease. Attention to microvascular pathology may also help in the management of these patients and should be regularly reported by pathologists. Histological evidence of pulmonary vascular remodeling in UIP/IPF is variable. CD31 immunostains can be helpful in estimating the degree to which the microvascular has been lost (Fig. 6.7A and B). The appropriate use of oxygen can help to reduce pulmonary vascular resistance in areas of hypoxic lung. Drugs used for the treatment of idiopathic pulmonary hypertension do not appear to be efficacious in this setting, likely due to different mechanistic pathways. The bronchial circulation is increased in UIP, as the lung is remodeled to produce bronchiectatic changes.

It appears that many patients do not decline gradually, but as a result of punctuated exacerbations. Whether these are part of the natural history of the disease, a complication of intercurrent airways infection, or microaspiration of gastric acid, is uncertain. The patient presents with increased dyspnea, hypoxemia, and ground glass infiltrates superimposed on the reticular pattern of fibrosis. The pathology of the lung in accelerated phase shows diffuse alveolar damage, fibrinous pneumonia, and/or organizing pneumonia (Fig. 6.8). A high ratio of fibrin and hyaline membranes to organizing pneumonia in the lung biopsy is a poor prognostic factor. Patients with primarily organizing pneumonia as part of the acceleration phase appear to respond better to corticosteroid intervention. There does not currently appear to be a role for anticoagulation, although this topic bears further investigation.

It is worthwhile noting that virtually all of the diffuse interstitial fibrotic disorders are capable of entering an accelerated phase. This has been documented for UIP, NSIP, and asbestosis. It appears that the once injured lung is at an increased risk for these events. An important question that remains to be answered is whether smoking-related fibrosis and emphysema (see chapter on smoking related disorders), commonly seen in heavy smokers, may also be susceptible to exacerbations that present as idiopathic acute lung injury.

NONSPECIFIC INTERSTITIAL PNEUMONITIS

This pathological syndrome was initially adopted as a generic term to refer to those nonpathognomonic cases of interstitial fibro-inflammatory disease which were *not* UIP. However, since the introduction of the term, it has been demonstrated to be a more specific pathological syndrome that includes a variety of potentially treatable disorders. The majority of cases of NSIP are associated with autoimmune disorders. It tends to affect younger patients than UIP, and there is a higher incidence among women. The prognosis depends primarily on whether the lungs are primarily cellular and inflamed, or scarred, with mixed categories being the most common (Table 6.7). Some have suggested that the presence of autoimmune disease is a favorable prognostic factor regardless of the pathological appearance but this remains an area of some controversy (Table 6.8).

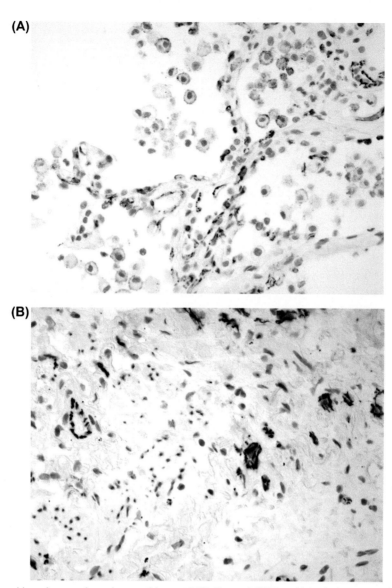

FIGURE 6.7 (A) CD31-positive microvasculature in normal lung and (B) disrupted CD31-positive microvascular endothelial cells (peroxidase).

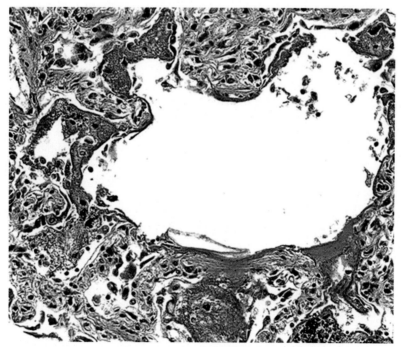

FIGURE 6.8 Hyaline membranes in accelerated usual interstitial pneumonitis.

TABLE 6.7 Features of Nonspecific Interstitial Fibrosis

Types
Cellular
Fibrotic
Mixed
Generally younger than patients with usual interstitial pneumonitis/idiopathic pulmonary fibrosis
Bibasilar distribution
Relative uniformity in appearance of age of lesions
Fibroblastic foci are rare
Minimal honeycombing but may show cysts lined by low cuboidal epithelium
Highly associated with autoimmune disorders and may be drug related or idiopathic
Comparable histology may be seen following recovery from acute respiratory distress syndrome
Cellular form has excellent prognosis
Pure fibrotic form has poor long-term prognosis
Fibrotic nonspecific interstitial pneumonitis changes can be seen in patients with classic features of usual interstitial pneumonitis seen in other
Biopsy fragments

TABLE 6.8 Autoimmune Interstitial Lung Disease

Virtually all patterns of lung injury can be seen in autoimmune disorders
Lung disease may precede systemic involvement or may persist as isolated disease
NSIP is most common followed by usual interstitial pneumonitis (RA, scleroderma)
Presence of frequent interstitial B-cell follicles and germinal center formation suggest an autoimmune disorder
Multicompartment involvement, i.e., lung, airways, vessels, pleura, favors autoimmune etiology
Vasculitis and pulmonary hemorrhage may be present
Obliterative vasculitis with pulmonary hypertension can be seen in scleroderma, SCL-70, and
Mixed connective tissue disorders

The distinction between NSIP and UIP is generally clear but there can be problematic overlaps, as has already been noted. Similar to UIP, most cases of NSIP show a basilar distribution of ground glass opacities. On HRCT scanning, one may see a small zone of lung sparing directly beneath the pleura, which is never present in UIP. Unlike UIP, the lung biopsy in NSIP shows a largely homogeneous appearance, in which virtually all areas of the biopsy show comparable changes and age. The presence of end-stage honeycomb lung is unusual as are large numbers of fibroblastic foci. When these findings are present, the diagnosis of UIP should be entertained and preferred, particularly in older patients. Not unexpectedly, patients with an indeterminate CT appearance are more likely to have histological changes that are likewise nondiagnostic.

Cellular NSIP is composed of infiltrates of lymphocytes, macrophages, and plasma cells (Fig. 6.9A). There may be focal areas of alveolar filling by nonpigmented macrophages. The cellular variant of idiopathic NSIP has an excellent response to corticosteroids and almost never leads to death. As might be expected, the mixed inflammatory/fibrotic cases (Fig. 6.9B) do somewhat worse and purely fibrotic forms have a 5- to 10-year survival that parallels that of UIP.

Virtually all of the autoimmune disorders can produce an NSIP pattern, but there are certain histological features that suggest an autoimmune etiology. These include follicular lymphocytic inflammation that includes large numbers of CD20+ B lymphocytes with cuffs of CD3+ T cells. CD20+ B cells are rarely found individually in the lung and instead are almost

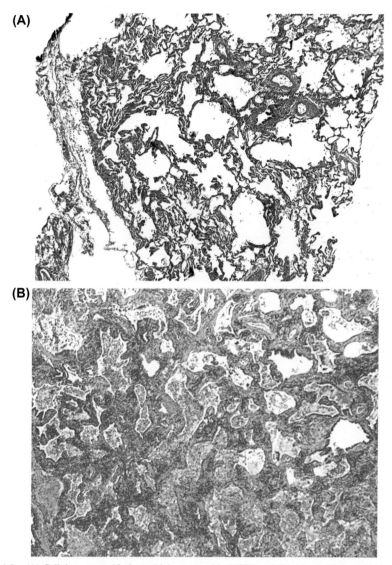

FIGURE 6.9   (A) Cellular nonspecific interstitial pneumonitis (NSIP) and (B) mixed cellular and fibrotic NSIP.

always seen within interstitial follicles (Fig. 6.10). Follicular germinal center formation may be present and this strongly suggests an autoimmune etiology.

In cases of rheumatoid disease and Sjogren's syndrome, the interstitium shows both lymphoid follicles and increased numbers plasma cells (Fig. 6.11). There is a propensity for the pulmonary microvasculature and the pleura to be involved by lymphoid inflammation in autoimmune disease. When multiple lung compartments are involved, one can be confident of an autoimmune etiology. Finally, as previously mentioned, cases with a UIP pattern and cystic spaces lined by cuboidal epithelium often prove to have an underlying autoimmune disease.

The diagnosis of an autoimmune disease is currently established noninvasively by the serological findings. Patients may show high titers of ANA or a positive rheumatoid factor or anti-CCP. In those with a positive ANA, further testing for ds-DNA, ENA, anti-CCP, SCL-70, anti-Ro, anti-La, anti-Jo, and other anti-muscle antibodies, including MDA-5, that suggest the "anti-synthetase" syndrome is indicated before establishing a diagnosis of IPF. In some cases, an elevated aldolase level can be present without a recognized anti-muscle serology. BAL should be sent for serologic analysis if it is performed, as the lung may be a primary source for autoantibody production.

It is noteworthy that some patients with anti-Ro, SCL-70, and myositis may show normal or low-level ANAs, so the absence of an elevated ANA does not exclude autoimmune disease. Recently, it has been recognized that pANCA mediated disease with MPO specificity can produce NSIP in the absence of vasculitis or pulmonary alveolar hemorrhage. Of course, many patients with autoimmune disease will also show extrapulmonary stigmata but an isolated pulmonary

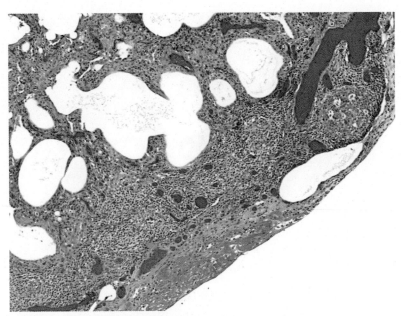

FIGURE 6.10 Lymphoid follicles in autoimmune lung disease.

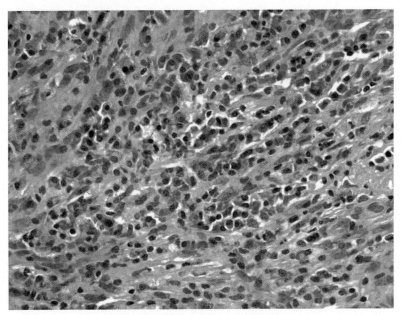

FIGURE 6.11 Plasma cells in rheumatoid lung.

presentation is sufficiently common to consider the spectrum of autoimmunity in the absence of associated extrapulmonary symptoms. On the other hand, a low-level positive ANA ($\leq$1:160) is highly nonspecific and should not be used to support a diagnosis of autoimmune disease. Other cases of nonautoimmune etiology may produce the pathologic findings of NSIP, including hypersensitivity pneumonia, drug hypersensitivity, and pneumoconiosis. A small percentage of cases ultimately prove to be idiopathic.

Based on this, some pulmonologists will argue that once the diagnosis of an autoimmune disease has been comfortably established there is no need for lung biopsy. However, an argument for lung biopsy can be made based on prognostic and therapeutic considerations, considering the current broader choice of therapeutic options. Finding a pattern of UIP or even substantial interstitial fibrosis suggests the possibility of a less robust response to immunosuppressive therapies. The presence of large numbers of CD20[+] lymphocytes or CD138[+] plasma cells in the biopsy may argue for a therapeutic approach that includes monoclonal antibodies directed against these cells (rituximab). However, these arguments although

TABLE 6.9 Rheumatoid Lung

Pleuritis with effusion (high WBC, low glucose, low pH)

Interstitial pneumonitis (nonspecific interstitial pneumonitis or usual interstitial pneumonitis)

Organizing pneumonia

Obliterative bronchiolitis

Rheumatoid nodules

Vasculitis

well-reasoned remain largely untested and appropriate clinical trials will be required to establish optimal therapeutic approaches based on the histology of lung disease.

Other patterns of lung disease are peculiar to certain collagen vascular diseases (Table 6.9). Patients with rheumatoid disease may develop pulmonary nodules that can range greatly in size and cavitate (Fig. 6.12). The pathology shows a serpiginous necrotizing process lined by granulomatous inflammation with large numbers of plasma cells. The lining of the necrotic center often shows epithelioid foamy macrophages lining up orthogonally with the center, a change that termed palisading granulomatous inflammation.

In patients with silicosis, silica and silicate crystals and hyalinizing fibrosis may be seen in the wall of the rheumatoid nodule, a phenomenon referred to as Caplan's syndrome. In this regard, it is noteworthy that silica has been implicated as an immune stimulating factor for a variety of autoimmune reactions. The pathology of rheumatoid nodules can also be seen in granulomatous polyangiitis and can closely resemble mycobacterial and fungal infections, all of which should be excluded.

Pulmonary hemorrhage is often seen in patients with systemic lupus erythematosus (Table 6.10). This may be due to alveolar immune complex deposition or to anti-phospholipid antibodies with a functional lupus anticoagulant. *Capillaritis* is most often seen with systemic lupus erythematosis (SLE) but can also complicate other autoimmune disorders (Fig. 6.13). Findings in capillaritis may be diffuse and obvious or focal and subtle. One looks for evidence of alveolar hemorrhage and exudate fibrin in the alveolar spaces, but the definitive diagnosis is established by identifying necrosis of the alveolar walls with neutrophilic inflammation and breakdown of granulocyte nuclei, i.e., *karyorrhexis*. In institutions prepared to perform immunofluorescence microscopy and ultrastructural examination, biopsies of lung tissue should be harvested and frozen fresh in cryomedium and/or placed in glutaraldehyde fixative for detailed studies, as these approaches will establish an immune complex etiology and argue in favor of an approach that includes rituximab, and/or plasmapheresis. Immunofluorescence shows granular deposits of Ig and C3 complement in the vessel walls and subendothelial granular complexes may be detected ultrastructurally.

The treatment of diffuse capillaritis constitutes a medical emergency as the disease can progress rapidly to respiratory failure. Systemic vasculitis may be present but not invariably so. Treatment includes immunosuppression and plasmapheresis if circulating immune complexes are implicated in the pathogenesis.

The use of bronchoalveolar lavage to establish the diagnosis of diffuse pulmonary alveolar hemorrhage (PAH) has become routine but has limitations. An increase in the DLCO is an insensitive test for diagnosing PAH. Cases of PAH have been seen in biopsies in which repeated lavagates have failed to demonstrate the characteristic progressive increase in blood, a phenomenon attributable to an initial washing out of anatomic *dead space*, followed by sampling of the distal source of alveolar hemorrhage. Furthermore, noninvasive approaches cannot distinguish pulmonary hemorrhage from capillaritis, which may be an important distinction for therapeutic purposes. The presence of increased neutrophils and nuclear "dust" with extravasated red blood cells and hemosiderin in the lavagate may be an indicator of inflammatory hemorrhage.

An unusual "ILD" seen in patients with SLE is so-called "disappearing lung." Here, the lung volumes are diminished with a relative preservation of the DLCO/VA ratio. The precise etiology is unknown; however, diaphragmatic weakness may play a role. Biopsies in such cases show chronic atelectasis with an otherwise near-normal lung.

Scleroderma exhibits interstitial scarring that is best described as pauci-inflamed (Fig. 6.14). UIP patterns are common, although the lining epithelium of "honeycomb" cysts is nonbronchiolar. In all cases of interstitial disease, and most particularly with those along the scleroderma spectrum, including anti-SCL 70 associated disorder, a careful examination of the pulmonary vasculature is indicated in parallel with estimations of pulmonary arterial pressures by ECHO cardiography or right heart catheterization. In patients with scleroderma, left heart disease due to myocardial fibrosis may be present.

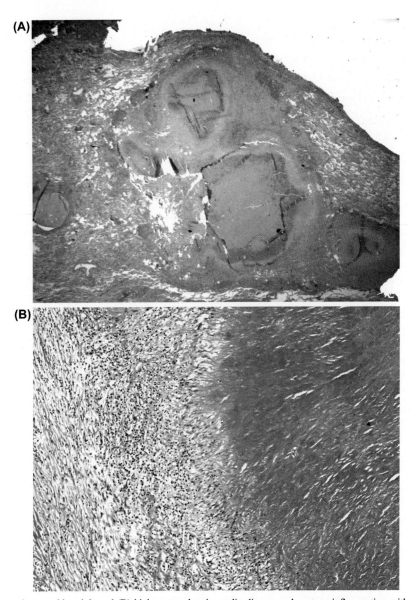

FIGURE 6.12 (A) Cavitary rheumatoid nodule and (B) high power showing palisading granulomatous inflammation with macrophages orthogonally arranged with respect to the necrotic center.

TABLE 6.10 Systemic Lupus Erythematosus
Pleuritis
Interstitial pneumonitis (nonspecific interstitial pneumonitis)
Capillaritis
Antiphospholipid syndrome
Disappearing lung disease

The pulmonary arterial bed is remodeled in interstitial disease primarily due to a loss of pulmonary alveolar vasculature. This can be determined in situ by virtue of immunostains directed at vascular antigens, e.g., CD31. However, in scleroderma and other collagen vascular diseases one may also see active remodeling of precapillary muscular arterial vessels. These small arteries show intimal onion skin—like proliferation of the intima with or without attendant vascular

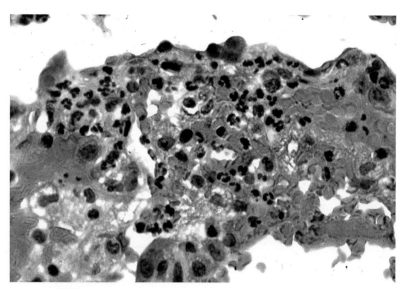

FIGURE 6.13   Pulmonary capillaritis in systemic lupus erythematosis.

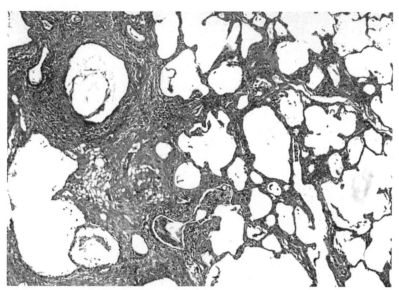

FIGURE 6.14   Pulmonary fibrosis in scleroderma.

wall chronic inflammation (Fig. 6.15). This finding can result in high-grade pulmonary hypertension clinically and is a poor prognostic sign. Many cases are associated with the syndrome of calcinosis, Raynaud's, esophageal dysmotility, sclerodactyly, and telangiectasia (CREST) with or without diffuse scleroderma, pulmonary fibrosis, and renal involvement.

Patients with systemic scleroderma have other extrapulmonary abnormalities that can complicate the evaluation of ILD. Myocardial fibrosis can lead to a reduced left ventricular ejection fraction and congestive failure. Esophageal fibrosis may reduce the tone of the lower esophageal sphincter leading to chronic aspiration. In general, it is well advised to treat all patients with chronic ILD with protein-pump inhibitors. If gastroesophageal incompetence is severe, fundoplication is indicated.

Secondary Sjogren's disease and ankylosing spondylitis show overlapping features with rheumatoid disease but can show distinct features. In addition to lymphocytic interstitial pneumonitis (LIP) that will be discussed below, Sjogren's frequently shows inflammation of the large and small airways that leads to a chronic atrophic bronchitis/bronchiectasis syndrome (Fig. 6.16). This reflects an autoimmune response directed at the minor salivary glands within the airways. Organizing pneumonia may be part of the spectrum of airways disease in Sjogren's or an indicator of reflux-induced injury. Cysts may develop in the lung.

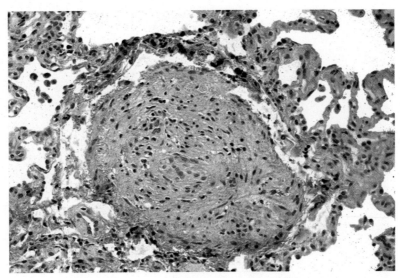

FIGURE 6.15   Obliterative intimal change in the CREST syndrome.

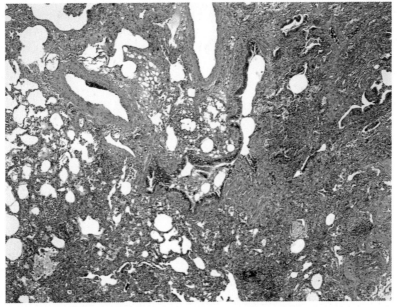

FIGURE 6.16   Airway changes in Sjogren's syndrome.

Ankylosing spondylitis has a propensity to produce fibrotic changes in the upper lobes near the apices (Fig. 6.17). The scarring is confluent and may be associated with emphysematous bullae. The pathogenesis is uncertain and whether it reflects mechanical-induced changes in ventilation due to the spinal deformity or direct pulmonary involvement by the autoimmune process is uncertain.

LYMPHOCYTIC INTERSTITIAL PNEUMONITIS

This is a disorder that has fallen out of favor with respect to the systemic classification of the ILDs. It is characterized by an exquisitely interstitial infiltrate of lymphocytes and plasma cells (Fig. 6.18). However, invasion of airways or blood vessels by lymphoid cells indicates a lymphoproliferative process that is likely malignant and should not be seen in LIP. Small microgranulomas occur seen in ~20% of cases.

The disorder is generally associated with an immune deficit, either secondary to HIV infection or in patients with dysgammaglobulinemias, and virtually any of the autoimmune disorders, but most frequently Sjogren's syndrome. The

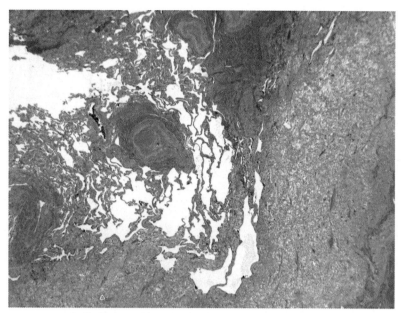

FIGURE 6.17 Upper lung fibrosis in ankylosing spondylitis.

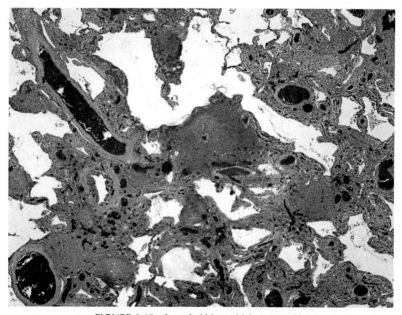

FIGURE 6.18 Lymphoid interstitial pneumonitis.

histological appearance can be difficult to distinguish from lymphoma, particularly mucosal associated lymphomas. In cases where monoclonality cannot be established but is suspected, PCR sequencing of Ig can at times identify a hidden clone that is neoplastic. The differential diagnosis includes cellular NSIP, sarcoidosis, and hypersensitivity pneumonitis. In cases that have been serially biopsied plasma cells can constitute the entire inflammatory infiltrate.

HYPERSENSITIVITY PNEUMONITIS

Hypersensitivity responses to inhaled antigens, classically the *thermophilic actinomyces* e.g., farmer's lung, and pigeon droppings or feather (Pigeon fancier's lung) can cause a variety of changes in the lung depending of the dose and frequency of exposures (Table 6.11). Clinically, patients can present with wheezes, dyspnea, and pulmonary infiltrates within hours of

TABLE 6.11 Causes of Hypersensitivity Pneumonitis

Name	Agent
Pigeon breeders/Bird fanciers	Avian proteins in feathers and droppings
Mold	Aspergillus and others
Humidifier	Thermophilic actinomycetes
Hot tubs	Mycobacterium avium
Bagassosis	Moldy sugar cane
Chemical workers	Isocyanates
Coffee workers	Coffee bean dust
Maple bark	Cryptospora corticale
Mushroom workers	Mushroom compost
Suberosis	Cork dust
Wood pulp	Pine and redwood molds

exposure to the offending antigen. The distribution of hypersensitivity pneumonitis (HP) (Table 6.12) is most marked in the upper lung zones radiographically and pathologically (Table 6.13). The histology of acute hypersensitivity pneumonitis is not well characterized but appears to be a fibrinous pneumonia (Fig. 6.19). However, the disease is rarely biopsied in this phase, as most symptoms will resolve once the offending antigen has been removed.

The most commonly observed and characteristic form of the disease is its subacute phase. This produces a microgranulomatous interstitial pneumonitis that is centrilobular in distribution reflecting the aerogenous deposition of antigen (Table 6.12). The small airways can show lymphohistiocytic inflammation or organizing pneumonia. There is virtually always a lymphocytic interstitial infiltrate admixed with macrophages that fans out from the airways (Fig. 6.20).

Intra-alveolar macrophages with foamy lipid-laden appearances are present indicating small airways obstruction (Fig. 6.21). A careful examination will show areas of distended alveoli consistent with air trapping, which is characteristically seen on expiratory HRCT scans. Lambertosis of the small bronchioles may be present. One can identify small aggregates of epithelioid macrophages, some of which contain cholesterol clefts in at least 85% of cases. This constellation of findings is considered pathologically to be diagnostic of hypersensitivity pneumonitis, although in nearly half of cases,

TABLE 6.12 Hypersensitivity Pneumonitis

Types
Acute
Subacute
Chronic
Upper lobe predominance
Centrilobular distribution
Air-trapping on expiratory phase of HRCT
Acute form rarely biopsied but shows fibrinous pneumonitis
Subacute form may show nonspecific interstitial pneumonitis or lymphohistiocytic inflammation with microgranulomas
Chronic hypersensitivity pneumonitis shows features that may be impossible to distinguish from UIP, except for rare
Microgranulomas, and air-trapping on HRCT

TABLE 6.13 Diseases With Upper Lobe Predominance

Ankylosing spondylitis

Centrilobular emphysema

Chronic mycobacterial and fungal infections

Cystic fibrosis

Hypersensitivity pneumonitis

Pleuropulmonary fibrosis

Respiratory bronchiolitis-interstitial lung disease

Sarcoidosis

Silicosis

Smoking-related fibrosis

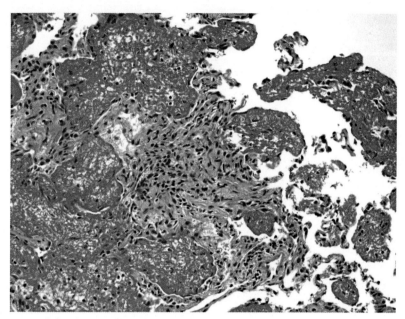

FIGURE 6.19   Fibrinous pneumonia in acute hypersensitivity pneumonitis.

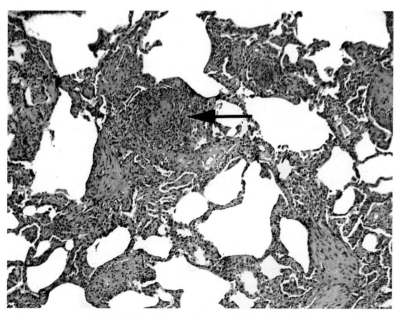

FIGURE 6.20   Subacute hypersensitivity pneumonitis.

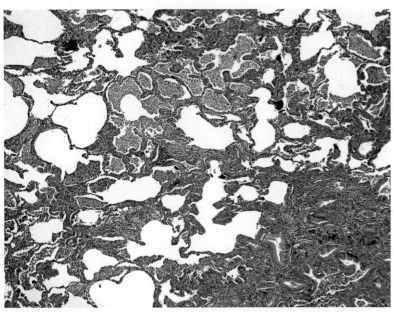

FIGURE 6.21 Intra-alveolar foamy macrophages in hypersensitivity pneumonitis are a marker of small airways obstruction.

an offending antigen is never determined by currently available serological panels, so that the diagnosis remains presumptive.

It is also worth noting that most cases do not show increased tissue eosinophilia, as the response is not IgE driven but instead appears to represent a combination of immune complex and cell-mediated immunity. When tissue eosinophils are present with the diagnostic features of HP, it suggests that the reaction may cause allergy to a hyphate mold, e.g., *Aspergillus spp*.

A subset of cases with hypersensitivity pneumonitis show either NSIP or LIP as their underlying pathologies and in such cases clinical and serological features are required to make the diagnosis.

Removal from the offending antigen is optimally required but may be difficult. Household contaminations, e.g., bird feather antigens, may persist even after what appears to be a thorough housecleaning, as hypersensitivity responses require only nanograms of antigen to evoke responses. Effective management at times may require the patient to change residences, which they understandably may be reticent to do.

The classical teaching was that patients with HP show increased numbers of $CD8^+$ lymphocytes in their bronchoalveolar lavagates. However, subsequent studies have shown that this is not invariably the case.

In its chronic phase, HP closely mimics the pathology of UIP although there may be (1) upper lobe predominance, (2) centrilobular predominance in areas, (3) evidence of air trapping on expiratory CT scans, and (4) residual microgranulomas seen microscopically (Fig. 6.22) (Table 6.14). The latter may be sparse or absent altogether. Although patients diagnosed with chronic HP should be removed from an environmental source if recognized, their long-term prognosis parallels that of patients with IPF with high 5-year mortality.

Finally, a cautionary note is worthy of mention. There is an increasing tendency among radiologists to suggest the presence of air trapping, in the presence of interstitial disease of *prima facie* evidence of hypersensitivity pneumonitis. This observation has some merit, particularly if there is clear upper lung zone predominance but also has limitations as other interstitial disorders that involve the small airways can produce comparable changes.

DIFFUSE ALVEOLAR DAMAGE/ACUTE INTERSTITIAL PNEUMONIA

Diffuse alveolar damage (DAD) is the pathological syndrome associated with the acute respiratory distress syndrome (ARDS). The term DAD is generally applied with two caveats: (1) the changes diffusely involve the lung, (2) an etiology can be identified, most commonly sepsis, infection, inhalational trauma, or hypovolemic shock. Currently, the term DAD should be avoided in the idiopathic setting. Patients present within hours to days of the etiological insult.

The pathologic changes may be divided into two phases. The exudative phase occurs within the first 72 h of the insult. It is characterized by a profound increase in capillary permeability (Fig. 6.23A). There is alveolar filling by proteinaceous

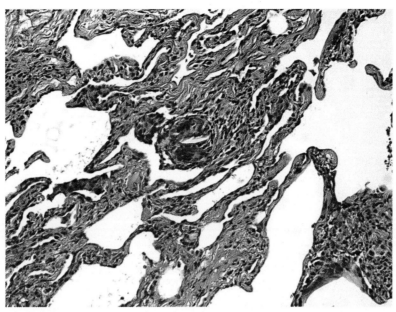

FIGURE 6.22   Chronic hypersensitivity pneumonitis showing rare microgranuloma.

TABLE 6.14 Diseases With Centrilobular Predominance and Tree-in-Bud* Radiographic Appearance

Airway-centered interstitial fibrosis
Acute and chronic bronchiolitis
Aspiration
Hypersensitivity pneumonitis
Langerhans' cell histiocytosis
Respiratory bronchiolitis-interstitial lung disease
Mycobacterial infection

edema as well as multifocal alveolar hemorrhage and neutrophil infiltration. The *sine qua non* of acute alveolar wall damage is the *hyaline membrane* (Fig. 6.23B). It is composed of necrotic alveolar lining cells and fibrin that remain tightly apposed to the injured alveolar surfaces. They are most prominent identified along the alveolar ducts where oxygen tension is the highest.

After 72 h, the lung shows a wide number of changes related to the healing process. All compartments of the lung are affected. A chronic inflammatory infiltrate composed predominantly of lymphocytes and macrophages is seen, together with a marked increase in interstitial mast cells and eosinophils that peaks around 2 weeks postinjury. Exudate monocytes enter into the alveolar space and play an important role in cross-linking the fibrin clot to fibronectin, type III collagen, and other matrix proteins via their surface membrane expression of Factor XIIIa. Macrophages release monokines that modulate the cellular and fibrotic response (Fig. 6.24).

The reparative epithelial phase appears to arise from two distinct compartments. In some cases, one sees squamous metaplasia in and around terminal bronchioles. These squamoid cells express p40 but also retain expression of the alveolar type II cell epithelial antigen thyroid transcription factor-1. The alveolar walls are repopulated by a population of cuboidal keratin-positive alveolar type II cells (Fig. 6.25). The source of these cells is uncertain but may be an epithelial stem cell or cells from the damaged small airways.

Critical to the optimal management of DAD/ARDS is the recognition that edema fluid is entrapped in the lung by a diffuse fibrin gel that cannot be easily mobilized by diuresis. For this reason, perfusion pressures should be maintained with near normal wedge pressures if possible and overdiuresis, which can potentially deplete intravascular volume and decrease right heart filling pressures, avoided.

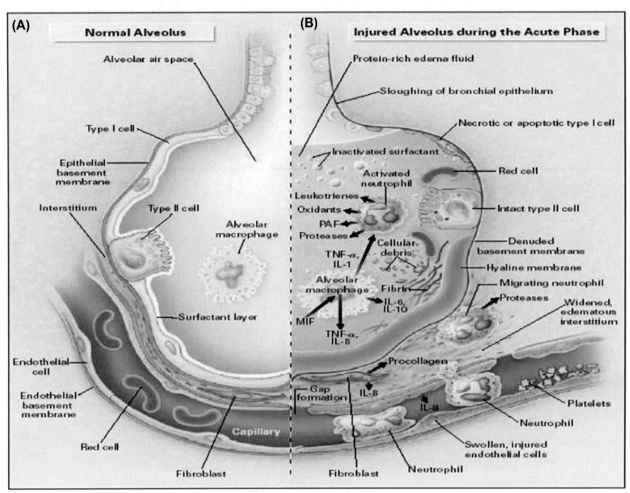

FIGURE 6.23 (A) Pathways of acute lung injury and (B) hyaline membranes in diffuse alveolar damage.

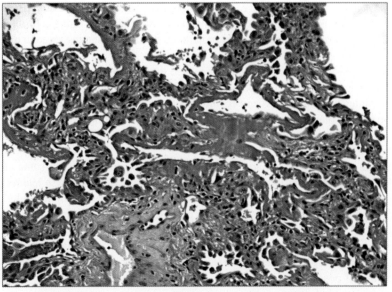

FIGURE 6.24 Early fibroproliferative phase of diffuse alveolar damage.

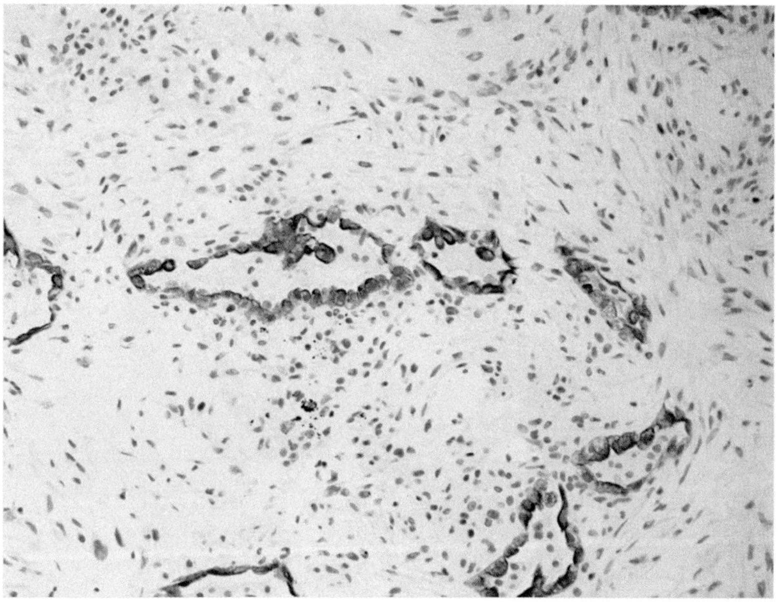

FIGURE 6.25   Keratin-positive alveolar type II cell hyperplasia in diffuse alveolar damage.

The ulcerated alveolar walls form *synechiae*, i.e., microadhesions that are subsequently lined by proliferating type II alveolar cells. These changes result in architectural simplification of the lung (Fig. 6.26). The pulmonary alveolar vascular bed is markedly reduced and can fail to fill due to intrapulmonary pressures due to the application of positive end-expiratory pressures. The proximal pulmonary vessels undergo intimal proliferation and show extension of the vascular smooth muscle into precapillary arterioles.

The massive release of tissue procoagulant due to cellular injury leads to intravascular and extravascular thrombus formation. Almost 50% of patients show in situ thrombosis of large pulmonary arteries by angiography and 20% of cases have disseminated intravascular coagulation with microthrombi in vessels (Fig. 6.27). There is a translocation of CD61[+] megakaryocytes from the bone marrow into the pulmonary microvasculature presumably due to stress and increased lung endothelial adhesiveness (Fig. 6.28).

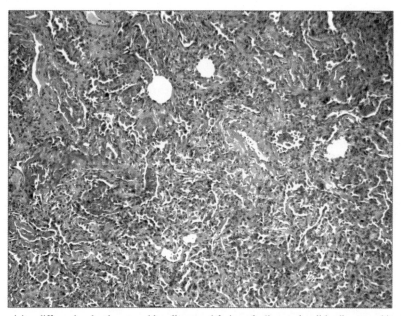

FIGURE 6.26   Organizing diffuse alveolar damage with collapse and fusion of adjacent alveoli leading to architectural simplification.

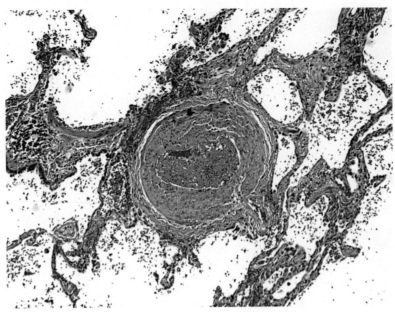

FIGURE 6.27 Intravascular thrombi in diffuse alveolar damage.

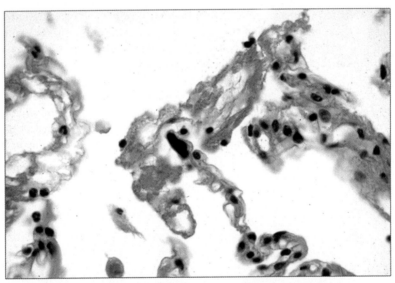

FIGURE 6.28 Megakaryocytes in alveolar capillaries in acute lung injury.

The proliferative phase of injury may proceed for weeks. Patients can die late in the disease from ventilatory complications with high peak airway pressures, increased dead space, and hypercarbia. Histological evidence of end-stage lung can be detected as early as 10 days after the onset of disease and must be distinguished from an exacerbation of an occult chronic interstitial disorder, at times with substantial difficulty.

Changes in ventilatory strategies have had a positive effect on DAD/ARDS. It has been convincingly demonstrated that low tidal volumes and permissive ventilation can reduce ventilator-induced stretch-induced injury in DAD. At the present time, mechanical ventilation and hemodynamic support remain the mainstays of treatment. However, extracorporeal membrane oxygenation may be used successfully to support these patients or as a possible bridge to lung transplantation.

Persistent controversy has surrounded the potential benefits of corticosteroids as an early treatment in ARDS. In a retrospective study at the MGH, we showed that a subset of patients with DAD who had a significant component (>25% of the biopsy) of associated organizing pneumonia may had a better overall outcome and apparent response to corticosteroids. Larger multicenter studies will be required to confirm this finding (Fig. 6.29).

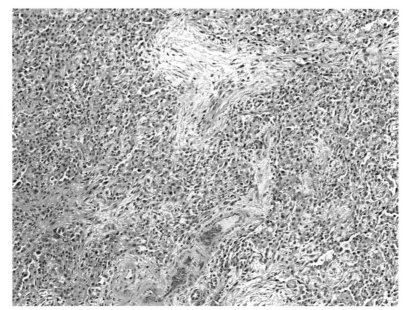

FIGURE 6.29   Diffuse alveolar damage with OP. Note presence of organizing pneumonia component.

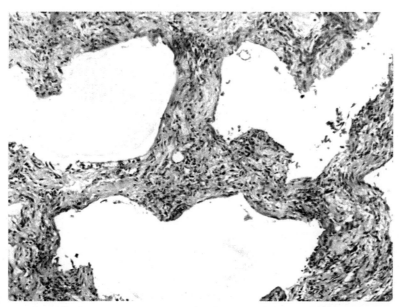

FIGURE 6.30   Post-diffuse alveolar damage showing the histology of fibrotic nonspecific interstitial pneumonitis.

Patients who survive ARDS can show near complete recovery of pulmonary flow rates and lung volumes, although radiographic changes suggesting interstitial scarring tend to persist and the DLCO may permanently diminished due to the loss of pulmonary microvasculature. The histology of the lung in such cases closely resembles the fibrotic form of NSIP (Fig. 6.30). Some patients post-ARDS can be maintained with low-dose corticosteroids with considerable success until their pulmonary function finally stabilizes, which may take more than a year.

ACUTE INTERSTITIAL PNEUMONIA

Acute interstitial pneumonia is the term given to lung disease that pathologically shows the histology of the fibroproliferative phase of DAD but with no identifiable etiology (Fig. 6.31). Typically, the clinical presentation includes a viral-type prodrome and weeks to months of progressive dyspnea on exertion with cough accompanied by diffuse lung infiltrates. In

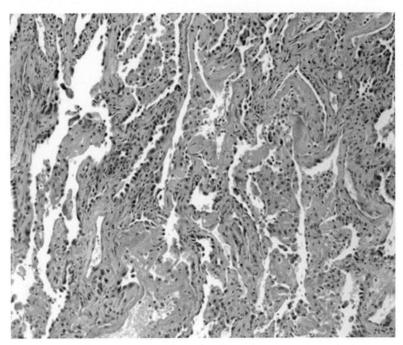

FIGURE 6.31   Acute interstitial pneumonitis is histologically equivalent to the fibroproliferative phase of diffuse alveolar damage.

the past, this was termed the Hamman–Rich syndrome. It is not known why these patients show a protracted course compared to the rapid progression seen in ARDS despite comparable histological changes. However, the prognosis is overall poor with the majority progressing to respiratory failure. Intravascular thrombi are common and should not be misinterpreted as thromboembolic disease.

DIFFUSE ALVEOLAR INJURY WITH DELAYED ALVEOLAR RE-EPITHELIALIZATION

An as yet idiopathic disorder has recently been recognized in two young patients with ARDS. Both presented with a "viral" syndrome, diffuse pulmonary infiltrates, and acute lung injury that ultimately required extracorporeal membrane oxygenation. Extensive efforts to exclude infection and toxic exposures were negative.

The biopsies in both cases showed diffuse alveolar injury (Fig. 6.32A) with a near complete loss of alveolar epithelial cells, edema, mild interstitial fibrosis, and sparse hyaline membranes. The peribronchiolar regions exhibited exuberant squamous metaplasia and Lambertosis (Fig. 6.32A, arrows). The alveolar walls showed a complete absence of keratin-positive regenerating epithelial lining cells, which are seen in the fibroproliferative phase of DAD. Alveoli were instead lined exclusively by CD68[+] macrophages with cuboidal and squamoid morphologies, as evidenced both by immuno-phenotyping and ultrastructural analysis (Fig. 6.32B). The explanted lung of one patient who subsequently received a lung transplant following 3 months on ECMO showed diffuse mild interstitial fibrosis and an exuberant proliferation of alveolar type II pneumocytes (Fig. 6.33). The cause of the delayed re-epithelialization is uncertain and is currently being investigated.

OTHER POORLY DEFINED ENTITIES

In Liebow's original description of the interstitial pneumonias, he referred to a disorder that he termed bronchiolitis-interstitial pneumonitis. Many of these cases would today be classified as BOOP. However, there are cases in which the dominant pathology is a cellular or mixed interstitial infiltrate that shows abundant small areas of either bronchiolar obliteration or fibroblastic foci (Fig. 6.34). Some cases have identifiable etiologies, whereas others are idiopathic. The overall prognosis and response to steroids is good. This pathological entity is not currently recognized as a separate entity by the American Thoracic or European Respiratory Societies.

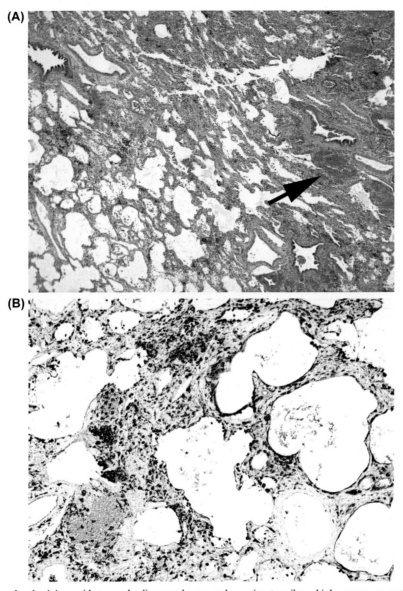

FIGURE 6.32 (A) Diffuse alveolar injury with sparse hyaline membranes and prominent peribronchiolar squamous metaplasia (*arrow*) in delayed alveolar re-epithelialization and (B) absence of keratin-positive alveolar cells with alveoli lined instead exclusively by CD68⁺ macrophages.

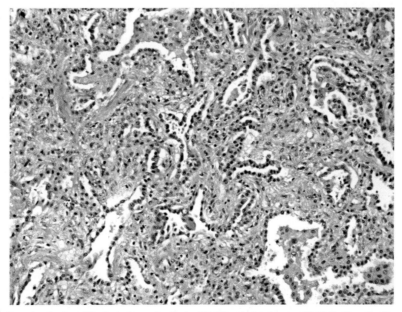

FIGURE 6.33 Keratin-positive alveolar type II cells in lung of patient with delayed alveolar re-epithelialization after four weeks of ECMO treatment suggesting delayed regeneration.

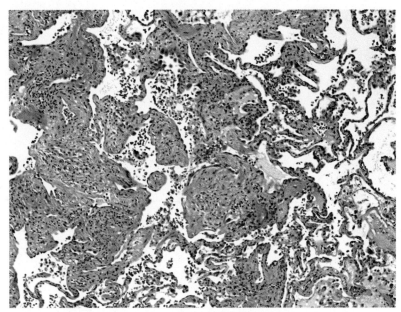

FIGURE 6.34 Fibroblastic foci with interstitial pneumonitis.

SARCOIDOSIS

Sarcoidosis is an idiopathic systemic disorder that tends to involve the lung and its regional lymph nodes but also affects extrathoracic sites and can present exclusively as extrathoracic disease (Table 6.15). Despite having a long and controversial history with respect to causation, it remains idiopathic. The most common presentation is a relatively young individual, more commonly women than men, with a high prevalence in African Americans and Scandinavians.

The disease presents as hilar and mediastinal adenopathy with lung involvement that may not be visible on normal chest radiographs but can frequently be identified by HRCT as nodules of various size with upper lung zone predominance. In 75% of cases, the disease resolves on its own. Currently, treatment is reserved for patients with symptomatic disease, lung function abnormalities, progressive radiographic findings, or extrapulmonary symptoms, including uveitis, hypercalcemia, congestive heart failure, cardiac arrhythmias, bony, neurological, and cutaneous disease.

Although considered a disease of exclusion, once mycobacterial and fungal infections have been excluded, there are histological features of sarcoidosis, both clinical and pathological, that suggest the diagnosis. Indeed, the incidence of positive stains for mycobacteria or fungi is vanishingly small, when the patient is not immunosuppressed for other reasons and there is no evidence of necrosis in the granulomas.

TABLE 6.15 Sarcoidosis
Non- or minimally necrotizing granulomatous inflammation
Lymphangitic patterns of infiltration
Upper lobe predominance
Hilar and mediastinal adenopathy but may affect any organ (skin, liver, posterior hypothalamus, heart, bone, spleen)
Peripheral blood lymphopenia with inverted $CD4^+/CD8^+$ ratio
Lymphocytic dominance in lavage with increased $CD4^+/CD8^+$ ratio
Pleural effusions (uncommon)
Diagnosis can be made by TBB in most cases
Most cases spontaneously regress
~20% progress
Therapy required only if symptomatic, decreasing pulmonary function (DLCO), or clear anatomic progression
Cardiac sarcoidosis diagnosis may be made by PET or MRI scanning
Hypercalcemia due to increased gut absorption through 1-hydroxyation of vitamin D precursor by macrophages

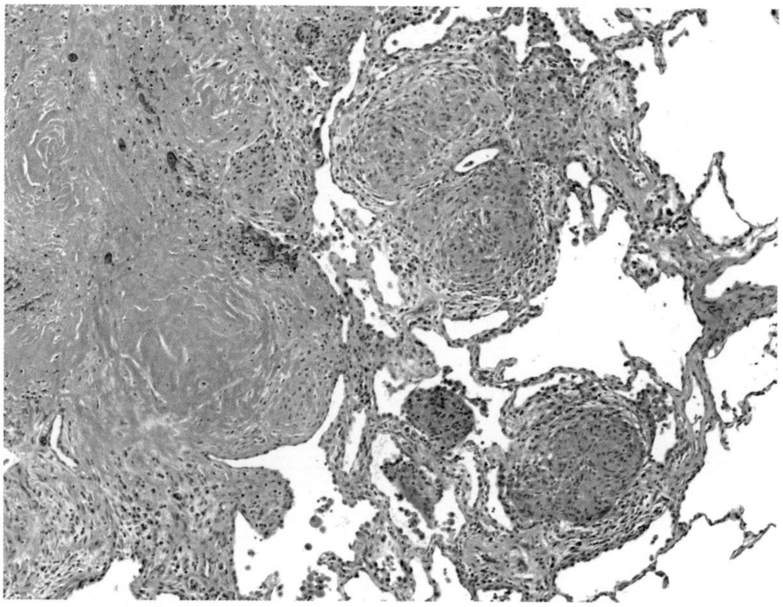

FIGURE 6.35   Lymphatic distribution of granulomas in sarcoidosis.

The *sine qua non* is the presence of nonnecrotizing granulomas in regional lymph nodes or distributed along lymphatic pathways in the lung (Fig. 6.35). This may be visualized by HRCT where one characteristically finds nodules along the pleura, bronchovascular bundles, and interlobular septa. Some patients show extensive disease in the airway mucosa leading to fixed obstruction (Fig. 6.36A and B). Vascular disease is unusual but can produce nodular pulmonary infarcts associated with granulomatous changes at the periphery, a disorder that has been termed "necrotizing sarcoidal granulomatosis" (Fig. 6.37A and B). Pulmonary venous obstruction rarely can cause pulmonary veno-occlusive disease. Pleural disease with effusions is uncommon but does occur.

Some patients develop distinct clinical syndromes. Fevers, arthralgias, and lower extremity nodular or diffuse areas of panniculitis dues to erythema nodosum are termed *Lofgren's* syndrome. Patients with this presentation can be treated

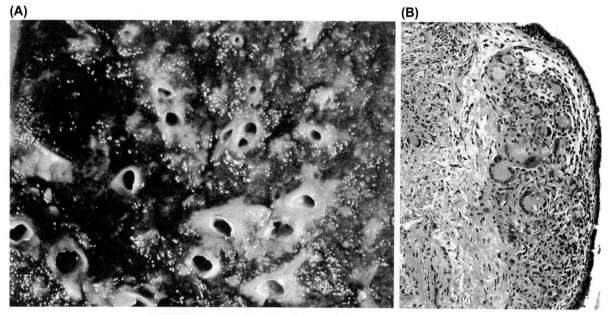

(A) **(B)**

FIGURE 6.36   (A) Sarcoid in airways and (B) granulomas in airways.

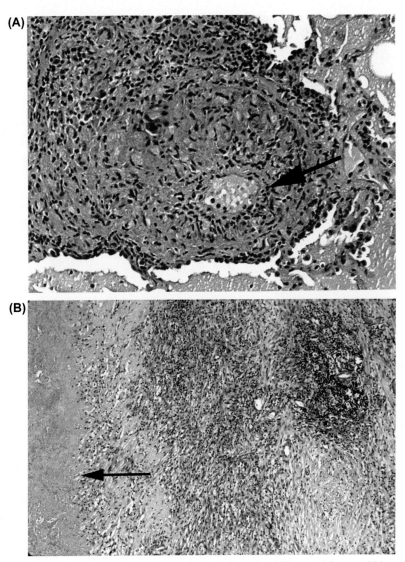

FIGURE 6.37 (A) Perivascular granulomas (*arrow* points to constricted vascular lumen) and (B) necrotizing sarcoidal granulomatosis showing area of ischemic necrosis (*arrow*).

symptomatically with nonsteroidal medications with an excellent response and do not require biopsy if symptoms rapidly resolve.

Extrathoracic disease is usually clinically silent. However, patients may develop uveitis and parotitis (*Heerfordt's syndrome*), uveitis alone, peripheral lymphopenia and anemia, hypercalcemia due to the 1-hydroxylation of 25-cholecalciferol within activated macrophages leading to increased gut absorption of calcium, congestive heart failure, or more ominously, atrio-ventricular heart block and ventricular tachyarrhythmias that can lead to sudden death if unrecognized. Positive-emission tomography is a sensitive technique for identifying granulomas in the myocardium and at other sites and more sensitive than magnetic resonance imaging (MRI).

One problem that invariably arises in some cases is whether the presence of necrosis within granulomas is inconsistent with the diagnosis of sarcoidosis and should trigger a more extensive work-up to exclude infection. Minor amounts of central fibrinoid necrosis in sarcoidosis are common (Fig. 6.38) and tend to accompany active disease with systemic symptoms. However, occasional cases show more extensive necrosis that suggests infection, but if diligent efforts fail to yield an infectious etiology, one is left to conclude that the changes most likely represent sarcoidosis, and the patient should be followed at regular intervals.

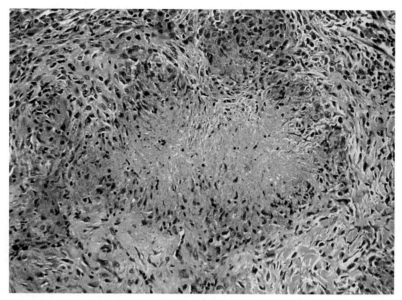

FIGURE 6.38 Central fibrinoid necrosis in sarcoidosis.

Other diseases besides infection can at times show nonnecrotizing granulomas as a feature and may be confused with sarcoidosis. These include malignancies in the lung, where regional lymph nodes and lung tissue can develop non-necrotizing granulomas that likely represent a hypersensitivity reaction to tumor. Generally, these granulomas are not extensive, confluent, or hyalinizing, but their presence can lead to the misdiagnosis of sarcoidosis.

Hodgkin's disease can produce a granulomatous reaction to the malignant lymphoid cells and this can complicate the correct diagnosis, as Hodgkin's can be focal with lymph nodes and easily missed in a background of granulomatous inflammation. Hypersensitivity pneumonitis generally shows microgranulomas that are loosely arranged but at times their frequency can produce confusion with sarcoidosis. LIP, granulomatous interstitial disease, drug-induced changes and Crohn's disease may mimic sarcoidosis in histological sections. On the other hand, some diseases that contain the term "granuloma" rarely show nonnecrotizing granulomas with any frequency and this includes Wegener's granulomatosis and lymphomatoid granulomatosis.

Bronchoalveolar lavage has been used to both diagnose sarcoidosis in the correct clinical setting and to monitor its progression. One sees a T-lymphocytic rich lavage fluid with a predominance of CD4$^+$ lymphocytes. Interestingly, this is the converse of what is seen in the peripheral blood where there is generally a CD3$^+$ lymphopenia and a predominance of CD8$^+$ lymphocytes. The utility of the BAL in practice is questionable.

Biopsies to establish the diagnosis may utilize one of two approaches. Mediastinoscopic biopsy is sensitive and specific, but relatively invasive. When there is concern for lymphoid malignancy or pulmonary tumor, one must interpret the biopsy with some caution based on what has already been discussed. The technique most widely adopted by the pulmonologist is the transbronchial biopsy. This is a sensitive and specific technique for sarcoidosis because granulomas tend to be distributed along the airway lymphatics in this disorder. The number of biopsies taken increases the yield and five interpretable biopsies should yield a diagnosis in >90% of cases. EBUS biopsies of the regional lymph nodes can increase the diagnostic yield.

The level of confidence for the histological diagnosis of sarcoidosis increases with (1) confluence of the nonnecrotizing granulomas and (2) hyalinizing fibrosis replacing the granulomas (Fig. 6.39). The latter is virtually never seen in malignancies and rarely with infections. Otherwise, in the absence of these findings, the diagnosis by the pathologist should be descriptive and the final diagnosis hinges on clinical and radiographic features.

It is furthermore worth noting that sarcoidal granulomas that are identical to sarcoidosis in appearance and even distribution may be seen in a variety of other autoimmune disorders, in response to certain drugs, including interferon-alpha, and there appears to be an increased incidence of sarcoidosis complicating a variety of malignancies.

Sarcoidosis can progress to produce extensive upper lobe scarring in a minority of cases. Areas of traction bronchiectasis, usually in the upper lobe, may be prone to colonization by *Aspergillus* and other hyphate molds. The lung shows irregular scarring and one may be hard-pressed to identify residual sarcoidal granulomas.

FIGURE 6.39   Hyalinizing fibrosis in sarcoidal granuloma.

GRANULOLYMPHOCYTIC INTERSTITIAL LUNG DISEASE

This is a poorly defined set of changes that occur in the lung, and in other organs, in patients with combined variable immunodeficiency disorder. In my experience the changes predominate around small airways that are bronchiectatic and show features of nodular lymphoid hyperplasia with poorly formed granulomas (Fig. 6.40). The disease may wax and wane naturally or in response to treatment. The chief concern is the development of a malignant lymphoma, although this is infrequent. Cases should be checked for evidence of monoclonal light chain production and patients monitored radiographically without repeated biopsies unless there has been a very significant change in the radiographic appearance or clinical course.

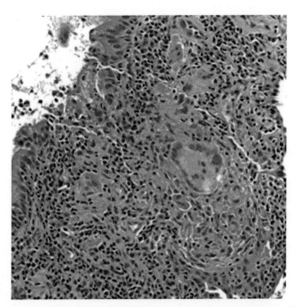

FIGURE 6.40   Granulomatous interstitial lung disease in patient with acquired immunodeficiency syndrome.

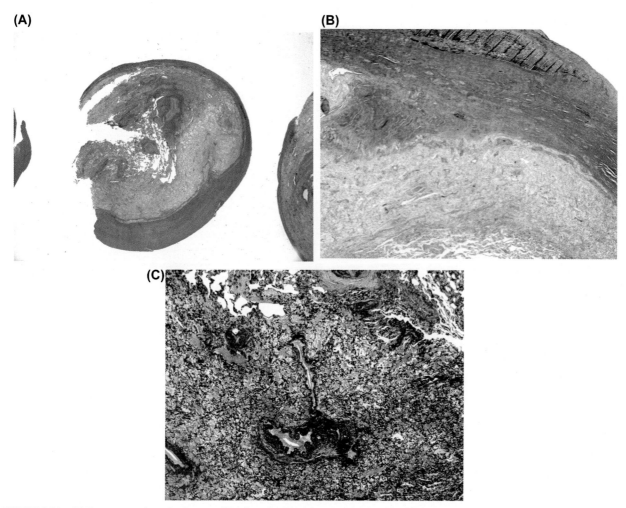

FIGURE 6.41 (A) Low power view of pulmonary fibroelastosis, (B) trichrome stain showing diffuse fibrosis, and (C) elastic stain showing extensive elastic fibers.

PLEURO-PULMONARY FIBROELASTOSIS

This disease has recently been recognized and included in the classification systems of ILD. Many excised upper lobes of lungs show apical fibroelastotic scars or so-called *apical caps*. In the past, these were generally attributed to old granulomatous infection, i.e., tuberculosis or histoplasmosis. In more recent years, as the prevalence of these infections has diminished, alternative explanations, including chronic apical ischemia, have been suggested to account for this change. One sees dense subpleural scars with large numbers of elastic fibers associated with distorted small airways, and with a small component of surrounding interstitial scar.

However, some patients show severe interstitial disease associated with subpleural scars. The radiographic and histopathological features are more extensive and not limited to the apices (Fig. 6.41A−C). The prognosis in these patients depends on the extent of disease, but it tends to be progressive and may require lung transplantation.

IGG4-MEDIATED LUNG DISEASE

The spectrum of diseases attributable to increased levels of IgG4 appears to be constantly expanding, since the disorder was first linked to autoimmune pancreatitis. In the lung, changes may occur within the proximal airways, in the lung parenchyma, and the pleura. Some of these cases were previously termed *plasma cell granuloma*.

Multiple patterns of IgG4[+] disease have been identified in the lung, including airway, nodular (Fig. 6.42A), and interstitial disease (Fig. 6.42B). They are characterized by active *storiform* fibrosis with a loose edematous stroma associated with dense lymphoplasmacytic inflammation. Many cases show associated fibro-obliterative phlebitis (Fig. 6.42C).

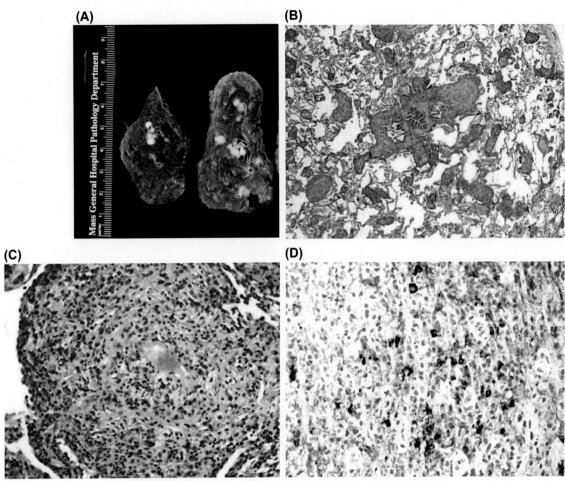

FIGURE 6.42 (A) Lung nodules in IgG4 disease, (B) fibrosis and plasma cells in IgG4 disease, (C) vein narrowed by fibrosis and inflammation, and (D) IgG4$^+$ plasma cells.

At least 30 infiltrating plasma cells per high-power field should show IgG4 expression, and a ratio of IgG4$^+$/IgG$^+$ plasma cells >40% is considered highly specific in the context of fibrosis, but no quantitative gold standard exists (Fig. 6.42D). Patients may respond to excision and rituximab therapy.

FURTHER READING

American Thoracic Society, European Thoracic Society, 2002. American Thoracic Society/European Respiratory Society international multidisciplinary consensus classification of the idiopathic interstitial pneumonias. Am J Respir Crit Care Med 165 (2), 277−304.

The current classification schema for interstitial lung diseases.

Deshpande, V., Zen, Y., Chan, J.K., et al., 2012. Consensus statement on the pathology of IgG4-related disease. Mod Pathol 25 (9), 1181−1192.

A recent review that explains the spectrum of diseases associated with this immune defect.

Kradin, R.L., 2015. Honeycomb lung: time for a change. Arch Pathol Lab Med 139 (11), 1398−1399.

The author's thoughts on how to interpret honeycomb lung with respect to its pathogenesis and role in the diagnosis of UIP.

Mandal, R., Mark, E., Kradin, R., 2008. Organizing pneumonia and pulmonary lymphatic architecture in diffuse alveolar damage. Hum Pathol 39 (8), 1234−1238.

A small but potentially important study suggesting that corticosteroids may benefit a subset of patients with ARDS who have elements of organizing pneumonia in their lung biopsies.

Vij, R., Noth, I., Strek, M.E., 2011. Autoimmune-featured interstitial lung disease: a distinct entity. Chest 140 (5), 1292−1299.

A recent review of the associations of autoimmune disease and interstitial lung disease.

Chapter 7

Nonneoplastic Smoking-Related Disorders

A variety of benign pulmonary disorders have been causatively linked to cigarette smoking. These range from those producing pulmonary nodules to scarring of the lung interstitium. Smoking consumption is an important aspect of the clinical history in pulmonary medicine. All of the smoking-related disorders, both neoplastic and nonneoplastic, are dose dependent, with the exception of acute eosinophilic pneumonia, which is a severe hypersensitivity reaction to a component in cigarette smoke with histological features of DAD. Radiographically, the smoking-related disorders, with the exception of desquamative interstitial pneumonitis (DIP), tend to have centrilobular predominance (Table 7.1).

Although it is unusual to encounter substantial pathological changes in patients who have less than a 20 pack-year cigarette consumption history, the epidemiological literature suggests that even low levels of consumption increase the risk of disease. The same may be said of "second-hand" smoke, which has been estimated to increase the risk of developing lung cancer by 20–30%.

The most common smoking-related disorders, i.e., emphysema and chronic bronchitis or as they are more commonly clinically termed, "COPD," have already been discussed in the chapter on airways disease.

RESPIRATORY BRONCHIOLITIS (INTERSTITIAL LUNG DISEASE)

Respiratory bronchiolitis (RB) represents an unusual form of inflammation, as neither neutrophils nor lymphocytes are prominent. RB shows the accumulation of pigmented macrophages within respiratory bronchioles and adjacent peri-bronchiolar alveolar spaces (Fig. 7.1). It is commonly observed in lung cancer resections from smokers and serves as a biological marker of smoking. The pigment in macrophages tends to be light brown and fine. Iron stains can be weakly positive but dense hemosiderin deposits should not be seen. Scarring of membranous bronchioles can extend into the juxta-bronchiolar alveoli. For reasons that are not evident, these changes generally produce neither symptoms nor radiographic abnormalities in most patients.

However, some patients with comparable histological changes develop centrilobular ground glass small opacities radiographically in the upper lung zones, and are *symptomatic* with dyspnea, decreased DLCO and mild restriction. The term RB-interstitial lung disease (ILD) should only be applied to symptomatic patients who have biopsy-proven RB. Pathologists should not apply the term *RB-ILD* in the absence of an appropriate clinical history and should instead indicate that in the proper setting the biopsy findings are consistent with that disorder.

TABLE 7.1 Smoking-Related Disorders
Chronic bronchitis
Emphysema
Respiratory Bronchiolitis-Interstitial Lung Disease
Desquamative Interstitial Pneumonitis
Langerhans' Cell Histiocytosis
Smoking-Related Interstitial Fibrosis
Usual Interstitial Pneumonitis

Understanding Pulmonary Pathology. http://dx.doi.org/10.1016/B978-0-12-801304-5.00007-1

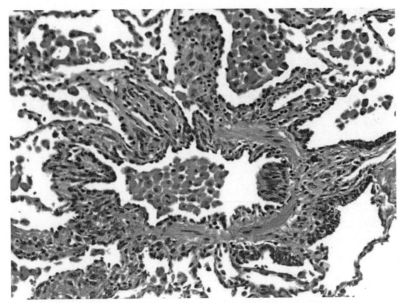

FIGURE 7.1   RB with macrophages in terminal bronchiole and surrounding alveoli.

It is noteworthy that ex-smokers may continue to show histological evidence of RB for more than 20 years after quitting smoking. If questions persist as to whether the patient is being disingenuous with respect to smoking cessation, cotinine level can be obtained for confirmation.

DESQUAMATIVE INTERSTITIAL PNEUMONITIS

Although first recognized by Liebow as a distinct disorder, within the last 25 years it has become evident that most cases of DIP are smoking related. Patients with DIP are heavy smokers who present with dyspnea, cough, and basilar crackles. Unlike other smoking-related disorders, radiographic ground glass opacities are accentuated in the lower lung zones as opposed to the upper lung zones. The pathology of DIP includes (1) diffuse alveolar filling by pigmented macrophages, (2) a lymphoplasmacytic infiltrate in the lung interstitium, and (3) mild interstitial fibrosis (Fig. 7.2A). All of these criteria should be present before confidently diagnosing DIP.

The differential diagnosis includes so-called "DIP-like" reactions, which can be seen in other pulmonary disorders, including smoking-related Langerhans' cell histiocytosis (LCH) (eosinophilic granuloma) and fibrotic interstitial lung

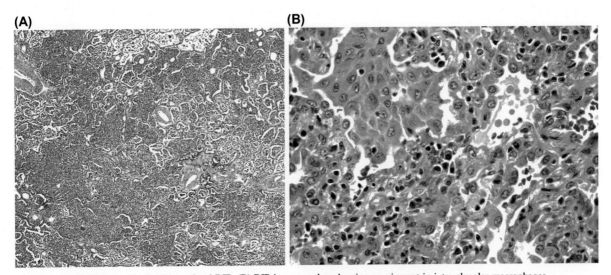

FIGURE 7.2   (A) Smoking-related DIP, (B) DIP in nonsmoker showing no pigment in intra-alveolar macrophages.

diseases (ILDs), in which the patient is either a smoker or has been occupationally exposed to inhaled dust. The distinction can be made based on the focality of the changes in "DIP-like" reactions, or to the presence of marked pulmonary fibrosis with architectural distortion, as the latter is not generally a feature of DIP.

Since the link to cigarette smoking was established, the first step in treatment has been smoking cessation. However, many cases of DIP will respond dramatically to corticosteroids and this approach should be considered either when the patient is markedly symptomatic or cannot quit smoking.

Approximately 15% of cases of DIP occur in nonsmokers. In these cases, the macrophages that accumulate in the alveolar spaces are not pigmented (Fig. 7.2B). Some cases are the result of drug toxicities, and a DIP-like condition with concomitant NSIP features has been reported in patients with myelodysplasia and it may augur acute leukemic transformation.

LCH (OR EOSINOPHILIC GRANULOMA)

LCH or eosinophilic granuloma is a disorder that leads to the nodular accumulation of Langerhans' cells, which are normally a small population of dendritic antigen-presenting cells in the airways. The disease tends to produce multiple small centrilobular nodules (Figs. 7.3A,B), some of which may cavitate. It has upper lobe predominance and a distinct propensity to spare the costophrenic angles.

The classical radiographic pattern of this disease is a centrilobular nodular and cystic disorder with upper lung zone accentuation. Although it is generally assumed that LCH is a benign disorder caused by smoking, questions have arisen as to whether it may be a low-grade neoplasm as clonal expansion of Langerhans' cells has been demonstrated in some cases.

Patients may be symptomatic with cough, dyspnea, pneumothorax, or hemoptysis, but most cases are incidentally discovered on chest imaging, which shows upper lobe centrilobular nodules with cyst formation performed for other reasons. Although the vast majority of patients have disease limited to the lung, the literature notes rib involvement, in 10−15% of cases, and rarely, the disease may involve visceral organs. Diabetes insipidus has been reported in patients with hypothalamic involvement.

Pathologically, the disease has different stages of progression. The earliest phase includes a nodular accumulation of Langerhans' cells admixed with eosinophils. The presence of the latter is variable and may be absent (Fig. 7.4A). In some cases, the nodules undergo central necrosis with cavitation. The surrounding lung frequently shows a prominent localized "DIP-like" reaction (Fig. 7.4B), and this can be a clue to the diagnosis.

There is a propensity for the lung in LCH to develop a cystic appearance which may represent emphysema or post-obstructive airspace dilatation due to air trapping. The end-stage of the disease is a centrilobular pauci-inflamed stellate scar (Fig. 7.5). The latter change is commonly seen in heavy smokers whose lungs are resected for neoplasm, and in the absence of areas of active nodular disease is clinically insignificant and should not prompt interventions that exceed smoking cessation.

(A)   **(B)**

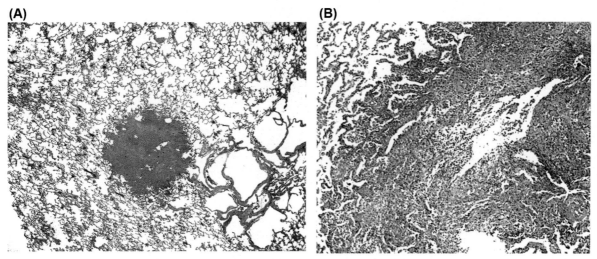

FIGURE 7.3   (A) Low power of nodule in LCH, (B) cavitary change in LCH.

(A)

(B)

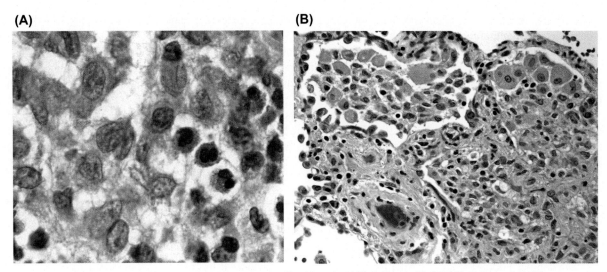

FIGURE 7.4   (A) Cleaved nuclei of Langerhans' cells with eosinophilic infiltrate, (B) "DIP-like" change in LCH.

The diagnosis is based on identifying Langerhans' cells in nodules. These cells are recognized by their characteristic nuclear membrane convolutions, but lineage is confirmed by expression of CD1a (Fig. 7.6A) and/or S-100, antigens by immunohistochemistry. Electron microscopy shows the presence of Birbeck granules (Fig. 7.6B), intracytoplasmic tennis racket-shaped microtubular inclusions that form a three-dimensional network within the cytoplasm of the cell, although an ultrastructural approach is rarely required.

Langerhans' cells accumulate nonspecifically within inflammatory lesions, so that the number of Langerhans' cells in a lesion should account for at least 15% of the cells in a nodule, and both clinical and radiographic criteria should be met before confidently diagnosing LCH. The cellularity of LCH tends to diminish with time so that the burnt-out lesions cannot be confidently diagnosed except in the presence of active nodules or by inference.

The behavior of LCH is variable. Patients should be encouraged to stop smoking. In most cases, the disease fails to progress and tends to resolve spontaneously. However, some patients develop progressive disease leading to respiratory failure with end-stage honeycomb lung. In such cases, cytotoxic therapy may be required, although there is little evidence to support its efficacy.

In the appropriate clinical setting, the diagnosis can be made with confidence based on the presence of increased numbers of CD1a/S100 + mononuclear cells in a bronchoalveolar lavagate and this approach is an alternative option to

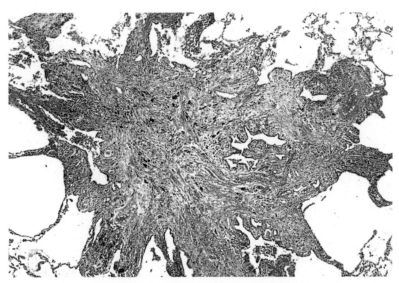

FIGURE 7.5   Stellate scar in "burnt-out" LCH.

(A)

(B)

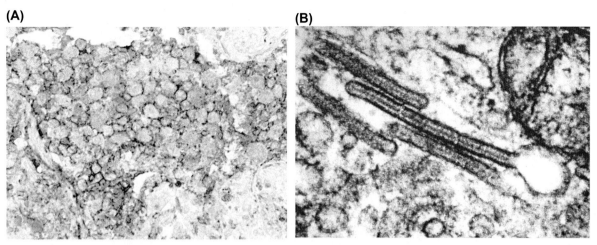

FIGURE 7.6 (A) CD1a immunostaining in LCH, (B) ultrastructural "tennis racket" appearance of Birbeck granule in LCH.

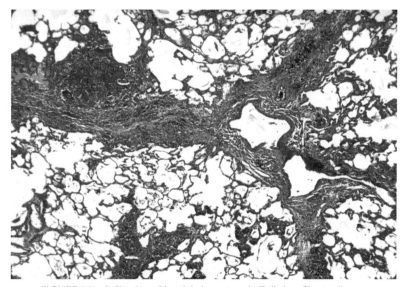

FIGURE 7.7 Infiltration of interlobular septum in Erdheim–Chester disease.

(A)

(B)

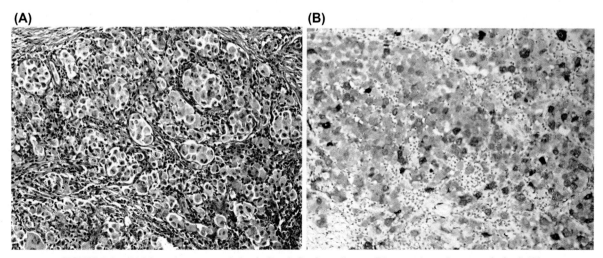

FIGURE 7.8 (A) Macrophage accumulation in Rosai–Dorfman disease, (B) macrophages immunostain for S-100.

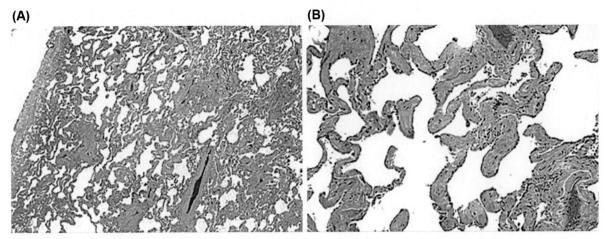

FIGURE 7.9 (A) Centrilobular accentuation and (B) paucicellular "ropey" collagen in SRIF.

biopsy. In addition, as the lesions tend to be centrilobular in distribution, transbronchial biopsy may suffice to establish the diagnosis but sensitivity of the approach is limited due to sampling error.

Other forms of LCH are rare in the lung. Letterer—Siwe disease is an acute disseminated form of Langerhans' cells generally limited to children under the age of three. Hand—Schuller—Christian disease is a systemic disorder that can at times affect the lung. Patients may exhibit exophthalmus, diabetes insipidus, and bone lesions. Smoking does not appear to play a role in this disease.

Erdheim—Chester disease is a non-LCH that affects the lung. Patients have systemic disease including ossifying bone lesions, skin, and central nervous system disease. About 50% of patients harbor point mutations in the BRAF gene at codon 600, which results in a substitute of the amino acid glutamine for valine. The infiltrating cell is an eosinophilic CD68 + macrophage that does not immunostain for S-100 or CD1a. The cells have a propensity to infiltrate the interlobular septae (Fig. 7.7), which are greatly expanded, a finding also seen by chest computed tomography.

Rosai—Dorfman disease rarely involves the lung. Although considered a benign entity, it can act aggressively. Common thoracic involvement includes the regional lymph nodes, pleura, and lung interstitium. The infiltrates are composed of CD68 + macrophages that are S-100 positive but negative for CD1a and Langerin (Figs. 7.8A,B).

SMOKING-RELATED INTERSTITIAL FIBROSIS

For years, smoking was not thought to be a cause of interstitial fibrosis. Patients with areas of bullous emphysema were recognized to have peri-bullous foci of scarring, and old eosinophilic granuloma could at times cause "bridging" forms of fibrosis linking centrilobular regions but interstitial disease was not recognized by most pathologists. In recent years, it has been noted by several groups that patients with heavy smoking histories show a form of subpleural interstitial fibrosis both radiographically and microscopically. The changes in some studies have been predominantly within the upper lobe with centrilobular accentuation and associated with RB. Most will have accompanying emphysema.

What appears to be characteristic is the pauci-inflammatory nature of the fibrosis, which has been described as showing "ropey collagen" (Fig. 7.9). It is important to distinguish this change from other forms of progressive fibrosis including NSIP-fibrotic type and UIP. It may be noteworthy that the latter disorders are also commonly seen in ex-smokers, suggesting that smoking-related injury may play a role in the pathogenesis of a spectrum of interstitial fibrotic disorders.

The natural history of this disorder is currently unknown. It is generally seen together with other smoking-related changes, e.g., emphysema. What if any role smoking-related interstitial fibrosis (SRIF) plays in the risk of developing lung carcinoma is also unknown, and to what extent the finding of SRIF contributes independently to pulmonary function abnormalities, especially the DLCO, is uncertain.

FURTHER READING

Ryu, J.H., Colby, T.V., Hartman, T.E., et al., 2001. Smoking-related interstitial lung diseases: a concise review. Eur Respir J 17 (1), 122—132.
As the title suggests, a concise review of the topic.
Bledsoe, J.R., Christiani, D.C., Kradin, R.L., 2014. Smoking-associated fibrosis and pulmonary asbestosis. Int J Chron Obstruct Pulmon Dis 10, 31—37.

A paper outlining the problems distinguishing smoking related fibrosis radiographically from pulmonary asbestosis in exposed workers.

Vassallo, R., Jensen, E.A., Colby, T.V., et al., 2003. The overlap between respiratory bronchiolitis and desquamative interstitial pneumonia in pulmonary Langerhans' cell histiocytosis: high-resolution CT, histologic, and functional correlations. Chest 124 (4), 1199–1205.

Discussion of two common smoking related disorders. http://journal.publications.chestnet.org/vassallo.robert@mayo.edu.

Chapter 8

Pulmonary Vascular Disorders

PULMONARY ARTERIAL HYPERTENSION

The pulmonary vasculature is normally a high-capacity low-pressure circulation. Disease in the pulmonary vessels can be caused by (1) intraluminal thrombosis or thromboemboli that increase vascular resistance and promote vascular remodeling, (2) inflammatory angiitis, (3) loss of microvasculature, as is seen in emphysema and other interstitial lung disorders, (4) chronic elevation of postcapillary pressures due to congenital heart disease or chronically elevated pulmonary venous pressures, (5) chronic hypoxemia, which is a stimulus for pulmonary vasoconstriction and vascular remodeling, (6) circulating vasoactive factors, e.g., serotonin in the carcinoid syndrome, (7) anorexigenic drugs that cause pulmonary hypertension, or (8) hepatocellular disease with portal hypertension. In many cases, multiple factors contribute, e.g., chronic hypoxemia in interstitial lung disease (Table 8.1).

TABLE 8.1 Pulmonary Arterial Hypertension

Group 1	Idiopathic
	Heritable • *ALK1* • *BMPR2*
	Drugs
	Autoimmune disease
	Congenital heart disease
	Chronic hemolytic anemia
	Portal hypertension
	Schistosomiasis
Group 1	PVOD
	PCH
Group 2	Left atrial hypertension
Group 3	COPD
	ILD
	Sleep apnea
	Alveolar hypoventilation
	Chronic exposure to high altitude
Group 4	Chronic thromboembolic disease
Group 5	Multiple or unclear mechanism
	Hematological disorders, splenectomy
	Vasculitis, sarcoidosis, LCH, LAM
	Tumor, fibrosing mediastinitis, renal failure on dialysis

LAM, lymphangioleiomyomatosis; *PVOD,* pulmonary veno-occlusive disease.

Understanding Pulmonary Pathology. http://dx.doi.org/10.1016/B978-0-12-801304-5.00008-3

(A)　　　　　　　　　　　　　　　　　　**(B)**

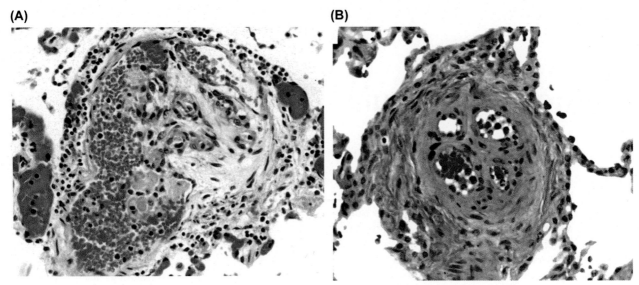

FIGURE 8.1　(A) Angiomatoid lesion in IPAH, (B) recanalization of occluded vessel in IPAH.

The changes in the pulmonary vascular bed that result from injury are largely stereotypic. One observes (1) thickening of the pulmonary vascular media, (2) thickening of the pulmonary intima, (3) reduplication of elastic fibers in the vessel wall, and (4) extension of muscular arterioles to the precapillary level. In congenital heart disease and primary pulmonary arterial hypertension, disorders that generate the highest level of pulmonary hypertension, one sees fibrinoid necrosis of the small arterial walls, with intraluminal thrombus formation and recanalization, the so-called *plexiform* lesion, which is associated with poststenotic dilation of capillary-sized vessels, the *angiomatoid* lesion (Fig. 8.1A and B).

One of the characteristic features of pulmonary vascular disease is its heterogeneity. The pathologist may have to search for a biopsy for some time before identifying a plexiform lesion and even changes due to low-grade pulmonary arterial hypertension (PAH) may be curiously patchy.

Another late stereotypic change complicating chronic pulmonary arterial hypertension is arterial dilatation and atherosclerosis in the proximal pulmonary arteries (Fig. 8.2). In the first three orders of arterial branching, atheromatous changes may be due to hyperlipidemia as in systemic arteries; however, when seen in the more distal pulmonary arteries, it virtually always indicates chronically elevated pulmonary arterial pressures. As in systemic atherosclerosis, luminal dilation and even aneurysmal changes can develop.

IDIOPATHIC PULMONARY ARTERIAL HYPERTENSION

Idiopathic PAH (IPAH) is predominantly a disease of young women. They present with dyspnea on exertion and late in the disease may succumb to right heart failure. The physical examination may be unremarkable or there may evidence of splitting of the second heart sound and a right ventricular heave. The chest radiograph shows increased caliber of the main pulmonary artery (>3.0 cm) and of lobar and segmental vessels. Electrocardiogram (ECG) can show right axis deviation, right bundle branch block, and right atrial hypertrophy. Pulmonary function tests routinely show a decreased DLCO and there is hypoxemia at rest. Patients should undergo ECHO cardiography to estimate PA pressures based on the degree of tricuspid regurgitation, together with an agitated bubble study to exclude intracardiac or intrapulmonary shunts. Right heart catheterization confirms the diagnosis and excludes elevated pulmonary capillary wedge pressures due to pulmonary venous hypertension.

A range of genetic mutations have been linked to IPAH. These include bone morphogenetic protein receptor type II (*BMPR2*), type I receptor *ACVRL1* and type III receptor *ENG*, caveolin-1 (*CAV1*), and a gene (*KCNK3*) that encode a two-pore potassium channel that yields an autosomal-dominant predisposition to idiopathic pulmonary hypertension (IPH) with incomplete penetrance. Scientists have also discovered mutations in eukaryotic initiation factor-2 alpha kinase 4 (*EIF2AK4*) in cases of pulmonary capillary hemangiomatosis and pulmonary veno-occlusive disease (PVOD).

Lung biopsy is rarely pursued to establish the diagnosis, as there is an elevated risk of anesthesia complications and death associated with lung biopsy in patients with severe pulmonary hypertension. Exceptions include atypical presentations and cases where pulmonary hypertension is observed only with exercise. If lung biopsy is performed, or lungs

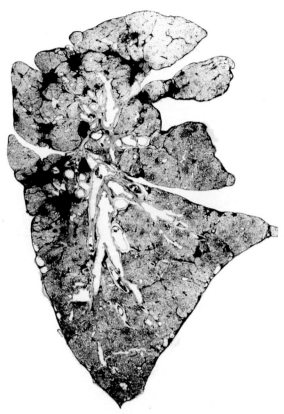

FIGURE 8.2   Generalized pulmonary arterial dilatation in IPAH.

are examined at the time of transplantation or autopsy, one finds plexiform lesions as well as remodeling of the small pulmonary arteries (Fig. 8.3). Foci of hemorrhage with hemosiderin deposition are also seen reflecting elevated micro-hemorrhage and fibrosis. The pulmonary venous system is normal.

A variety of new intravenous and oral medications that target endothelin receptors and prostaglandin activity have improved the prognosis, although the long-term outcome remains poor. Lung transplantation is potentially curative and combined heart–lung transplantation may be indicated in advanced cases where right ventricular failure has developed.

PULMONARY VENO-OCCLUSIVE DISEASE

PVOD results from narrowing and remodeling of the pulmonary veins. Most cases are idiopathic and reflect small venular obliteration, although large caliber veins can also be involved. Patients present with dyspnea on exertion and hypoxemia, and less often with hemoptysis. The diffusing capacity is diminished. The pulmonary capillary wedge is generally normal or minimally elevated. Early diagnosis can be fostered by exercise testing, as some patients with early pulmonary venular disease show near normal pulmonary arterial pressures at rest that become elevated with exercise. The diagnosis is most often made noninvasively.

If lung biopsy is pursued, it reveals venular stenosis due to intimal obliteration or extrinsic compression by venous adventitial scarring (Fig. 8.4A). Pulmonary veins run in the interlobular septa and that is where the primary pathology should be sought. Deposits of hemosiderin and iron-laden macrophages highlight the affected interlobular septae and this "endogenous hemosiderosis" is a clue to the diagnosis (Fig. 8.4B). The downstream pulmonary arteries show intimal and medial thickening but high-grade plexiform lesions are unusual.

Comparable histological changes can be seen in long-standing pulmonary hypertension due to chronic left atrial hypertension due to mitral valvular disease, cor triatriatum, or left ventricular failure, so that pulmonary artery catheterization is essential to exclude elevated capillary wedge pressures. Bush tea alkaloids and chemotherapeutic agents, as well as compression of the major pulmonary veins by tumor, sarcoidal granulomas, or mediastinal fibrosis can mimic the findings of PVOD, although these pathologies often produce asymmetric findings. Pulmonary venous ablation in the treatment of cardiac arrhythmias is a recognized iatrogenic cause of PVOD.

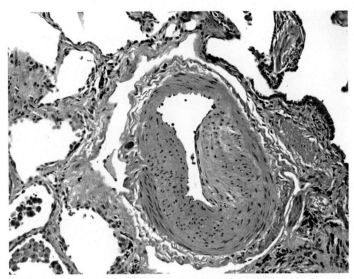

FIGURE 8.3   Remodeled pulmonary artery with thickened tunica media and intima.

(A)   **(B)**

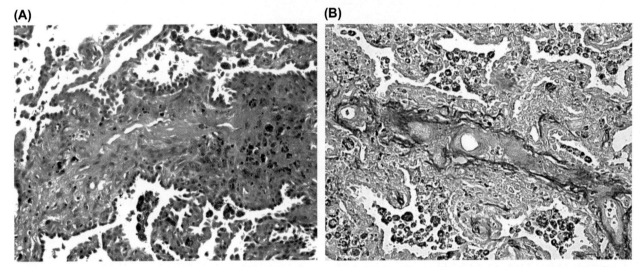

FIGURE 8.4   (A) Venous obliteration and hemosiderosis in pulmonary veno-occlusive disease, (B) elastic stain showing venous stenosis.

There is currently no effective treatment for PVOD other than lung transplantation. Pulmonary vasodilators are contraindicated, as they can produce flash pulmonary edema when the increased pulmonary blood flow encounters a fixed resistance in the pulmonary venous circulation.

PULMONARY CAPILLARY HEMANGIOMATOSIS

There is controversy as to whether pulmonary capillary hemangiomatosis (PCH) is a disease *sui generis* or a variant of PVOD. Most patients with PCH show histological features of PVOD and vice versa. The current classification is largely based on the preponderance of one component versus the other. The clinical presentation is virtually identical to that of PVOD and only biopsy can distinguish these two pulmonary microvascular disorders.

The lung biopsy shows patchy proliferations of capillary-like vessels that appear to "invade" interlobular septae (Fig. 8.5). There is duplication of capillaries in the alveolar walls, and this is the *sine qua non* for diagnosis. The remaining pulmonary arteries may show hypertensive remodeling. The findings of PCH are nonspecific and have been reported in patients with congenital heart disease and genetic vascular malformations. Also, severe pulmonary congestion from heart failure can mimic PCH and should not be overinterpreted. As in PVOD, the treatment is lung transplantation. Antiangiogenic factors have not been consistently effective in treating this disorder and neither has interferon-α.

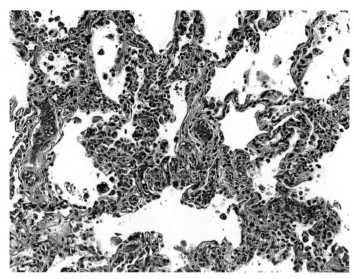

FIGURE 8.5   Duplication of capillary vessels in PCH.

OTHER CAUSES OF PULMONARY VASCULAR DISEASE

A number of estrogen and progesterone substances have been demonstrated to cause pulmonary hypertensive remodeling that is irreversible. Drugs with amphetamine-like activities that stimulate catecholamine release, including cocaine, or those that suppress appetite, e.g., aminorex and phentermine (Phen-fen), can yield high levels of pulmonary arterial hypertension.

Possibly related to the diminished catabolism of estrogen-like moieties by a cirrhotic liver, the hepato-pulmonary syndrome may include changes that mimic primary pulmonary hypertension. As there are other reasons for elevated pulmonary arterial pressure in patients with hepatic failure, including pulmonary venous hypertension these must be excluded by right heart catheterization.

Patients with sleep apnea can develop pulmonary hypertension due to nocturnal O_2 desaturations and hypoxic-mediated vascular remodeling. Finally, patients with both chronic airways disease and ILD develop pulmonary hypertensive complications due to loss of pulmonary capillary bed and persistent hypoxic-mediated vasoconstriction.

PULMONARY THROMBOEMBOLIC DISEASE

Pulmonary thromboembolic disease is a common and potentially lethal disorder, as well as a complex topic with respect to both diagnosis and management. The incidence of pulmonary thromboembolic disease is substantially higher than recognized based on clinical symptomatology, and small pulmonary emboli are commonly detected at autopsy in the lungs of bed-ridden patients with chronic disease. For the most part, small embolic events cause neither symptoms nor physiological deficits. However, there is a syndrome of multiple small pulmonary emboli that presents with nonspecific symptoms or fever of unknown origin.

The risk factors for the development of pulmonary thromboembolic disease have been recognized since Virchow's description in the 19th century. They include stasis, vascular wall injury, and prothrombotic factors, including abnormalities in the clotting cascade and procoagulant factors produced by malignancies, especially those that are mucin producing. Estrogens, progestational agents, obesity, immobility, and traumatic tissue injury are all recognized risk factors for the development of thromboembolic disease.

Symptomatology is dependent on the size of the pulmonary embolic burden and the presence of underlying cardiovascular or pulmonary disease. Young patients may survive a massive pulmonary embolus, whereas an elderly patient with chronic left ventricular failure may succumb to a much smaller embolic clot burden.

Most vascular thromboemboli that produce symptoms and life-threatening disease arise from the deep venous iliofemoral system of the lower extremities. However, clinically relevant thromboemboli also arise in the vena cava, pelvic veins, right heart, or upper extremities. Deep venous thrombi in small distal veins can propagate to larger caliber veins if left untreated.

As the lung is supplied by both the pulmonary and bronchial vasculature, sufficient collateral circulation protects the lung against infarction. The earliest response to pulmonary embolus is stasis within the distal pulmonary microvasculature.

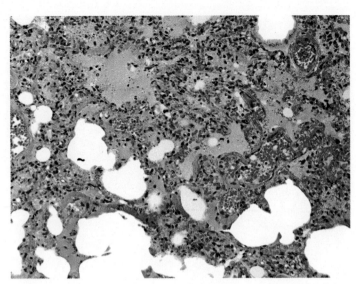

FIGURE 8.6   Severe pulmonary vascular congestion in incipient infarction.

This is referred to as "incipient" pulmonary infarction (Fig. 8.6), in which the lung shows severe pulmonary vascular congestion and alveolar hemorrhage.

True lung infarction with necrosis of the lung parenchyma (Fig. 8.7A, B, and C) is often associated with concomitant elevated pulmonary venous pressures, as seen in patients with congestive heart failure (CHF), or in patients who have the underlying structural lung disease. The infarcted lung is nonviable and prone to superinfection by bacteria or fungi, often *Aspergillus spp*. Radiographically, the classic pulmonary infarction is a subpleural wedged-shaped defect in the lower lung zones producing what has been termed *Hampton's hump.*

Early symptomatology following a pulmonary embolus includes hypoxemia, tachycardia, and localized wheezing due to the release of serotonergic factors released by platelets in the thrombus. Evidence of right ventricular strain on electrocardiograms and elevated right ventricular pressures with dilation and reduced ejection fraction are ominous signs, and when these are accompanied by systemic hypotension, thrombolytic therapy, either systematic or via a pulmonary artery catheter, or surgical thromboarterectomy may be indicated, if there are no preexisting contraindications.

The natural history of thromboembolic disease is variable. Some patients have a clinical prodrome of dyspnea that reflects small pulmonary emboli from a source in the venous system. This may be followed by a submassive or massive and fatal pulmonary embolic event. In patients who do not survive, and whose lungs are examined at autopsy, one discerns characteristic changes with respect to the pulmonary embolus. Massive emboli usually show a saddle configuration that straddles the bifurcation of the main pulmonary arteries. There may be coiling of a large embolus in the right ventricle that obstructs the pulmonary arterial outflow track. In hyperacute embolic events, a distal pulmonary artery may be dilated in the area where the embolus has lodged under pressure, with perithrombotic hemorrhage due to rupture of the feeding *vasa vasorum* in the adventitia of the affected vessel.

It can be difficult to discern the difference between a fresh thromboembolus and a *postmortem* clot. The latter occurs frequently and can fill the vascular bed and mimic embolic disease. Even after examining a postmortem clot microscopically, it may still be challenging to distinguish it from a fresh thromboembolus. What pathologists are trained to look for is evidence of *organization* of the thrombus.

Unfortunately, the term *organization* is used in two distinct ways that are pertinent to pulmonary thromboembolic disease and can cause confusion. Premortem organization of the *thrombus* is evidenced by the presence of strands of fibrin and platelets interspersed with red blood cells throughout the clot (Fig. 8.8A). These are termed *lines of Zahn,* and they are *prima facie* evidence that the clot is formed prior to death and is a real thrombus/embolus.

The second mode of organization concerns the relationship of the thrombus to the *pulmonary arterial wall.* The early response to an embolus is activation of the adjacent vascular endothelial cells, followed within hours by the slow ingrowth of capillary-like vessels into the thrombus (Fig. 8.8B). One can roughly date the age of the thrombus by virtue of the depth of endothelialization into the thrombus.

With time, the thrombus is incorporated by the in growth and replacement of fibrin by fibroblasts secreting collagens, endothelial cells, factor XIIIa + macrophages, and mast cells. Eventually, the organized thrombus becomes part of a

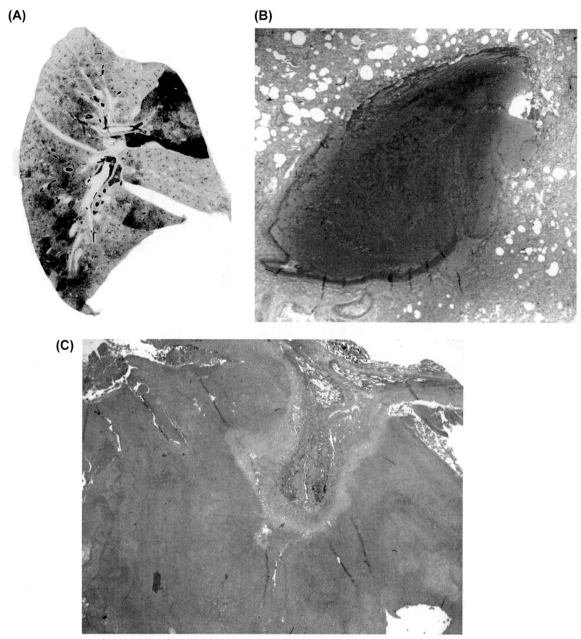

(A)

(B)

(C)

FIGURE 8.7   (A) Multiple areas of subpleural pulmonary infarctions, (B) microscopic area of hemorrhagic infarction, and (C) organizing area of infarction.

vascular *neointima*. The lumen of the vessel now shows multiple vascular channels, evidence of previous thrombosis and organization (Fig. 8.8C).

Long-term management of thromboembolic disease requires a commonsensical approach. Chronic anticoagulation carries an increased risk for the elderly, those who are prone to trauma, ethanol abusers, and patients receiving medications that can promote bleeding. However, patients with persistent risk factors, e.g., stasis and venous injury should be continued on anticoagulation as long as they are at risk, as long as the risk of thrombus formation outweighs the potential benefits of withdrawing anticoagulation. Although patients with no risk factors can have anticoagulation withdrawn as early as 3 months postembolic disease, there is no magic time for withdrawing therapy and each case must be judged on its own merits. Failure to adopt this approach may lead to recurrent thromboembolic disease. A determination of D-dimer formation may help in the assessment of persistent systemic procoagulant activity.

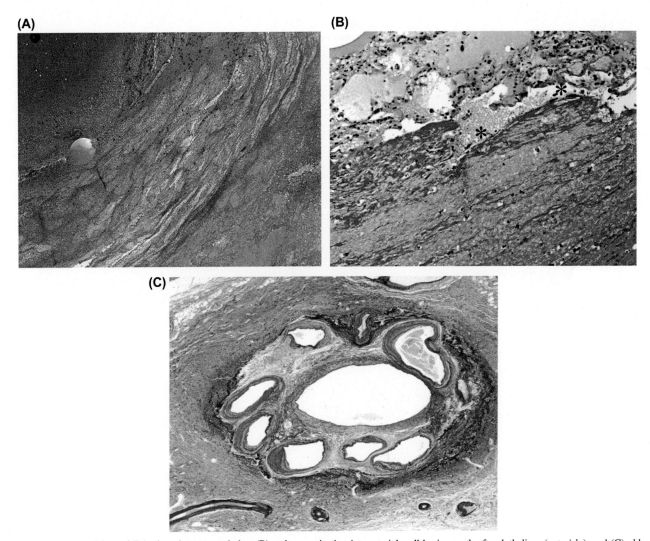

FIGURE 8.8 (A) Lines of Zahn in pulmonary embolus, (B) early organization into arterial wall by ingrowth of endothelium (*asterisks*), and (C) old recanalized thrombus.

The question of fibrinolytic therapy in submassive thromboembolic disease is controversial and most centers continue to adopt a conservative approach that includes heparin with a bridge to warfarin or a novel oral anticoagulant that targets factor Xa.

CHRONIC THROMBOEMBOLIC PULMONARY HYPERTENSION

Patients with chronic thromboembolic pulmonary hypertension (CTEPH) present late in the course of their disease with dyspnea and advanced pulmonary hypertension. Some have sustained previously documented pulmonary thromboembolic events but this is not invariably the case. ECHO cardiography and right heart catheterization establish the presence and degree of pulmonary hypertension. Ventilation/perfusion scanning is helpful in establishing the diagnosis, as it usually shows multiple areas of V/Q mismatch in a pattern that suggests regional embolic events. Dual energy computed tomography (CT) scanning appears to show some promise in diagnosing this disorder. The risk factors include those already mentioned for acute thromboembolic disease.

Patients with CTEPH may be divided somewhat arbitrarily into those who have predominant disease in the proximal pulmonary vascular bed and those who have mostly distal disease. The first group can benefit from surgical thromboendarterectomy, whereas the second generally does not. Pathological examination of endarterectomy specimens in these patients shows an extensively remodeled and recanalized neointima (Fig. 8.9A and B). Many cases show fresh thrombi within the vascular lumen, which most likely represents in situ thrombosis rather than repeated episodes

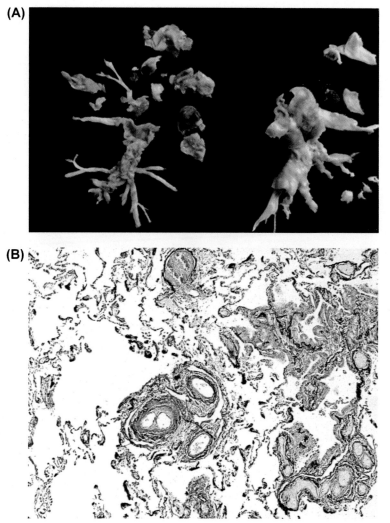

FIGURE 8.9 (A) Endarterectomy specimen from patient with CTEPH, (B) recanalized arteries in CTEPH.

of thromboembolic disease, especially as thrombosis in the peripheral venous circulation is not detected in most of these patients. The presence of atherosclerotic change in endarterectomy specimens may be a poor prognostic indicator.

The disorder is almost certainly a syndrome. It is likely that some patients with CTEPH have suffered previous embolic disease leaving an abnormal and prothrombotic vascular intimal surface. But primary disorders of the pulmonary endothelium and procoagulant factors may exist that have not yet been identified. Treatment includes chronic anticoagulation and when the disease is amenable, surgical endarterectomy.

OTHER FORMS OF PULMONARY THROMBOEMBOLIC DISEASE

Virtually any foreign substance that gains access to the systemic venous system can lead to thromboembolic disease. Following trauma to bones, fat emboli are common. However, the clinical syndrome of fat emboli, which includes dyspnea, neurological, and cardiovascular abnormalities, is experienced by only a fraction of patients with histological evidence of fat emboli in the lung. The presence of intravascular fat in small caliber vessels is best visualized with Oil Red O stains applied to frozen or nonalcohol treated sections of lung tissue (Fig. 8.10A and B), as alcohol dissolves fat so that it cannot be identified except as empty vacuoles in small vessels.

Patients who have undergone bony trauma, surgical resection of a rib, or most commonly, cardiac resuscitation with rib fractures often show small bone marrow emboli in their pulmonary microvessels. In most instances, these are asymptomatic and of no clinical consequence (Fig. 8.11).

(A)

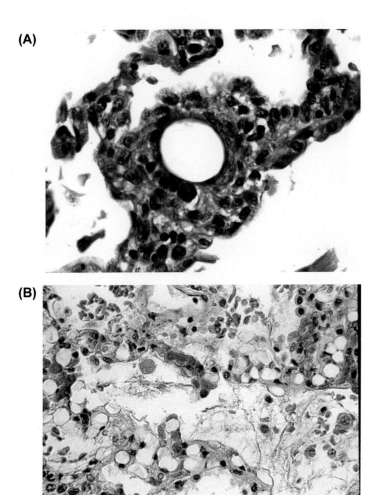

(B)

FIGURE 8.10 (A) Fat emboli, (B) Oil Red O staining of fat globules in vessels.

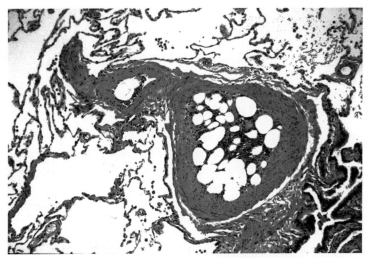

FIGURE 8.11 Bone marrow embolus postresuscitation.

Air emboli occur primarily in the setting of intravenous introduction of a bolus of air in patients receiving intravenous medications. Symptoms, including sudden death, generally require a bolus of air greater than 40 cc in volume. Symptoms are immediate. When air embolus leads to death, it is generally due to obstruction of blood flow in the pulmonary artery outflow tract. The autopsy pathologist, therefore, must carefully incise the main pulmonary artery under water to document the release of air bubbles from the vessel.

Amniotic fluid embolus in the setting of labor and delivery can be a devastating event, associated with ARDS and disseminated intravascular coagulation (DIC). The diagnosis is made clinically or at autopsy due to the presence of amniotic epithelium or "squames" in thrombosed vessels (Fig. 8.12). Other causes of thromboembolic disease include fragments of catheters or injected foreign materials.

Intravenous drug users often inject drugs "cut" with talc that can be seen in the pulmonary microvasculature. Talc is birefringent when examined with polarized light and it evokes a foreign body macrophage and giant cell response

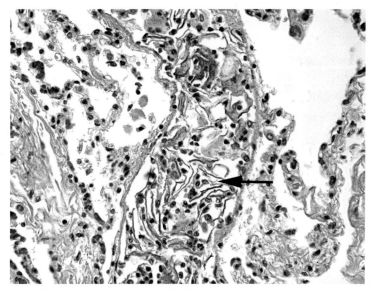

FIGURE 8.12 Amniotic fluid emboli with "squames" (*arrow*) in microvasculature.

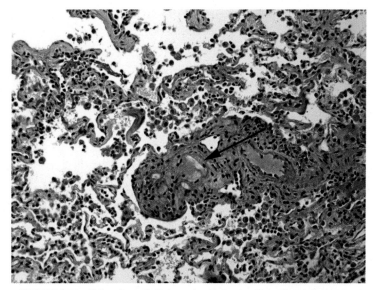

FIGURE 8.13 Talc microemboli (*arrow*) with foreign body giant cell response.

(Fig. 8.13). Over time, the talc crystals work their way out of the vascular lumen and into the surrounding lung interstitium, where they evoke a fibrosing foreign body reaction with radiographic interstitial disease, pulmonary hypertensive changes, decreased DLCO, and hypoxemia.

Silicone, increasingly used for cosmetic procedures, can be inadvertently injected into systemic veins. The silicone lodges in the microvasculature of the lung evoking a similar reaction as that described for talc (Fig. 8.14). Silicone is weakly birefringent and may be confused with oil droplets in the absence of a detailed history.

The nematode *Dirofilaria immitis*, the canine heart worm, can gain entry to the systemic veins of the skin and embolize to the human lung, where it forms a single coin-shaped pulmonary infarction (Fig. 8.15). As humans are not the definitive host, the worm dies in the lung and its remnants can be identified with the assistance of trichrome stains (see chapter on infection).

Schistosomules and schistosome ova can embolize and migrate from the liver to the lung to produce granulomatous pulmonary arterial hypertension (Fig. 8.16), and this is the most common cause of pulmonary hypertension worldwide.

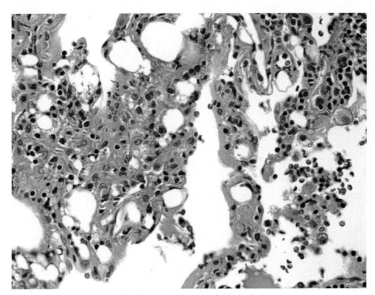

FIGURE 8.14   Silicone microemboli.

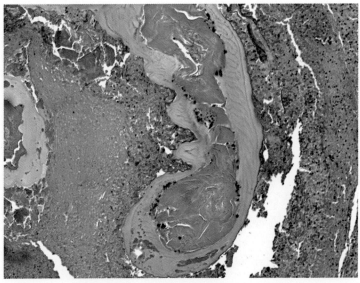

FIGURE 8.15   Dirofilaria in vascular infarction.

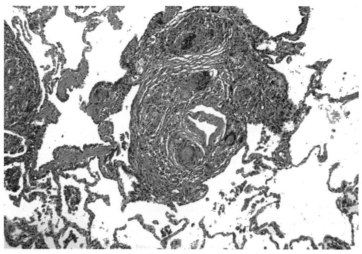

FIGURE 8.16   Granulomatous pulmonary angiitis in pulmonary hypertension due to schistosomiasis.

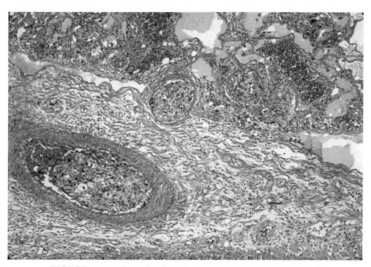

FIGURE 8.17   Embolic disease due to gastric adenocarcinoma.

Most cases are attributable to *Schistosoma mansoni* or less frequently *Schistosoma japonicum*, which cause disease associated with the portal venous system.

TUMOR EMBOLI

A variety of tumors can enter the venous circulation or a proximal pulmonary artery and embolize to the pulmonary microvascular bed leading to hypoxemia (Fig. 8.17). Perhaps the most common cause of this syndrome is seen in patients with gastric carcinoma. Lung cancers that erode into a major pulmonary artery can also cause a shower of pulmonary emboli. All sarcomas tend to metastasize via the systemic circulation to produce pulmonary embolic disease. Among the carcinomas that generally metastasize via lymphatics, a notable exception is follicular carcinoma of the thyroid that metastasizes through the blood circulation.

PULMONARY ANGIITIS

Pulmonary angiitis is a complex and diagnostically challenging form of pulmonary artery disease. Large caliber pulmonary vessels may be involved in giant-cell arteritis, Takayasu' arteritis, and granulomatosis with polyangiitis (GPA), formerly known as Wegener's granulomatosis.

GPA—WEGENER'S GRANULOMATOSIS

Wegener's granulomatosis is generally referred to as an angiitis, although it frequently occurs in the lung in the absence of definitive vascular inflammation. It has a variety of clinical presentations. In its most classic form, GPA is a systemic disorder characterized by pan-sinusitis, lung disease, and rapidly progressive glomerulonephritis. However, it can affect virtually any organ, and pulmonary involvement in the absence of sinusitis and renal disease is well recognized. In the lung, one may see nodular or pneumonic infiltrates or a syndrome of diffuse pulmonary hemorrhage associated with "capillaritis."

The diagnosis of GPA is usually established by the presence of anticytoplasmic antibodies (ANCA) in an appropriate clinical setting. However, ~15% of cases with characteristic histology will be ANCA negative, at least at the time of presentation.

The pathology of GPA is diverse. One virtually always sees "granulomatous inflammation," i.e., lymphohistiocytic infiltration with giant cells, many of which have a Teuton-like appearance with polarization of nuclei to one side of the giant cell (Fig. 8.18). Necrosis tends to develop in a geographic distribution with what has been termed *pathergic* necrosis, a term that refers to collagenolysis that produces a distinct fibrinoid appearance (Fig. 8.19). Small collections of neutrophils

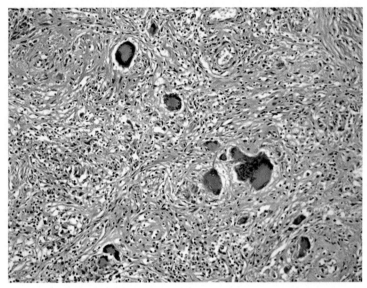

FIGURE 8.18   Granulomatous inflammation with giant cells in GPA.

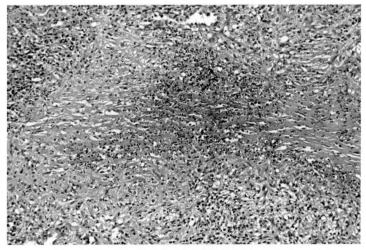

FIGURE 8.19   "Pathergic" necrosis in GPA with dissolution of collagen.

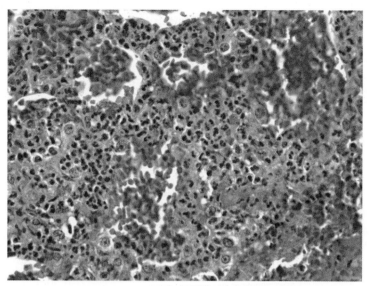

FIGURE 8.20   Focal microabscess formation in GPA.

can be seen to produce microabscesses that resemble the Pautrier's abscesses of mycosis fungoides of the skin (Fig. 8.20). Cavitary nodules that are indistinguishable from rheumatoid nodules may develop. There is a form of protracted disease that exclusively affects the upper airways and tracheobronchial tree.

When all of the above changes are present one may confidently *suggest* the presence of GPA, and this can be confirmed by ANCA testing. However, to definitively diagnose the disorder histologically, one needs to find evidence of granulomatous angiitis, defined as fibrinoid necrosis in the wall of a vessel, together with infiltration by giant cells, lymphocytes, and macrophages (Fig. 8.21). When (1) all of these changes are present and (2) there is no evidence of infection by mycobacteria or fungi, a diagnosis of GPA can be established even in the absence of ANCA confirmation.

The treatment of GPA includes corticosteroids and cytotoxic agents. Recent evidence suggests that rituximab may have beneficial effects in the treatment of this disorder. However, the long-term prognosis of the disease is poor as most patients relapse. Acute and chronic renal failure adversely affects the prognosis.

Polyarteritis nodosum that is often associated with chronic hepatitis B virus infection rarely involves the lung for reasons that are obscure.

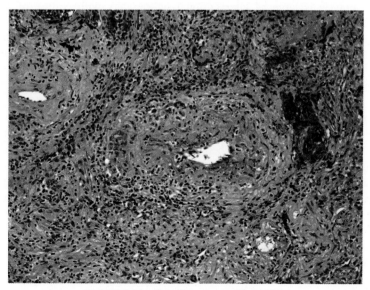

FIGURE 8.21   Granulomatous angiitis in GPA.

TABLE 8.2 Pulmonary Hemorrhage

Immune-mediated	Anti-GBM disease
	ANCA-mediated disorders
	Autoimmune disease
	Immune complex mediated disease
	Cryoglobulinemia
	Antiphospholipid antibody syndrome
	Celiac disease
	Vasculitis
Pauci-immune	Acute bronchitis
	Bronchiectasis
	Coagulopathy
	Infection
	Arteriovenous malformation
	Neoplasm
	Trauma
	Lymphangioleiomyomatosis
	Pulmonary veno-occlusive disease
	Behçet's syndrome
	Idiopathic pulmonary hemosiderosis

PULMONARY ALVEOLAR HEMORRHAGE SYNDROMES

Pulmonary hemorrhage can be seen in a wide spectrum of disorders (Table 8.2). The most common cause of intrapulmonary bleeding is localized bronchitis and bronchiectasis. Localizing the site of bleeding is important when the volume of hemoptysis is large as interventional radiography with coiling of a feeding bronchial artery may be required. Pulmonary contusion, coagulopathy, bleeding into an inflamed bulla, and the invasion of a blood vessel by tumor or infection are all potential causes of pulmonary bleeding. Determining the site of active bleeding in the lung radiographically and bronchoscopically may be complicated by aspiration of blood into dependent lung zones.

Other causes of diffuse pulmonary hemorrhage may be subcategorized based on the presence or absence of inflammation in the vessels, or by the presence or absence of antibody-mediated disease.

The diagnosis of diffuse PAH is confirmed by the progressive retrieval of blood from serial lavages. This reflects the initial washing out of the pulmonary dead space prior to observing dominant contributions from the alveolated lung. But this approach cannot establish the cause of alveolar hemorrhage and it is not without its pitfalls so that the diagnosis of PAH cannot be dismissed based on negative serial lavages.

Histologically, the diagnosis is evident when there is evidence of pulmonary hemorrhage and hemosiderin-laden macrophages (Fig. 8.22A and B). In acute disease, it may be difficult to discern the distinction between surgical hemorrhage and disease. This diagnosis of acute pulmonary hemorrhage is established by the concomitant presence of alveolar fibrin enmeshing alveolar macrophages, many of which will show early hemosiderin formation within their cytoplasm when a histochemical stain for iron is applied. This is a diagnosis that can easily be missed by pathologists who should be prompted to search for these findings, if PAH is a serious diagnostic consideration.

(A) **(B)**

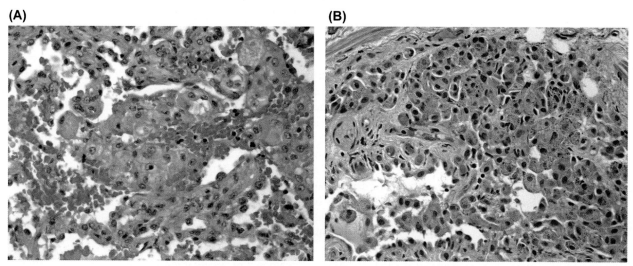

FIGURE 8.22 (A) Pulmonary hemorrhage, (B) hemosiderin deposition.

ANTIALVEOLAR BASEMENT MEMBRANE DISEASE OR GOODPASTURE'S DISEASE

Antialveolar basement membrane disease (ABM) or Goodpasture's disease is mediated by the deposition of immuno-globulin and complement along the alveolar basement membrane. Most cases have concomitant renal involvement of the glomerular basement membrane (GBM). The diagnosis is most often established serologically and the immunoassay is both sensitive and specific. Biopsies that are rarely performed when the serology is positive tend to show minimal inflammation of the alveolar wall.

If the disease is suspected, direct immunofluorescence for the detection of IgG or less frequently IgA along the basement membrane (Fig. 8.23) should be performed on frozen tissue as the test cannot be performed accurately with formalin-fixed tissue. Electron microscopy shows electron-dense linear deposits along the ABM in 40% of cases of anti-GBM disease. However, the serological tests for anti-GBM have a >90% sensitivity and specificity and suffice to confirm or exclude the disease.

Inflammatory causes of diffuse alveolar hemorrhage (DAH) include ANCA-mediated diseases, both Wegener's and microscopic polyangiitis-myeloperoxidase disorders. The latter includes a variety of microangiopathic diseases including

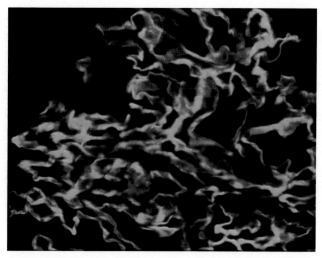

FIGURE 8.23 Immunofluorescence showing linear IgG staining along the alveolar basement membrane.

Churg–Strauss disorder and drug-mediated vasculitides. The criteria for the diagnosis of Churg–Strauss disease have undergone modification in recent years. Churg and Strauss originally described a granulomatous inflammatory condition similar to Wegener's disease, associated with asthma and marked tissue and peripheral eosinophilia. Most cases had multiorgan involvement.

Currently, it is clinically defined by the presence of asthma, pulmonary infiltrates, and the presence of tissue and peripheral eosinophilia. Biopsies may show eosinophilic angiitis and parenchymal eosinophilia with granulomatous inflammation, but granulomatous changes are not required to establish the diagnosis as it is currently conceived (Fig. 8.24A and B). Mononeuritis and paranasal sinus abnormalities are currently included as diagnostic criteria. The leukotriene inhibitor montelukast (Singulair) used to treat asthma can evoke Churg–Strauss disease in some patients.

Collagen vascular disorders, including SLE, RA, and cryoglobulinemia, can yield microvascular injury and hemorrhage (Fig. 8.25). Serological confirmation is generally required as the biopsy findings are rarely specific. Ultrastructural examination may show vascular immune complex deposition. A rare autoimmune cause of DAH is celiac disease. The diagnosis is established by appropriate serological testing but may be confirmed by small bowel biopsy. There may be cases, as in rheumatoid lung disease, that do not involve the bowel, but this question is being investigated.

Antiphospholipid disease is an inflammatory coagulopathy that can complicate other immune disorders, e.g., SLE, or may occur independently. Antiphospholipid disease in the lung can lead to DAH, ARDS, and organizing pneumonia. Biopsies are nonspecific but the diagnosis is suggested by the concomitant presence of microthrombi associated with acute and lymphocytic vascular inflammation. The diagnosis is established by serological evidence of elevated titers of acute and chronic titers of anticardiolipin antibodies together with the presence of a "lupus" anticoagulant.

(A)   **(B)**

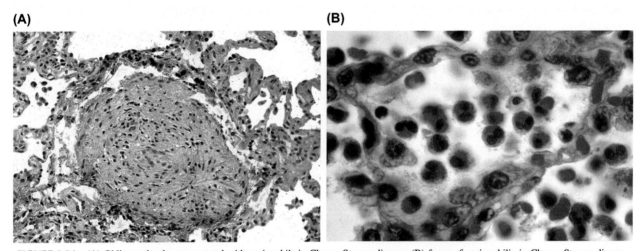

FIGURE 8.24   (A) Obliterated pulmonary vessel with eosinophils in Churg–Strauss disease, (B) focus of eosinophilia in Churg–Strauss disease.

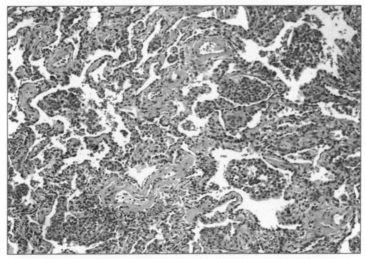

FIGURE 8.25   Inflammatory pulmonary hemorrhage in cryoglobulinemia.

Behçet's syndrome is a poorly defined entity that presents with a complex of sign of systemic angiitis, uveitis, cutaneous disease, mucous membrane ulcerations, and a spectrum of other pathologies. Cases may involve the lungs with DAH or localized bleeding. The pathology is nonspecific but includes acute and chronic inflammation of small arteries and veins (Fig. 8.26).

Idiopathic pulmonary hemosiderosis (IPH) is generally a disease of children and one of exclusion. However, adult cases of chronic DAH that cannot be characterized are referred to as idiopathic hemosiderosis. Some patients have a long history of intrapulmonary bleeding, sufficient to cause an iron-deficiency anemia and pulmonary fibrosis. The lung biopsy shows stigmata of chronic pulmonary bleeding, including the deposition of hemosiderin with encrustation of the elastic fibers of the alveolar wall (Fig. 8.27). The prognosis is variable, although many patients survive for greater than 5 years following diagnosis. The efficacy of anti-inflammatory medications is uncertain.

DRUG-INDUCED ANGIITIS

Drug-induced angiitis topic is addressed in the chapter on drug-related pulmonary disease. Changes due to drugs may include the cuffing of small vessels with lymphocytes or eosinophils. Frank vasculitis with necrosis of the vessel wall is uncommon. The diagnosis is one of exclusion and requires clinical correlation.

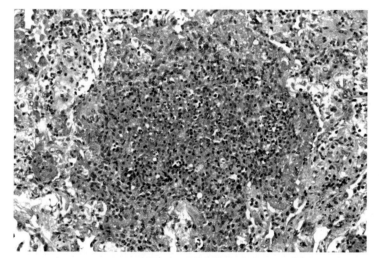

FIGURE 8.26   Pulmonary angiitis in Behçet's disease. Patient had multiple mucosal ulcerations.

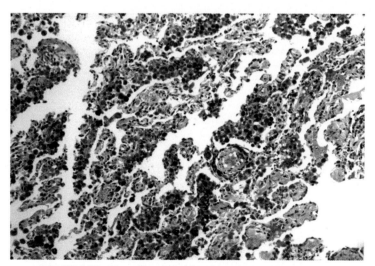

FIGURE 8.27   Hemosiderosis in idiopathic pulmonary hemosiderosis.

(A)
(B)

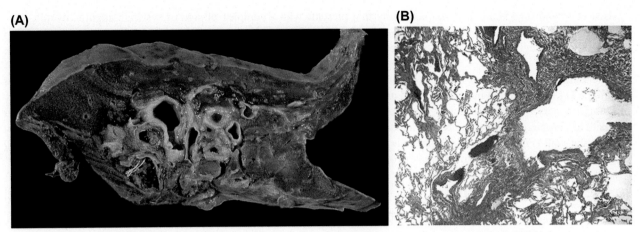

FIGURE 8.28 (A) Large pulmonary arteriovenous malformation (AVM), (B) small microscopic AVM.

ARTERIOVENOUS MALFORMATIONS

Arteriovenous malformations (AVMs) occur idiopathically but are increasingly recognized to reflect underlying genetic abnormalities such as hereditary hemorrhagic telangiectasia (Rendu—Osler—Weber) syndrome. The diagnosis may be spurred by symptoms due to hemoptysis or by arteriovenous shunting with hypoxemia. The lesions may be large enough (Fig. 8.28A) to be identified radiographically or microscopic and too small to detect (Fig. 8.28B). ECHO cardiography with agitated bubbles shows late appearance of bubbles in the left heart, as opposed to their early appearance when there is an intracardiac shunt communication. Pathological detection is not easy when the lesions are small as pulmonary vessels normally tend to collapse when the lung is removed and may falsely leave the impression of abnormal organization.

Clusters of malformed vessels seen best with elastic stains are required to establish the pathological diagnosis with confidence (Fig. 8.28). The pathologist should be informed clearly when the diagnosis is being considered so that he or she can do due diligence in his or her search for the abnormal vessels.

FURTHER RADING

Simonneau, G., Robbins, I.M., Beghetti, M., et al., 2009. Updated clinical classification of pulmonary hypertension. J Am Coll Cardiol 54 (Suppl. 1), S43—S54.

A recent clinical classification of pulmonary hypertensive disorders.

Wagenvoort, C.A., Wagenvoort, N., 1970. Primary pulmonary hypertension. A pathologic study of the lung Vessels in 156 clinically diagnosed cases. Circulation 42, 1163—1184.

The classic and still relevant study of the pathology of primary pulmonary hypertension.

Jennette, J.C., Falk, R.J., Andrassy, K., et al., 1994. Nomenclature of systemic Vasculitides Arthritis. Proposal of an international consensus conference. Arthritis Rheum 37 (2), 187—192.

The proposal of a consensus conference of experts on Wegener's and other forms of vasculitis that can affect the lung and other organs.

Stein, P.D., Woodward, P.K., Weg, J.G., 2006. Diagnostic pathways in acute pulmonary embolism: recommendations of the PIOPED II investigators. Am J Med 119 (12), 1048—1055.

A detailed approach to the clinical diagnosis of pulmonary thromboembolic disease.

Chapter 9

Pulmonary Infection

As the lung is a portal between the ambient environment and the internal milieu, it is the most frequent site of serious infections. A number of factors predispose to pulmonary infection, and these include distortions in lung anatomy, decreased mucociliary clearance, as well as abnormal cellular and humoral immune responses.

The optimal treatment of infection requires diagnosing its cause. As a large variety of microbes can infect the lung, and as the histopathology of noninfectious conditions frequently mimics infection, the differential diagnosis of pulmonary infection is often broad. While in many cases, the clinical history, radiographic findings, and the noninvasive sampling of secretions can establish the cause of infection, at times, a lung biopsy will be required.

THE APPROACH TO SAMPLING FOR INFECTION

The optimal approach to sampling the lung for infection depends on whether disease is localized or diffuse (Table 9.1). In immunosuppressed patients, diffuse pulmonary infiltrates due to infection are often diagnosed by sputum induction or by bronchoalveolar lavage (BAL). This is particularly the case when the microbial burden is large. However, noninvasive approaches are less sensitive than biopsy in diagnosing localized infections, and they cannot distinguish a colonizing commensal from an invasive pathogen. Lung biopsy is also often required to exclude infection and establish a noninfective diagnosis, e.g., acute lung injury due to chemotherapy.

Pathologists prefer the opportunity to examine generous samplings of lung because diagnoses based on larger biopsies are generally more accurate, yield more information with respect to the host immune response, and sometimes reveal additional potentially treatable disorders. For this reason, and based on the specific details of a case, the diagnostic pathologist should be prepared to educate clinicians with respect to the limits of minimally invasive lung sampling, thereby sparing patients the unnecessary discomforts and delays that can attend nondiagnostic procedures.

TRANSBRONCHIAL BIOPSY

The lung has roughly the surface area of a tennis court and so sampling error is an unavoidable potential pitfall in diagnostic pulmonary pathology. The transbronchial biopsy (TBB) preferentially samples peribronchiolar lung tissue, yielding tissue fragments of 1−3 mm in diameter. However, many peripherally located lung lesions cannot be accessed by this approach.

TBB is usually adequate for diagnosing diffuse pulmonary infections and peribronchiolar granulomatous diseases, such as sarcoidosis and lymphangitic spread of tumor. But at times the findings of a TBB can be nonspecific and misleading. For example, "organizing pneumonia" in a TBB may represent a nonspecific reaction adjacent to an adjacent focus of infection or malignancy, a nonspecific manifestation of chemotherapy effect, aspiration, or cryptogenic disease. For this reason, the findings gleaned from a TBB must always be thoughtfully correlated with clinical and radiographic findings.

FINE NEEDLE ASPIRATION BIOPSY

Computed tomography (CT)-guided fine needle aspiration biopsies have a high yield in the diagnosis of peripheral nodular infiltrates. Biopsies can be semiliquid, or include a 1 mm core of tissue. When performed with the assistance of a cyto-technologist, rapid diagnoses can be proffered by preparing and examining stained smears directly at the bedside. Fine needle aspirates are useful in diagnosing localized infections and cytopathologists can suggest the pattern of inflammation based on the types of inflammatory cell subsets in the sample and the presence or absence of necrosis.

TRANSBRONCHIAL NEEDLE ASPIRATION BIOPSY

Transbronchial needle aspiration biopsy of regional lymph node groups is often a low-yield procedure because nonspecific reactive regional lymphadenitis is common in the presence of pulmonary infection. The procedure is prone to artifacts that

Understanding Pulmonary Pathology. http://dx.doi.org/10.1016/B978-0-12-801304-5.00009-5

TABLE 9.1 Approach to the Isolation of Pulmonary Microorganisms

Expectorated sputum
Induced sputum
BAL
Fine-needle aspirate (1 mm)
Bronchial biopsy (1—3 mm)
Transbronchial biopsy (1—3 mm)
Transbronchial needle biopsy (1 mm)
Video-assisted thoracoscopic biopsy (2—3 cm)
Open-lung biopsy (2—3 cm)
Surgical lobectomy
Autopsy

may present diagnostic difficulties for the surgical pathologist. However, when adopted judiciously, this approach may be adequate for the diagnosis of infection, as in one series in which ~50% of cases of tuberculous lymphadenitis were accurately diagnosed.

VIDEO-ASSISTED AND OPEN THORACOSCOPIC BIOPSY

Video-assisted thoracoscopic (VATS) lung biopsy has largely replaced open thoracotomy biopsy as the optimal approach for obtaining large samples of lung. The procedure is associated with modest and acceptable morbidity, has the advantage of allowing direct access to widely separated lung segments, and provides generously sized wedge biopsies of 2—3 cm. Consequently, VATS should be considered a first-line approach when a timely accurate diagnosis is essential.

HANDLING LUNG BIOPSIES

Appropriate handling of the lung biopsy is essential for obtaining the highest diagnostic yield (Fig. 9.1). Sampling the lung for microbiological culture should ideally take place under sterile conditions in the operating room but the pathologist processing the biopsy is ultimately responsible for ascertaining that all necessary diagnostic tests have been ordered and be prepared to harvest additional samples for testing that may have been overlooked. When preparing tissue for microbiological isolation, the lung should be minced rather than crushed, as hyphate fungi, e.g., *Zygomycoses spp.*, may fail to grow in culture following maceration. It is substandard care for a pathologist to place a lung biopsy directly into fixative, without first considering a diagnosis of infection. If questions arise as to which tests to order, or how best to transport the specimen to the laboratory, discussions with the hospital microbiology laboratory staff or a hospital infectious disease specialist will generally answer them.

The examination of touch imprints of lung tissue is a simple and rapid way of identifying pathogens. Touch imprints can be prepared from foci of pulmonary consolidation, necrosis, or suppuration, and rapidly stained for bacteria, mycobacteria, and fungi in the surgical pathology suite or the microbiology laboratory. Concomitantly biopsies may be harvested for ultrastructural analysis, polymerase chain reaction (PCR) assays, or research purposes. For large biopsies, it may be possible to inflate the lung with 5% formalin via a small (23—25) gauge needle to optimize subsequent histological examination.

PULMONARY INJURY IN INFECTION

Pulmonary Host Response

The diagnosis of infection requires both an interpretation of the morphological changes evoked by the pathogen and the identification of a pathogen in situ. The pattern of pulmonary inflammation often suggests the route of entry of an

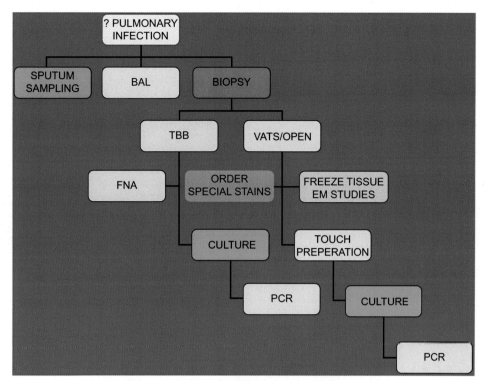

FIGURE 9.1   Approach to the handling of lung biopsies in infection.

infectious agent and may help to narrow the diagnostic possibilities. It is necessary to be familiar with the multiplicity of response patterns evoked by infection and to recognize that these can vary depending on the route of entry, pathogen load, and the competence of host defenses. For example, although *herpesvirus-1* can produce a miliary pattern of fibrinoid necrosis in an immunosuppressed patient with viremia, it can also cause ulceration of the tracheobronchial mucosa in a chronically intubated patient.

Microbes are rarely identified randomly or diffusely in infections; rather, they tend to be compartmentalized, so that substantial effort may be wasted in searching for them where they are not likely to be found. Mycobacteria and fungi are virtually always localized in areas of necrosis; *Rickettsia spp.* and *Bartonella spp.* largely target the microvasculature; viruses tend to attack the airways and so the surgical pathologist must both be acquainted with the pulmonary micro-anatomy and with the preferential localization of microbes in pulmonary tissues.

Pulmonary Defenses

Most microbes are small ($<5 \mu M$) and can penetrate to the distal gas-exchanging surfaces of the lung, although the majority are excluded by the defenses of the upper airways or deposit along the conducting airways to be cleared by the mucociliary escalator. Humoral factors, including IgA and defensins, released by airway cells limit microbial penetration into tissues. Airway mucosal dendritic cells trap microbial antigens and transport them to regional lymph nodes, where they are processed and present to both T and B lymphocytes, evoking adaptive immunity (Fig. 9.2).

Ulceration or thickening of the gas exchange surface limits diffusion of oxygen and carbon dioxide. For this reason, the alveolus is under normal conditions maintained sterile by resident macrophages that scavenge inhaled particulates and secrete monokines, including interleukin-10 and transforming growth factor-β, that locally suppress inflammation and promote immunotolerance.

When the alveolar lining is injured, or when the number of invading organisms exceeds the phagocytotic capacities of resident macrophages, neutrophils and exudate monocytes are recruited to sites of lung infection. Even small numbers of virulent pathogens can greatly amplify inflammation via the release of chemokines, cytokines, and complement, by host immune cells. These defenses promote the clearance of infection, but can also damage the lung. Lung biopsies afford the pathologist a unique opportunity to assess these dynamic responses, directly, in addition to identifying a causative pathogen.

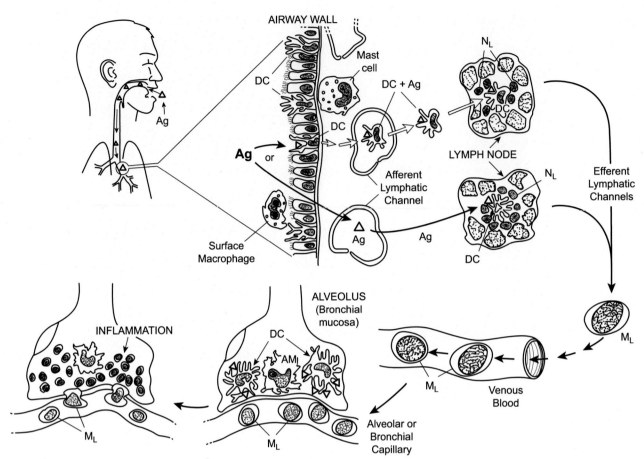

FIGURE 9.2 Pulmonary immune anatomy.

Patterns of Lung Injury due to Infection

A number of generic patterns of inflammation may be evoked by infection but how they are distributed is specific to the involved tissue (Table 9.2).

Clinicians and radiologists have developed classification systems that are distinct from those of pathologists with respect to pulmonary infection. For example, a variety of infectious agents yield a radiographic picture that clinicians term "atypical" interstitial pneumonia, which differentiate them from the "typical" bacterial pneumonias. However, the histopathology of an "atypical pneumonia" may be centered on the lung interstitium, small airways, or the alveolar spaces. As this text is aimed primarily at surgical pathologists, pathological schemas of classification will be adopted primarily with reference to their clinical counterparts when appropriate.

Tracheobronchitis/Bronchiolitis and Miliary Infection

Many pathogens target the conducting airways to produce tracheobronchitis and bronchiolitis. Pathologic changes range from superficial erosion of the lining respiratory epithelium, to ulceration and repair. The type of inflammation will vary from intraluminal neutrophilic exudates (Fig. 9.3) to airway cuffing by lymphocytes and histiocytes (Fig. 9.4), depending on the offending pathogen.

Diffuse Alveolar Damage

Disease of the gas-exchange alveolar surfaces can show a spectrum of changes, including acute ulceration and septal infiltration by chronic inflammatory cells. Diffuse alveolar damage (DAD) represents a global injury to the gas-exchange surfaces due to disruption of the blood—air barrier leading to exudative edema and fibrosis, resulting in severely impaired blood and tissue oxygenation (Fig. 9.5). The *sine qua non* of DAD is the *hyaline membrane* that is composed of necrotic alveolar lining

TABLE 9.2 Injury Patterns Seen in Lung Infection

Tracheobronchitis
Bronchiolitis (acute, chronic, necrotizing)
Bronchiectasis
Bronchopneumonia (acute, chronic, necrotizing)
Eosinophilic pneumonia
Pulmonary hemorrhage
Pulmonary edema
Diffuse alveolar damage
Pulmonary nodules and micronodules
Cavitary pneumonia
Vasculitis
Capillary dissemination
Lymphatic dissemination
Pulmonary hypertension
Pleuritis

cell debris and an extravascular fibrin coagulum apposed to an ulcerated alveolar wall, which yields a gel that entraps lung water (Fig. 9.6). Although DAD is the most frequent pathological cause of the clinical entity, the adult respiratory distress syndrome (ARDS), other diseases, including extensive bronchopneumonia and acute pulmonary hemorrhage can also lead to ARDS (Table 9.3). The pathology of the exudative phase of DAD can focally mimic acute bacterial infection and one must maintain a high threshold for making the diagnosis of acute infection in this setting (Fig. 9.7).

Viruses are the most common infectious cause of DAD, although bacteria, fungi, and parasites also produce diffuse lung injury. DAD can result from sepsis that complicates either pulmonary or extrapulmonary infection. Whenever DAD is present, the surgical pathologist must examine the lung for evidence of viral-induced cytopathic changes. These vary with the type of viral infection, and some viruses do not produce cytopathic changes, so that viral infection is always in the differential diagnosis of DAD (Table 9.4). A common pitfall in diagnosis is to mistake the hyperplastic reparative alveolar type II cells of DAD with viral infected cells, particularly when examining rapidly frozen sections, in which these changes may be especially prominent (Fig. 9.8).

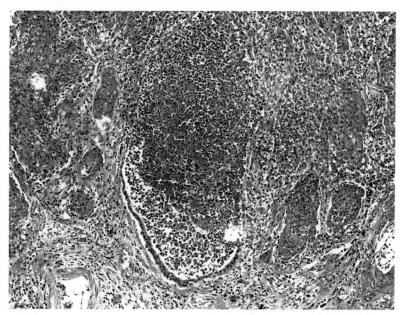

FIGURE 9.3   Acute bronchiolitis showing neutrophilic exudate in the lumen of a small airway.

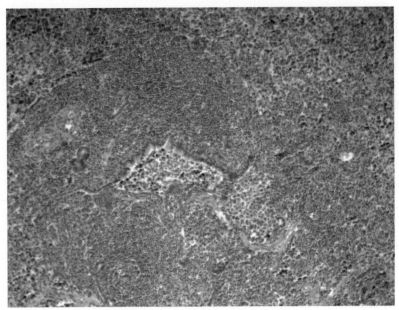

FIGURE 9.4   Chronic lymphocytic bronchiolitis.

RNA VIRUSES

Influenza

Influenza is a rod-shaped RNA virus that can cause either bronchiolitis or DAD without cytopathic changes. Influenza infection recurs each year due to a high incidence of mutation of its hemagglutinin (H) and neuraminidase (N) antigens,

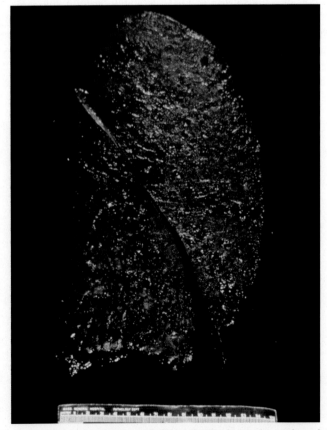

FIGURE 9.5   Consolidated lung with the beefy red appearance of diffuse alveolar damage in a patient who died of influenzal pneumonia.

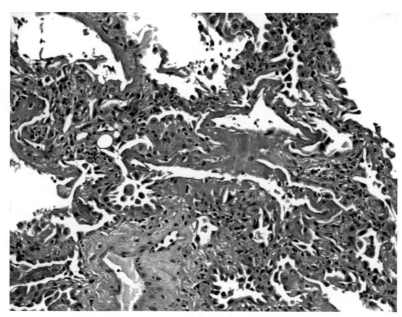

FIGURE 9.6   Hyaline membrane lining an alveolar duct in DAD.

and these determine its virulence. When mutations occur concomitantly in both the H and N antigens, pandemics with potentially high degrees of morbidity due to the lack of immunity may ensue (Fig. 9.9). Currently, epidemiologists are carefully monitoring the evolution of an avian influenza in southeast Asia for evidence of spread to man.

Influenza is the most common cause of viral pneumonia, although most cases are subclinical. The virus most commonly causes a diffuse tracheobronchitis/bronchiolitis in which the normal ciliated respiratory epithelium is sloughed. But when DAD develops, it carries a high mortality even in the absence of acute bacterial superinfection (Fig. 9.10). The lungs in DAD due to influenza of patients with prolonged survival often develop prominent squamous metaplasia of bronchial-alveolar lining cells (Fig. 9.11A). Although these findings are characteristic, they are also nonspecific, so that immunostains, in situ hybridization, electron microscopy, or viral antigen detection, may be required to establish the diagnosis (Fig. 9.11B). Superinfection by pyogenic bacteria, including *H. influenza, Group A Streptococcus,* and *Staphylococcus,* is a well-recognized complication and may mask evidence of a healing influenza infection.

Serious Acute Respiratory Syndrome

The recent epidemic of the zoonotic *coronavirus* infection termed serious acute respiratory syndrome (SARS) fortunately has not recurred, as the virus led to acute respiratory distress with high mortality. The lungs at autopsy showed DAD with scattered multinucleated giant cells of uncertain diagnostic significance. Otherwise, the virus otherwise produced no cytopathic changes and was essentially histologically indistinguishable from DAD due to influenza.

TABLE 9.3 Causes of Diffuse Alveolar Damage
Pulmonary Infection
Systemic infection with sepsis
Vasodilatory shock
Aspiration
Drugs
Radiation
Trauma
Accelerated phase of chronic interstitial pneumonia
Idiopathic

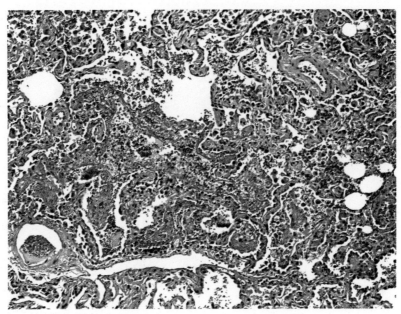

FIGURE 9.7   Lung in exudative phase of DAD with hemorrhage and inflammation. This finding should not be over interpreted as evidence of infection.

Middle Eastern Respiratory Syndrome

Like SARS, middle eastern respiratory syndrome (MERS) is caused by a coronavirus. The first cases were reported in Saudi Arabia in the fall of 2012, but the first recognized cases occurred in Jordan earlier that year. All cases of MERS to date have been linked through travel to or residence in countries in and near the Arabian Peninsula. The largest known outbreak of MERS outside the Arabian Peninsula occurred in the Republic of Korea in 2015 and was associated with a traveler returning from the Arabian Peninsula. MERS-CoV spreads from ill people to others via close contact, such as caring for or living with an infected person, and patients have ranged in age from infants to nonagenarians.

Patients present with fever, cough, and dyspnea. Almost 75% of reported patients with MERS have died and most have had an underlying medical condition. Some infected people have had mild symptoms of a upper respiratory infection (URI) or even no symptoms at all, and they all recovered. Incubation time ranges from 2 days to 2 weeks. Diagnosis is confirmed by RT-PCR at the Center for Disease Control for the implicated coronavirus. The pathology of the disorder has not been established but radiographically the pictures appear to be that of a severe organizing pneumonia.

TABLE 9.4 Changes Seen in Viral Infected Lung Cells

Organism	Cytopathic Change
Influenza	No cytopathic change
SARS (coronavirus)	No cytopathic change
Respiratory syncytial virus	Polykaryons, inconspicuous cytoplasmic inclusions
Parainfluenza	Polykaryons, intracytoplasmic inclusions
Measles	Polykaryons, intranuclear inclusions
Adenovirus	Intranuclear inclusions (smudge cells)
Herpesvirus	Intranuclear inclusions, polykaryons
Cytomegalovirus	Intranuclear and cytoplasmic inclusions
Varicella zoster	Intranuclear inclusions
EBV	No cytopathic change

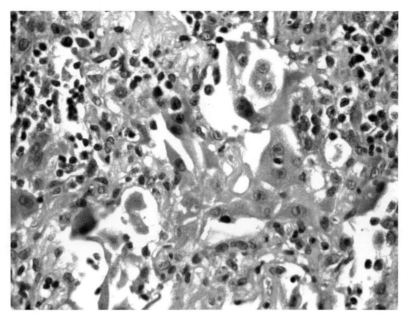

FIGURE 9.8   Lung in reparative phase of acute lung injury showing highly atypical alveolar lining cells with changes that mimic viral infection.

Respiratory Syncytial Virus

Respiratory syncytial virus (RSV) causes a benign respiratory infection in older children and has been recognized as a cause of adult community-acquired pneumonia, acute bronchiolitis, and DAD in the immunosuppressed host. The infection targets the respiratory lining epithelium producing syncytial giant cells with nonprominent eosinophilic inclusions (Fig. 9.12A and B). Human metapneumovirus produces changes comparable to respiratory syncytial virus (RSV) and must be included in its differential diagnosis.

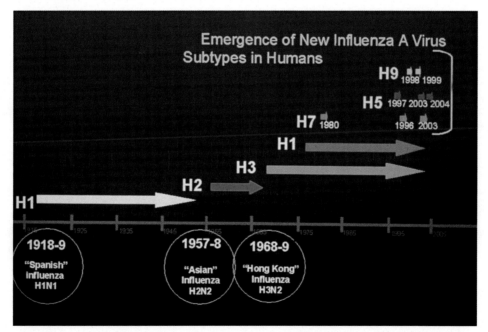

FIGURE 9.9   Neuraminidase and hemagglutinin expression of *influenza* correlates with epidemic outbreaks.

FIGURE 9.10   Lung from patient who died in the 1918 influenza epidemic showing DAD with no cytopathic changes.

Parainfluenza

Parainfluenza causes a benign URI in children that rarely progresses to DAD, although severe disease may develop in the immunosuppressed host. Like RSV, parainfluenza produces bronchiolitis and DAD with syncytial giant cells and epithelial cell intracytoplasmic inclusions. However, the latter are both more frequent and larger than those seen in RSV (Fig. 9.13A and B).

Measles

Measles pneumonia is a rare and serious complication of the childhood viral exanthem. The pathology of pulmonary measles infection ranges from bronchiolitis (Fig. 9.14A) to DAD. The virus produces multikaryons with prominent glassy eosinophilic nuclear Cowdry type A inclusions (Fig. 9.12B). The differential diagnosis of giant cell pneumonia includes RSV and hard-metal pneumoconiosis; however, the giant cells in the latter disorders lack intranuclear inclusions and the giant cells in hard-metal pneumoconiosis specifically lack the exudative features of an acute infection.

(A)   **(B)**

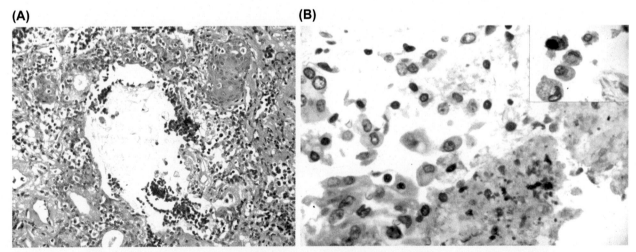

FIGURE 9.11   (A) Lung in patient with DAD due to influenza showing prominent squamous metaplasia of terminal airways; (B) immunostain confirms the presence of influenza A.

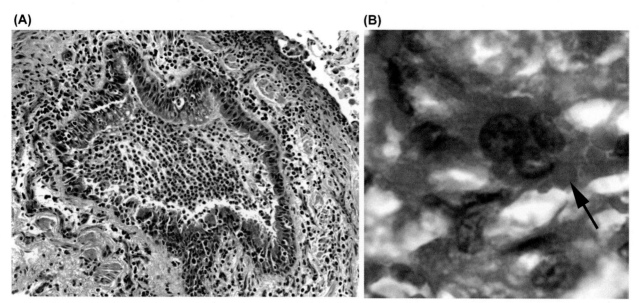

FIGURE 9.12 (A) Acute bronchiolitis in RSV; (B) multinucleated epithelial cells in RSV contain inconspicuous eosinophilic cytoplasmic inclusions (*arrow*).

DNA VIRUSES

Adenovirus

Adenovirus primarily affects the immunosuppressed host but can produce outbreaks in healthy subjects living at close quarters, e.g., military recruits. *Adenovirus* typically produces (1) ulcerative bronchiolitis with karyorrhexis (Fig. 9.15A), (2) neutrophilic pneumonia (Fig. 9.15B) (3) acute intrapulmonary necrosis with *hemorrhage* (Fig. 9.15C), or (4) DAD. Infected cells can exhibit amphophilic intranuclear inclusions with perinuclear clearing that mimic *herpesvirus* infection but more characteristically produce "smudge cells" (Fig. 9.15D), showing hyperchromatic nuclei extruding beyond the confines of their nuclear membranes.

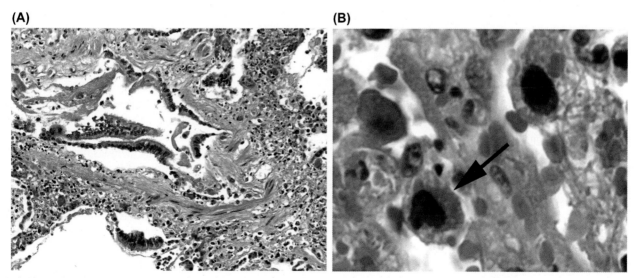

FIGURE 9.13 (A) Bronchiolitis in parainfluenza infection; (B) epithelial cell showing eosinophilic inclusions that are both larger and more frequent than in RSV (*arrow*).

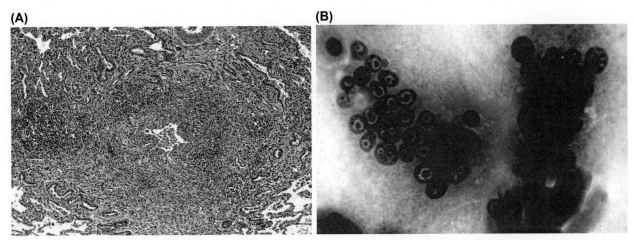

FIGURE 9.14 (A) Chronic inflammation involving a small airway in patient with measles. (B) Multinucleated cell showing glassy nuclear Cowdry type A inclusions.

The appearance of "smudge cells" can be mimicked by pulmonary cytotoxic drug injury or by epithelial repair in the early proliferative phase of DAD. For this reason and because of the potential overlap with *herpesvirus*-induced cytopathic changes, the diagnosis of *adenovirus* infection should always be confirmed by immunohistochemical staining, ultrastructural examination, or viral isolation.

Cytomegalovirus

Cytomegalovirus occurs at the extremes of age, or as a result of immunosuppression, and it is a common infection in HIV/ acquired immunodeficiency syndrome (AIDS). The number of cells showing cytopathic changes can vary considerably and parallels the severity of infection. Cytomegalovirus (CMV) primarily targets pulmonary macrophages and endothelial cells, but virtually any cell can show cytopathic features. The most common distribution is blood-borne miliary disease (Fig. 9.16) but bronchiolitis and DAD also occur. The diagnostic features of infection are (1) cytomegaly, (2) intranuclear inclusions with characteristic Cowdry Type B inclusions (Fig. 9.17A), (3) ill-defined amphophilic intracytoplasmic inclusions that are seen with hematoxylin and eosin (H&E), PAS, and GMS stains (Fig. 9.17B). In patients receiving prophylactic treatment with antivirals, CMV infection may fail to exhibit cytopathic changes. However, immunostains and in situ hybridization will continue to identify intracellular CMV antigens (Fig. 9.17C).

CMV is a frequent copathogen in the immunosuppressed patient and a cause of immunosuppression in its own right. CMV infection can be seen together with other viral infections, *Pneumocystis jirovecii*, or opportunistic fungal infections (Fig. 9.18).

Herpesvirus

Herpesvirus types 1 and 2 both infect the lung. The incidence of herpesvirus infection increases with immunosuppression and when mucosal barrier defenses have been breached, *herpesvirus* characteristically elicits a prominent neutrophilic response that mimics pyogenic bacterial infection, and foci of necrosis, cell karyorrhexis, and piled up viral infected cells with amphophilic nuclei confirm the diagnosis (Fig. 9.19). Cytopathic diagnosis includes type A or type B Cowdry nuclear inclusions showing molding of adjacent cells with multikaryon formation (Fig. 9.20). Immunosuppressed patients with *herpesvirus* viremia develop miliary foci of hemorrhagic necrosis with prominent fibrinous exudates (Fig. 9.21A and B).

Herpesvirus pulmonary infections also develop in patients with structural abnormalities of the airways or as complications of primary infections of the oropharynx and esophagus. Intubated patients receiving chronic ventilatory support and are at increased risk due to local barotrauma from inflated endotracheal tubes. The respiratory mucosa is the primary target (Fig. 9.22). At times, extensive necrosis of an ulcerated airway suggests the diagnosis but immunostaining for herpes viral antigen can demonstrate high background staining obscuring the diagnosis. When this is the case, examining paraffin-embedded tissues by electron microscopy can reveal diagnostic virions (Fig. 9.23A−C).

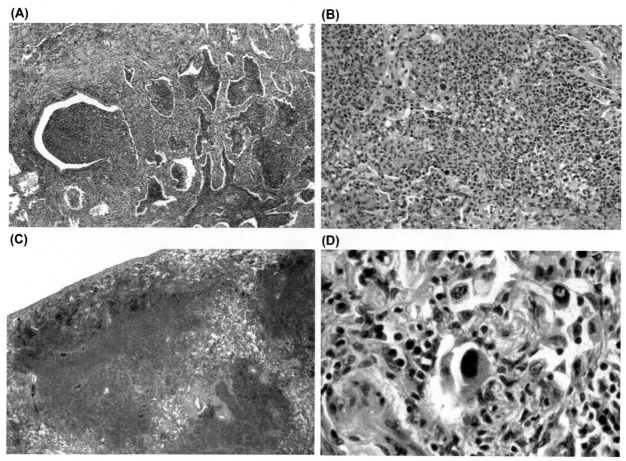

FIGURE 9.15 (A) Ulcerative bronchiolitis in adenovirus infection. (B) Neutrophilic pneumonia due to adenovirus. (C) Necrotizing hemorrhagic pneumonitis. (D) "Smudge cell" showing extrusion of nuclear contents beyond the confines of the nuclear membrane.

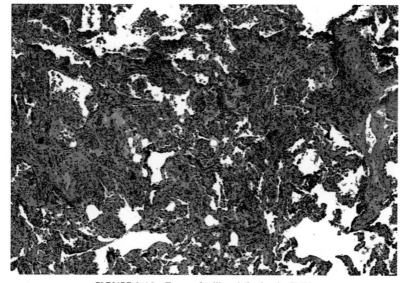

FIGURE 9.16 Focus of miliary infection in CMV.

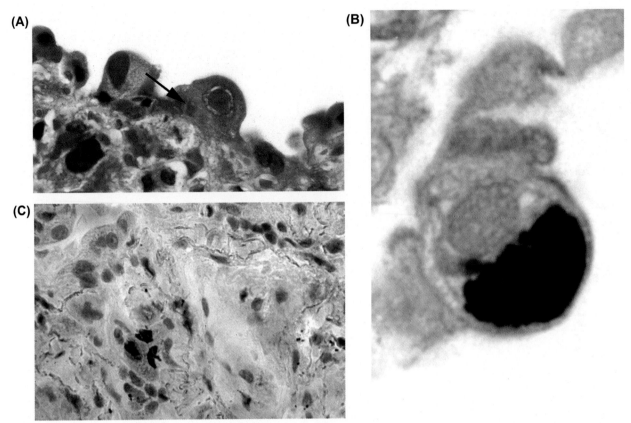

FIGURE 9.17 (A) Alveolar type II lining cells with prominent cytomegaly, Cowdry type B inclusions, and cytoplasmic inclusions (*arrow*). (B) Cytoplasmic inclusions stain positive with GMS. (C) Immunostain demonstrates CMV antigen in patient treated with ganciclovir.

Varicella Zoster

Varicella zoster pneumonia is a rare complication of the childhood chickenpox and it is more commonly encountered as the consequence of reactivated virus in the immunocompromised host. Following nonlethal pulmonary infections in childhood, the lung shows multiple calcified miliary lesions. Cases coming to biopsy or autopsy show miliary nodules of

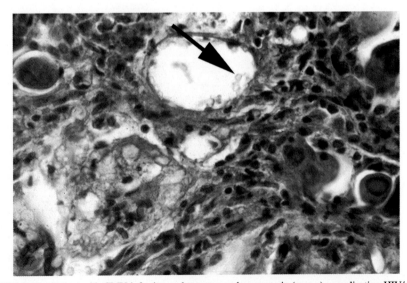

FIGURE 9.18 Patient with CMV infection and cryptococcal pneumonia (*arrow*) complicating HIV/AIDS.

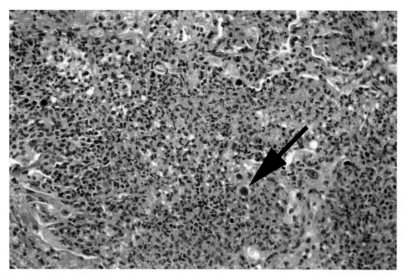

FIGURE 9.19   Neutrophilic bronchopneumonia due to *herpesvirus-1*. *Arrow* points to viral infected cell.

hemorrhagic necrosis in lung and pleura or DAD (Fig. 9.24). Infected cells with primarily Cowdry type A inclusions may be seen at the edges of the lesion but they are harder to identify than in *herpesvirus* pneumonia.

Hantavirus

This virus produced an epidemic in the four corners region of the southwestern United States in 1993. The infection is a zoonosis transmitted by infected rodent feces. The most common radiographic presentation is diffuse pulmonary edema with pleural effusions mimicking congestive heart failure. Histologically, the lung shows pulmonary edema with scant poorly formed hyaline membranes (Fig. 9.25A) with atypical lymphocytes circulating within the pulmonary vasculature (Fig. 9.25B). Confirmation of the diagnosis requires specific immunohistochemistry, serological evidence of hantavirus-specific IgM and PCR or ultrastructural identification of the causative virions.

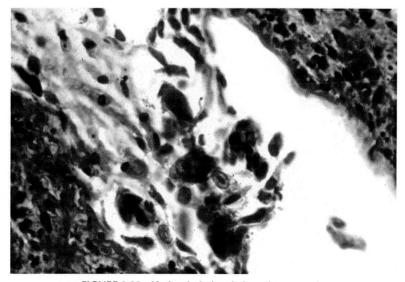

FIGURE 9.20   Nuclear inclusions in herpetic pneumonia.

(A)   **(B)**

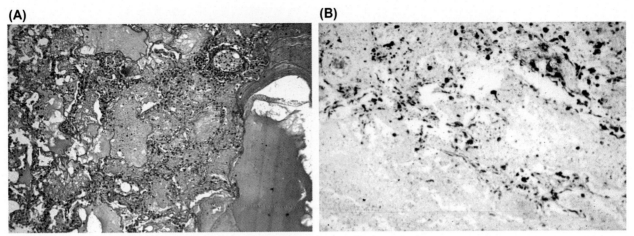

FIGURE 9.21   (A) Hemorrhagic necrotizing pneumonia in immunosuppressed patient with herpesvirus-1 viremia. (B) Multiple infected cells immunostaining for *herpesvirus-1*.

OTHER "ATYPICAL PNEUMONIAS"

Mycoplasma Pneumonia

Mycoplasma are the smallest (0.2–0.8 μM) free-living bacteria but they lack a true cell wall. They are facultative anaerobes, except for *Mycoplasma pneumoniae*, the most common pulmonary pathogen, which is a strict aerobe. Mycoplasma pneumonia occurs worldwide with no increased seasonal activity, but epidemics predictably occur every 4–8 years. Although primarily an infection of young adults, it can attack the elderly. The most common clinical syndrome is tracheobronchitis, with one-third of patients developing a mild but persistent pneumonia.

Mycoplasma pneumonia is rarely biopsied, as positive cold agglutinin and specific complement fixation antigen assays will establish the diagnosis. Biopsied cases show lymphocytic or neutrophilic bronchiolitis with alveolar wall inflammation and fibrinous exudates (Fig. 9.26). Similar changes are seen in both *Chlamydia* (Fig. 9.27) and *Coxiella* pneumonia.

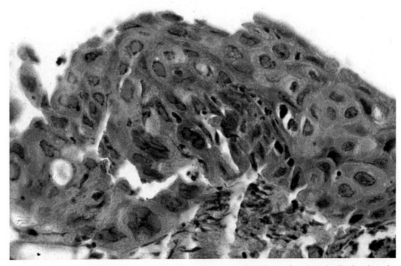

FIGURE 9.22   Herpetic inclusions in squamous respiratory epithelium of a chronically intubated patient.

(A)

(B)

(C)

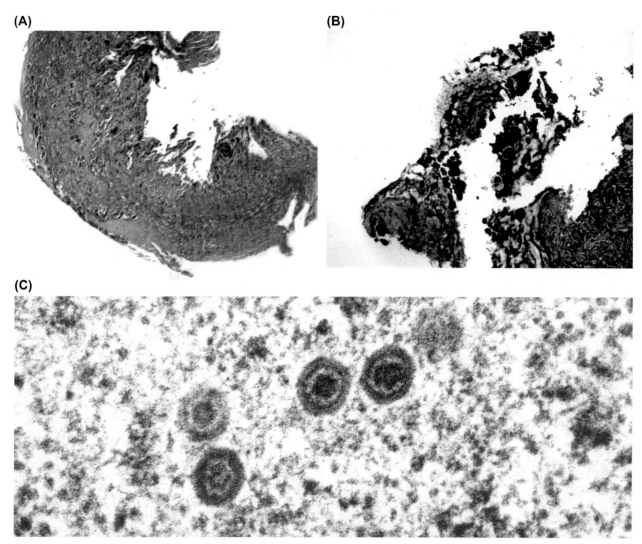

FIGURE 9.23   (A) Ulcerated tracheal lesion showing; (B) intense immunostaining for *herpesvirus-1* with diagnosis; (C) confirmed by ultrastructural examination demonstrating diagnostic virions.

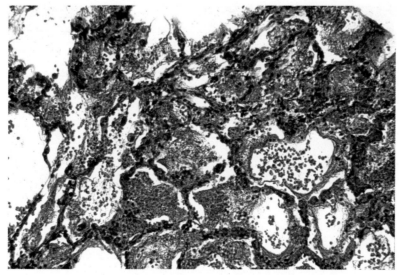

FIGURE 9.24   Hemorrhagic pneumonia due to varicella zoster.

(A)   **(B)**

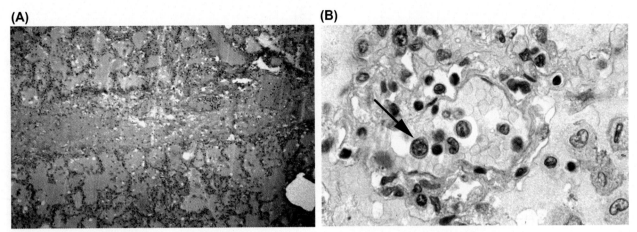

FIGURE 9.25 (A) Pulmonary edema in patient with *hantavirus* infection. (B) Pulmonary vessel with intraluminal atypical lymphocytes (*arrow*).

Epstein–Barr Virus

Epstein–Barr virus (EBV) has been implicated in disorders ranging from the mononucleosis syndrome to malignant lymphoid neoplasia. The mononucleosis syndrome includes pharyngitis, lymphadenitis, and hepatosplenomegaly. Pulmonary involvement can occur as part of the syndrome but it is unusual and rarely biopsied. EBV pneumonia shows patchy peribronchiolar and interstitial polyclonal lymphoid infiltrates with scant interstitial and intra-alveolar fibrin exudates (Fig. 9.28). The diagnosis is generally established serologically by EBV antigen titers but can be confirmed by in situ hybridization.

Pneumocystis jirovecii

For years, the organism formerly known as *Pneumocystic carinii* was thought to be a protozoon; however, it is now confidently classified as a fungus. Originally described as a plasma-cell interstitial pneumonia in malnourished children, it was subsequently seen in patients with hematological malignancies, and in those receiving chemotherapy, or chronic corticosteroids. In the 1980s, *Pneumocystis* pneumonia became a signal infection in establishing the diagnosis of AIDS. It was encountered in epidemic proportions, until prophylactic use of trimethoprim-sulfa (*Bactrim*) became a routine aspect of HIV/AIDS management. *P. jirovecii* is currently most often diagnosed by sputum induction, and lung biopsy is reserved for diagnostic challenges.

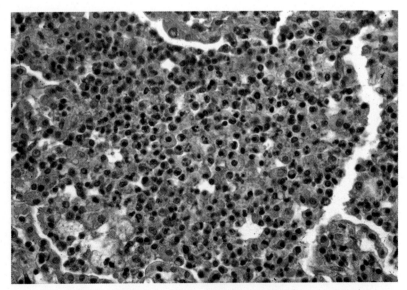

FIGURE 9.26 Histiocytic and fibrinous exudates in *mycoplasma* pneumonia.

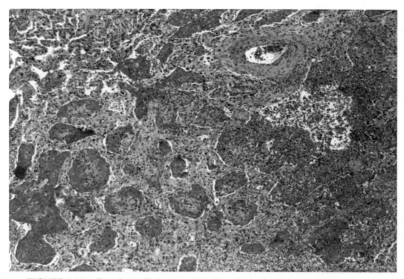

FIGURE 9.27   Prominent fibrinous and histiocytic exudates due to *Chlamydia* spp.

Although the disease presents radiographically as an atypical "interstitial" pneumonia, interstitial inflammation is not its most striking pathological feature. In pneumocystis pneumonia, the alveoli are filled with a frothy eosinophilic exudate that mimics exudative pulmonary edema or alveolar lipoproteinosis (Fig. 9.29A). The pulmonary interstitium shows a mild plasma cell-rich pneumonitis and prominent alveolar type II cell hyperplasia.

The diagnosis is confirmed by the presence of oval, helmet, or crescenteric-shaped GMS+ "cysts," 4−6 μM in greatest dimension, within the alveolar froth (Fig. 9.29B). Pneumocystis is distinguished from GMS+ fungal yeast forms by the absence of budding, pericapsular accentuation, and a so-called intracytoplasmic "dot," in the former. However, when these features are absent and when the organism load in the biopsy is low, it may be difficult to confidently exclude *histoplasma* or *cryptococcus*. In these cases, specific immunohistochemistry can confirm the diagnosis.

Multiple patterns of unusual host reactions to *P. jiroveci* have been recognized. These include DAD (Fig. 9.30A), solitary necrotizing granulomas, (Fig. 9.30B), miliary infection, lymphoid interstitial pneumonia, and regional lymphadenitis. Microcalcifications may be seen in areas of infection and thin-walled cyst formation may develop.

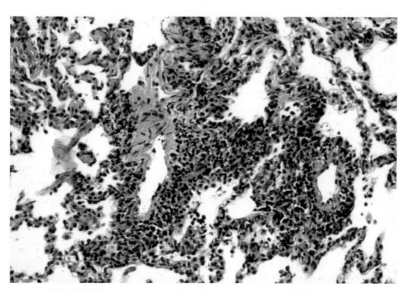

FIGURE 9.28   Dense lymphoid infiltrate in patient with EBV pneumonia.

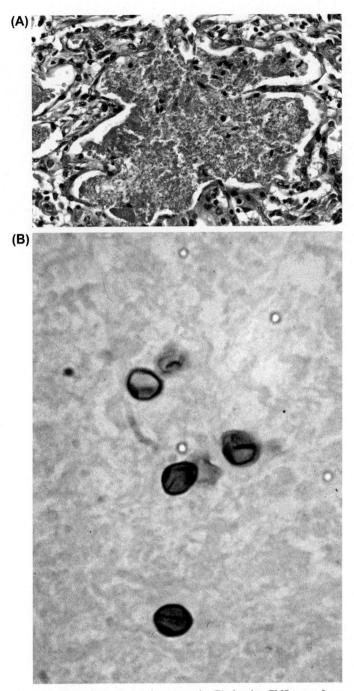

FIGURE 9.29 (A) Alveolar eosinophilic exudate in pneumocystis pneumonia, (B) showing GMS+ cyst forms with pericapsular accentuation.

Bronchiectasis

Bronchiectasis is one of the major manifestations of pulmonary infection. Dilatation and anatomical distortion of conducting airways can result from acute and chronic airway injury from airway infection (wet bronchiectasis) or from adjacent parenchymal scarring ("dry" or traction bronchiectasis). A number of infections can damage the large airways and leading bronchiectasis. Reversible cylindrical bronchiectasis may appear radiographically in patients following an acute bacterial pneumonia. As this change often resolves, biopsies of suspicious bronchiectatic regions should be deferred until a period of weeks past postacute infection. As regions of chronic bronchiectasis are fed by varicose bronchial arteries that often course directly beneath the airways surface, endoscopic biopsies of bronchiectatic areas are contraindicated.

(A)　　　　　　　　　　　　　　　　**(B)**

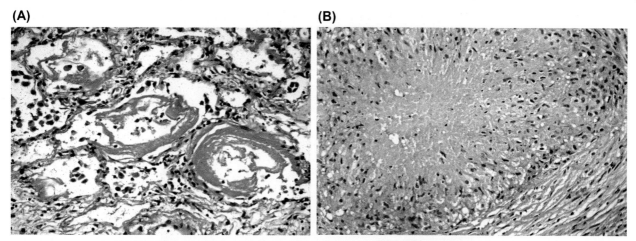

FIGURE 9.30　(A) DAD and (B) necrotizing granulomatous inflammation due to pneumocystis.

Prior to the development of childhood vaccination, viral exanthems and *Bordetella pertussis* were common causes of bronchiectasis (Fig. 9.31). Cystic fibrosis, a genetic disorder of chloride transport, is currently a leading cause of severe bronchiectasis in young adults, as advances in management have resulted in patients surviving beyond childhood (Fig. 9.32).

In bronchiectasis, the airways lose their cartilaginous support due to chronic inflammation and become dilated and prone to repeated bouts of mucoid impaction and infection. The airways develop a spectrum of gross changes ranging from cystic, varicose, to cylindrical dilatation, associated with distal peribronchiolar inflammation (Fig. 9.33A) and increasing degrees of parenchymal scarring with loss of gas-exchanging alveoli (Fig. 9.33B). Although the large airways are invariably ectatic, the distal airways in bronchiectasis are generally narrowed by constrictive bronchiolitis.

Recurrent polymicrobial infections with necrotizing bronchopneumonia and abscess formation complicate bronchiectasis. Patients are also prone to develop infections due to antibiotic-resistant mucoid forms of *Pseudomonas* and *Burkholderia spp.* (Fig. 9.33C). and *Staphylococcal* spp. Some patients with infective bronchiectasis develop allergic bronchopulmonary aspergillosis (ABPA) and superinfection with atypical mycobacteria.

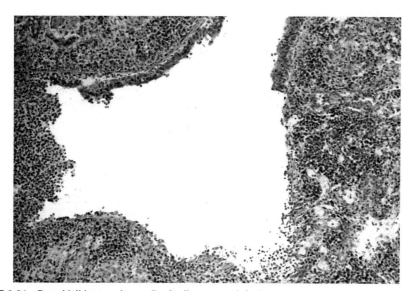

FIGURE 9.31　Bronchiolitis secondary to *Bordetella pertussis* infection, once a common cause of bronchiectasis.

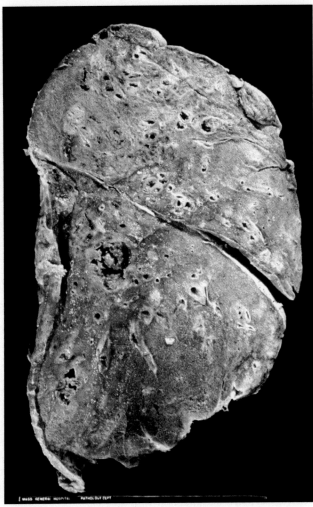

FIGURE 9.32   The lung in cystic fibrosis showing bronchiectasis and a cavitary abscess.

Acute Bronchopneumonia

Acute bronchopneumonia is the most common distribution of pulmonary infection. Gram-positive and Gram-negative bacteria, as well some viruses, including *herpesvirus* and *adenovirus*, elicit primarily an exudation of neutrophils, whereas the cellular responses to other viruses, fungi, *mycoplasma*, and *chlamydia* are primarily lymphohistiocytic. Bronchopneumonia generally results from microaspiration of pathogens that have colonized the oropharynx (Fig. 9.34A). At times, the aspiration of colonized food particles can carry bacteria and fungi into the lung (Fig. 9.34B). Terminal episodes of aspiration often show colonies of Gram-positive *Aerococci*, previously referred to as "Gaffkya," (Fig. 9.34C) and their appearance is both common and characteristic in autopsy lungs.

Bacterial Infections

The pyogenic bacteria are distinguished by their propensity to evoke acute neutrophilic inflammation and "pus." Pyogenic infections as well as other acute necrotizing bronchopneumonias progress to *organizing pneumonia*, characterized by a fibrohistiocytic response that obliterates small airways along with inflammation of the surrounding alveolar interstitium (Fig. 9.35). This reaction is nonspecific, and "organizing pneumonia" or "bronchiolitis-obliterans-organizing-pneumonia" is simply a rubric for the generic lesion that may be due to infection, noninfective inflammatory disorders, or may be idiopathic. It is important for the surgical pathologist to convey clearly to the treating clinicians that a diagnosis of "organizing pneumonia" does not indicate a specific etiology.

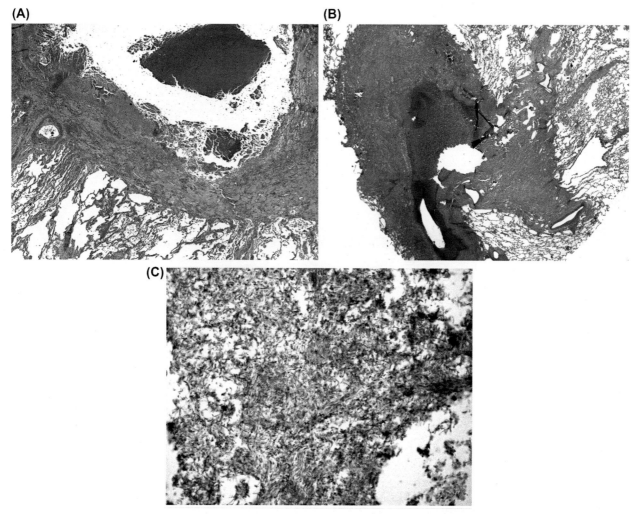

FIGURE 9.33 (A) Bronchiectasis in CF showing dilatation and destruction of airway wall, (B) dense peribronchiolar scar in bronchiectasis, (C) bronchiectatic airway with colonies of Gram-negative bacilli identified by culture as *Burkholderia cepacia*.

Pneumococcal Pneumonia

Streptococcal pneumonia (pneumococcal pneumonia) is a community-acquired pneumonia that classically produces a lobar pneumonia healing by resolution, i.e., without necrosis or scarring. In some cases, pneumococcal pneumonia complicates a resolving viral tracheobronchitis or influenza. In the age of antibiotic treatment, most cases do not progress to lobar involvement and are limited to acute nonnecrotizing bronchopneumonia. A large number of serotypes of *S. pneumonia* have been isolated, and type 3 can be virulent, producing necrotizing pneumonia, bacteremia, and death, despite prompt antibiotic treatment.

Although rarely biopsied, lobar pneumococcal pneumonia is still seen at autopsy. The early phase of the disease, the *red hepatization*, shows exudation of edema fluid with the diapedesis of red blood cells (Fig. 9.36A). The exudate spreads to fill an entire lung lobe via the pores of Kohn, interalveolar potential channels within the normal lung (Fig. 9.36B). The acute cellular response is neutrophilic, but this is followed within days by alveolar filling with exudate macrophages that ingest the infected exudate, the *gray hepatization* (Fig. 9.36C). Despite the extensive inflammatory changes, alveolar necrosis is absent—a finding best assessed by elastic stains—and the lung heals by *resolution* with minimal sequelae. The offending organisms are Gram-positive lancet-shaped cocci growing in pairs (diplococci) and in short chains, and they can be identified by both tissue Gram and GMS stains (Fig. 9.36D).

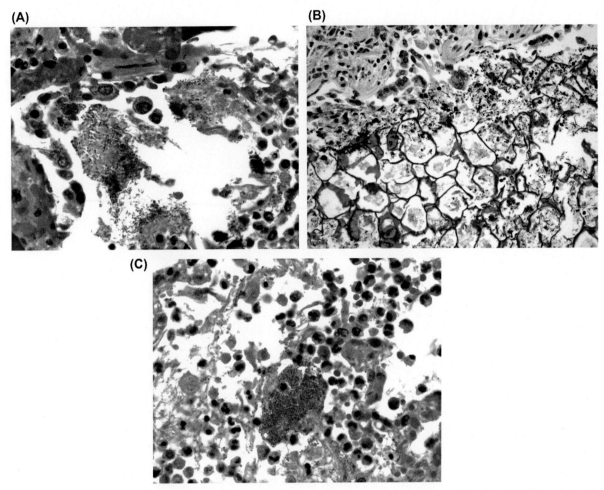

FIGURE 9.34 (A) Acute bronchiolitis due to *Staphylococcus,* (B) Lentil aspiration with colonies of Gram-positive bacteria; (C) terminal aspiration showing intraluminal colony of Gram-positive *Aerococcus* spp. (Gaffkya).

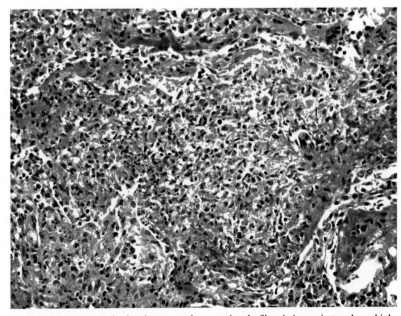

FIGURE 9.35 Organizing pneumonia showing macrophages and early fibrosis in respiratory bronchioles and alveoli.

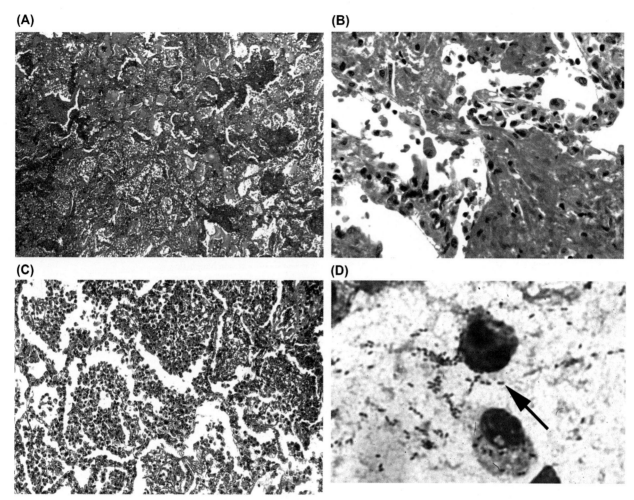

FIGURE 9.36 (A) In the red hepatization phase of pneumococcal lobar pneumonia, alveoli are filled by fibrin with diapedesis of red blood cells. (B) The infected exudate extends throughout the lobe via the pores of Kohn, potential channels between adjacent alveoli. (C) In the gray hepatization phase, alveoli are filled by leukocytes that scavenge bacteria and detritus prior to resolution of the infection. (D) The pneumococcus is a lancet-shaped Gram-positive coccus that grows in pairs (diplococci) (*arrow*) and chains.

Group A Streptococci

Group A streptococcal pneumonia occurs at the extremes of life or as a complication of resolving influenza infection. It produces a rapidly progressing and life-threatening pneumonia with edema, hemorrhage, abscess formation, empyema, and septicemia. *Group A streptococci* evoke a brisk increase in pulmonary capillary permeability, yielding what can be mistaken for cardiogenic pulmonary edema on low-power microscopic examination (Fig. 9.37A). But further scrutiny reveals necrotic macrophages and innumerable Gram-positive bacteria in chains (Fig. 9.37B). Pulmonary hemorrhage and abscess formation are frequently seen. Streptococci have a predilection to course along pulmonary lymphatic channels, recapitulating erysipeloid spread in the skin to produce an early empyema (Fig. 9.38).

Staphylococcal aureus

Staphylococcal aureus has emerged as frequently encountered life-threatening pulmonary pathogen. Previously seen as a community-acquired complication of influenza, staphylococcal infection is now a primary cause of nosocomial pneumonia, and its evolving drug resistance accounts for methicillin-resistant strains that are of particular concern for hospital infection control epidemiologists.

   S. aureus spp. produce a necrotizing pyogenic pneumonia with abscess formation (Fig. 9.39A). The organisms grow in clusters as Gram-positive microcolonies (Fig. 9.39B). Infection heals by organization with scarring, and cystic

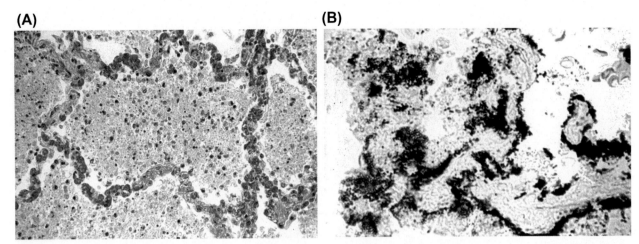

FIGURE 9.37 (A) The alveolar spaces in streptococcal pneumonia show an eosinophilic exudate that mimics proteinaceous pulmonary edema at low-power. (B) Innumerable Gram-positive cocci in chains are seen in the exudates and tend to invade lymphatics.

pneumatoceles may develop. Some phage-infected strains of the organism produce an exotoxin that can activate the CD3 receptor to promote the release of the T-lymphocyte's complement of lymphokines, leading to a *toxic-shock syndrome* with sepsis physiology, DAD, and disseminated intravascular coagulation that leads to death if not treated promptly.

Gram-negative Bacteria

Most Gram-negative bacilli produce a necrotizing bronchopneumonia with hemorrhage and abscess formation. Certain virulent Gram-negative species, including *Klebsiella, Pseudomonas, Acinetobacter*, and *Burkholderia spp.*, have a propensity to infect the pulmonary microvasculature leading to necrosis, bacteremia, and septic shock. The lung shows fibrinoid necrosis with colonies of Gram-negative bacilli streaming along the vessel walls, where they create an ill-defined purplish hue in H&E-stained sections (Fig. 9.40A and B).

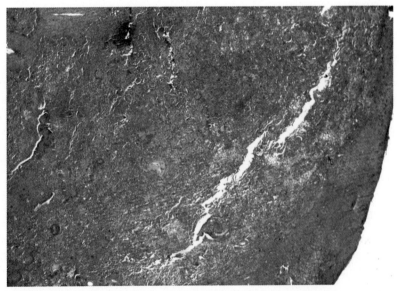

FIGURE 9.38 Abscess formation and empyema occur early in streptococcal pneumonia.

(A)

(B)

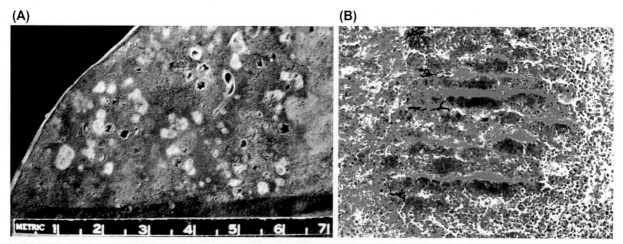

FIGURE 9.39 (A) Lung in staphylococcal pneumonia showing bronchopneumonia with abscess formation. (B) Microabscess with Gram-positive *Staphylococcal aureus* in clusters.

Klebsiella

Klebsiella pneumonia generally occurs in patients who are immunocompromised due to age, ethanol abuse, or diabetes mellitus. It is a common of ventilator-associated pneumonia. Like the pneumococcus, *Klebsiella spp.* classically produce a lobar pneumonia, with an unexplained predilection for the upper lobes (Fig. 9.41A). Infection produces hemorrhagic necrosis, microabscesses, and cavity formation. The organisms are short Gram-negative bacilli that can be demonstrated with both tissue Gram stain (Fig. 9.41B) and GMS by virtue of their capsules. Less commonly, *Klebsiella spp.* produce a chronic necrotizing pneumonia with scarring and distortion of the pulmonary anatomy (Fig. 9.41C).

Lung Abscess due to Oropharyngeal Aspiration

Lung abscess complicates the aspiration of polymicrobial oropharyngeal bacteria. The organisms isolated from a lung abscess include mixed Gram-positive and -negative aerobic flora, as well as anaerobes. Patients with poor oral hygiene due to dental caries, those with gingival and tonsillar disease, and those with disorders that impair either consciousness or normal swallowing, e.g., ethanolism, seizure disorder, and cerebrovascular accidents, are at increased risk for aspiration pneumonia and lung abscess. The disorder begins as a necrotizing bronchopneumonia in a dependent segment of the lung and progresses to produce a cavitary abscess communicating with a feeding adjacent airway. Most lung abscesses are diagnosed and treated noninvasively; however, failure to respond to medical treatment due to poor drainage and closure may prompt surgical excision.

Microscopically, the wall of the lung abscess is irregular with a shaggy fibrinous lining (Fig. 9.42A) that may be difficult to distinguish from a necrobiotic rheumatoid nodule or Wegener's granulomatosis. However, palisading granulomatous inflammation is not a prominent finding in lung abscess. The area around the cavity shows acute and organizing bronchopneumonia (Fig. 9.42B). The activity of a lung abscess may be determined microscopically by examining its lining, as actively infected cavities show a squamous epithelium that indicates complete healing, and may be mistaken for a region of primary bronchiectasis.

Actinomycosis

Actinomyces spp. are Gram-positive filamentous bacteria that cause a chronic distal necrotizing pneumonia with a proclivity toward penetrating into the adjacent soft tissues of the chest wall *Actinomyces* are aspirated from the oropharynx, where they are commonly part of the tonsillar flora in younger patients or a pathogen related to poor oral hygiene and gingivitis in the elderly. The risk factors for actinomycosis are similar to those of lung abscess. The infection extends into the pleura and then forms sinuses within the soft tissues of the chest wall that ultimately exit at the skin surface. The indurated pulmonary lesion may be mistaken clinically for an aggressive peripheral lung malignancy but its gross appearance at surgery is usually distinct (Fig. 9.43).

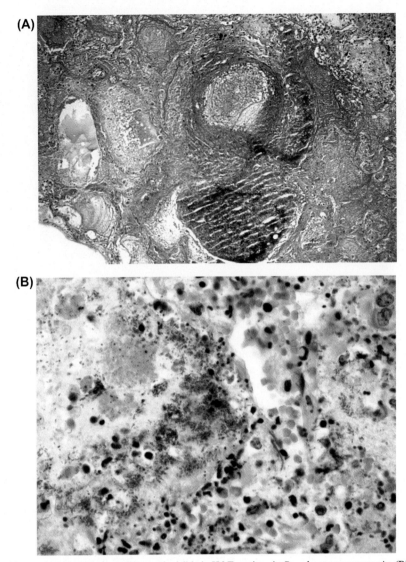

FIGURE 9.40 (A) Fibrinoid necrosis with dense bacterial growth visible in H&E sections in *Pseudomonas pneumonia*. (B) Clusters of Gram-negative bacilli are identified.

The histological response evoked by actinomycosis is variegate, with microabscesses, polymorphous infiltrates of lymphocytes, histiocytes, plasma cells, giant cells, and fibrosis. At times, the extent of the fibroinflammatory response can lead one to consider a diagnosis of "inflammatory pseudotumor." The correct diagnosis is established by the presence of "sulfur granules" (Fig. 9.44A)—bright yellow specks seen with the naked eye or with the aid of a hand lens that microscopically represent colonies of tangled Gram-positive, GMS-positive, beaded filamentous bacilli (Fig. 9.44B and C), coated by an eosinophilic matrix of exudate plasma proteins termed the *Splendore−Hoeppli* reaction. Treatment includes long-term antibiotics and surgical resection.

The differential diagnosis of "sulfur granules" in the lung includes *botryomycosis* a term used to describe a variety infections by colonies of Gram-positive cocci, either *Streptococcus* or *Staphylococcus* that grossly appear comparable to the sulfur granules caused by *Actinomyces*. Distinguishing these entities depends on the morphology of the bacteria in the granules, i.e., cocci in botryomycosis versus filamentous bacteria in actinomycosis (Fig. 9.45A and B).

Nocardia

Nocardia spp. produces pneumonia primarily in the immunosuppressed host. The organism shares morphological and histochemical staining features with *Actinomyces*, but its clinical and histological responses are usually easily

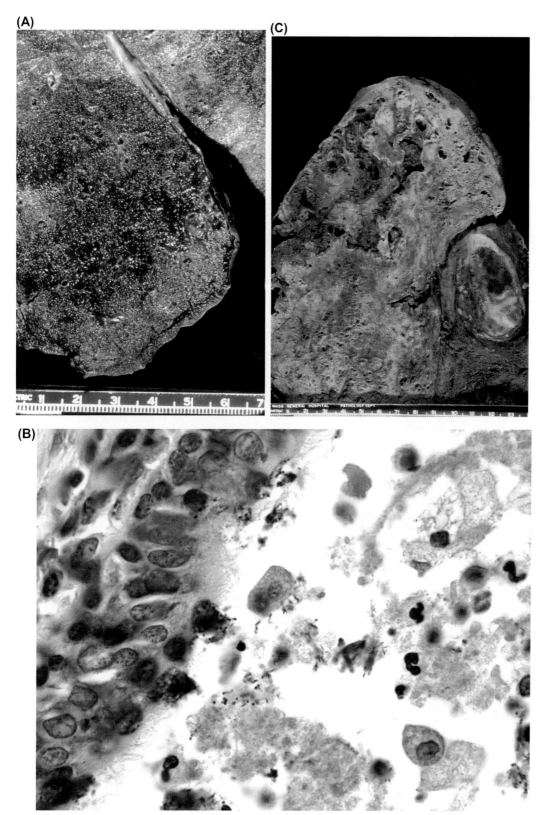

FIGURE 9.41 (A) Hemorrhagic lobar pneumonia due to *Klebsiella pneumonia*. The organism has a predilection for the upper lung lobes; (B) *Klebsiella* is a small nonmotile Gram-negative rod that also stains with GMS; (C) chronic necrotizing pneumonia due to *Klebsiella* spp. with extensive scarification of necrotic lung.

(A)

(B)

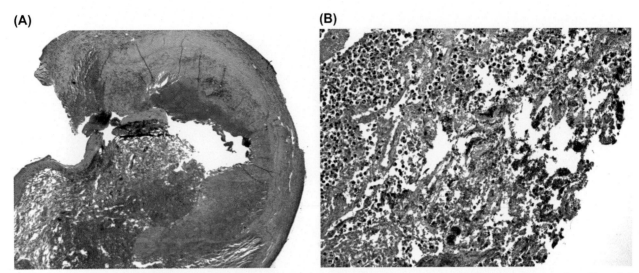

FIGURE 9.42 (A) Shaggy fibrinous exudate lining wall of lung abscess due to aspiration of oropharyngeal mixed flora; (B) necrosis and bacteria in lung abscess.

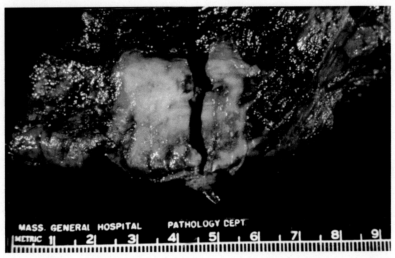

FIGURE 9.43 Subpleural nodule of actinomycosis resected for suspicion of carcinoma. Note the yellow-tan appearance that suggests inflammation rather than malignancy.

(A)

(B)

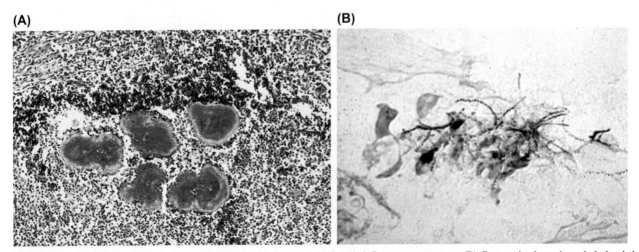

FIGURE 9.44 (A) Multiple actinomycotic sulfur granules within granulohistiocytic inflammatory response; (B) Gram stain shows irregularly beaded filaments of *Actinomyces* spp. that also stain with GMS. Recall that GMS stains all Gram-positive organisms and with no specificity for actinomycosis.

(A)

(B)

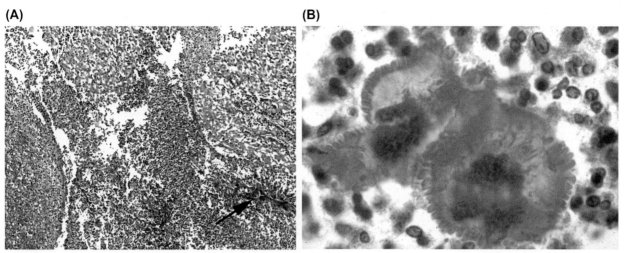

FIGURE 9.45 (A) Low-power view of botryomycosis in an infected airway (*arrow*); (B) high power of botryomycotic granule due to *Staphylococcus aureus*.

(A)

(B)

(C)

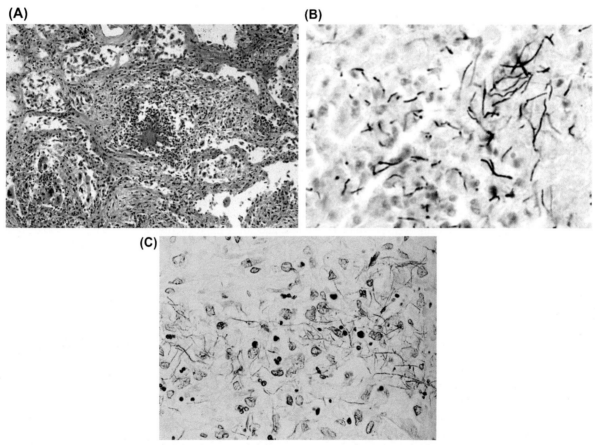

FIGURE 9.46 (A) Necrotizing pneumonia due to *Nocardia asteroides* shows granulohistiocytic response; (B) like *Actinomyces*, *Nocardia* is a Gram-positive filamentous bacillary actinomycetes that also stains with GMS. (C) Unlike *Actinomyces*, *Nocardia* is weakly acid-fast, and can be demonstrated with the Fite–Ferraco stain.

distinguished. *Nocardia* produces a necrotizing pneumonia with granulohistiocytic inflammation (Fig. 9.46A) and special stains highlight tangles of Gram-positive, GMS-positive filamentous organisms showing less beading than *Actinomyces* (Fig. 9.46B). Sulfur granules are rarely seen in pulmonary infections. As distinct from *Actinomyces*, *Nocardia* are weakly acid fast and can be highlighted with modified Ziehl–Neelsen stains, e.g., Fite–Ferraco (Fig. 9.46C). Nocardiosis is a

(A)   **(B)**

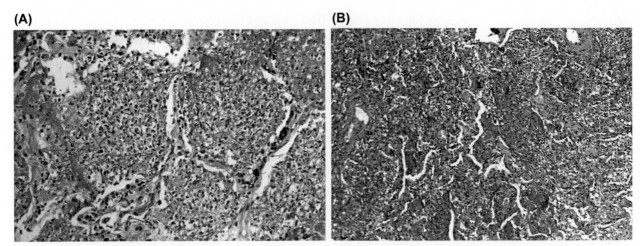

FIGURE 9.47   (A) *Legionella* spp. characteristically produce a necrotizing bronchopneumonia with alveolar filling by fibrin and histiocytes; (B) This appearance must be distinguished from AFOP, in which the alveolar spaces are filled with fibrin, but this disorder is not due to infection.

recognized complication of pulmonary alveolar lipoproteinosis and has recently been observed in patients receiving therapies that interfere with the activities of tumor necrosis factor-α.

Legionella

Legionella spp. produce pulmonary infections ranging in severity from a mild respiratory illness to life-threatening pneumonia. Patients may suffer from a modest degree of immunosuppression due to diabetes and ethanolism. The organism is water-borne, and infected water and air conditioning sources have caused point outbreaks of infection.

The diagnosis of *Legionella* infection is currently established noninvasively by immunoassays; however, the disease may be identified histologically in biopsy or autopsy tissues. The infected lung characteristically shows a fibrinohistiocytic response with alveolar filing by fibrin and macrophages with a sparse neutrophilic component (Fig. 9.47A), although its histology may also be indistinguishable from that of pyogenic infections. It can be mimicked by acute fibrinous organizing pneumonia (AFOP), a recently recognized noninfective pattern of pulmonary injury (Fig. 9.47B). Although *Legionella* is a Gram-negative coccobacillus, it stains weakly with Gram stains, and silver impregnation stains are required to blacken the organisms, which are often abundant in situ in the absence of prior treatment (Fig. 9.48A). *Legionella micdadei* is distinguished by its staining with modified Ziehl–Neelsen stains (Fig. 9.48B).

(A)   **(B)**

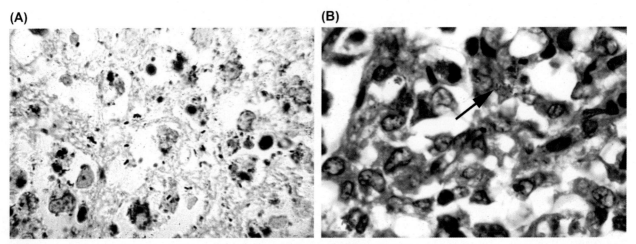

FIGURE 9.48   (A) *Legionella* spp. are Gram-negative coccobacilli but must be demonstrated by silver impregnation. (B) *Legionella micdadei*, the Pittsburgh pneumonia agent, also stains with modified acid-fast bacteria stains (*arrow*).

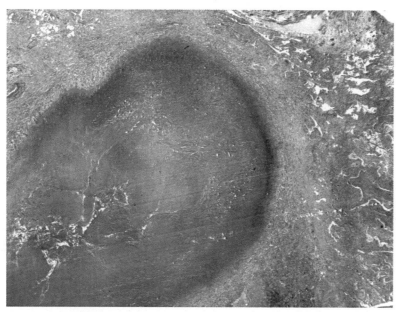

FIGURE 9.49   *Rhodococcus equi* causes a cavitary nodular pneumonia.

Rhodococcus equi

Rhodococcus pneumonia is a zoonotic infection that causes a nodular histiocytic and cavitary pneumonia in immunosuppressed patients, most commonly with HIV/AIDS (Fig. 9.49). The causative Gram-positive cocci are easily identified (Fig. 9.50A) and also stain positive with the modified Ziehl–Neelsen stain (Fig. 9.50B). The inflammatory response shows *malakoplakia* (Fig. 9.51A) with formation of intracellular calcific concretions termed *Michaelis–Gutman* bodies that although nonspecific are characteristically seen in the infection and can be highlighted by both PAS and iron stains (Fig. 9.51B).

Tropheryma whippelii (Whipple's Disease)

Whipple's disease is a rare disorder caused by the actinomycete *Tropheryma whippelii*. Whipple's disease most commonly causes intestinal malabsorption but pulmonary and neurological disease also occur. In the lung, Whipple's disease can present as interstitial infiltrates, pleural effusions, or as pulmonary hypertension. The characteristic change in the disorder is the accumulation of foamy macrophages that show intense staining with PAS (Fig. 9.52). In some cases, microgranulomas

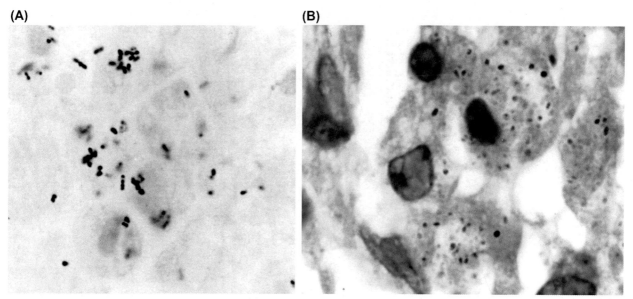

FIGURE 9.50   (A) The causative Gram-positive cocci also; (B) stain with modified acid-fast bacteria preparations.

(A)

(B)

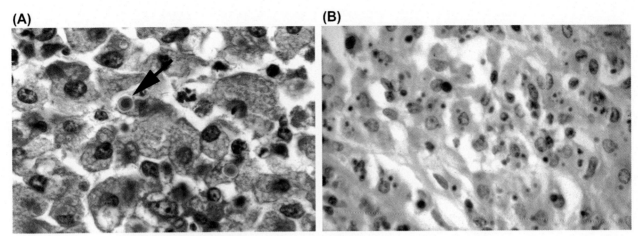

FIGURE 9.51 (A) Malakoplakia with *Michaelis—Gutman* bodies, calcific concretions that are also positive in PAS and (B) iron-stained sections.

that are histologically comparable to those seen in sarcoidosis may confuse the diagnosis (Fig. 9.53), together with the fact that the PAS staining in these granulomas may be equivocal. It is uncertain whether sarcoidosis is associated with Whipple's disease or whether the sarcoidal granulomas are evidence of early infection. In these cases, a small bowel biopsy will generally establish the diagnosis. Further confirmation can be achieved via ultrastructural examination demonstrating the bacillary organisms (Fig. 9.54) or by specific PCR.

Granulomatous Pneumonia

Pathologists apply the term "granuloma" to a variety of histological responses, including micronodular collections of tightly knit epithelioid macrophages (tuberculoid granulomas), necrotizing histiocytic reactions (necrotizing granuloma), and diffuse polymorphic infiltrates composed of lymphocytes, histiocytes, and plasma cells (granulomatous inflammation). The organisms that evoke these cellular responses are limited, and include the actinomycetes, mycobacteria, fungi, and helminths.

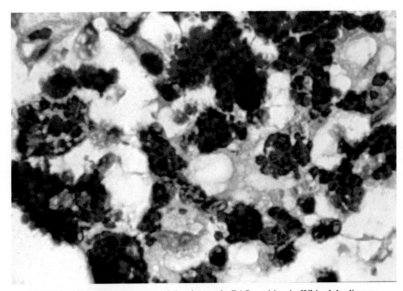

FIGURE 9.52 Macrophages staining intensely PAS-positive in Whipple's disease.

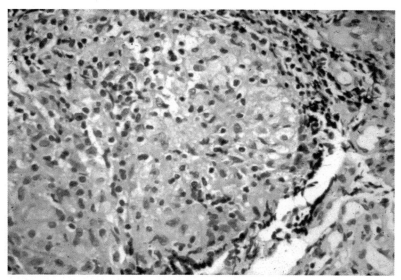

FIGURE 9.53 Nonnecrotizing granulomas mimicking sarcoidosis in Whipple's disease.

Mycobacterial Infection

Pulmonary necrotizing granulomatous inflammation is most commonly caused by mycobacterial or fungal infections. Tuberculosis is an ancient disease caused by a Gram-positive soil actinomycete, and it continues to represent a major source of global morbidity and mortality, largely due to its recrudescence in the setting of HIV/AIDS. The term *tuberculosis* is properly limited to infections caused by *Mycobacterium tuberculosis* and its genetically related congeners, e.g., *Mycobacterium bovis*, and is not the appropriate appellation for infections due to nontuberculous or "atypical" mycobacteria.

M. tuberculosis can affect the pulmonary airways, parenchyma, or pleura. Inhaled mycobacteria proliferate in the alveolar spaces and are then transported via lymphatics to regional hilar lymph nodes, from which they can enter the systemic circulation to spread to other organs. Progression of infection is limited by the acquisition of effective cell-mediated immunity. The initial pulmonary focus of infection (*Ghon focus*) subsequently heals by fibrosis and may show dystrophic calcification. This Ghon focus together with the accompanying calcified site of infection in a hilar lymph node is termed the *Ranke complex* (Fig. 9.55), and the foci of primary disseminated infection are termed *Simon's foci*. At all sites, the cellular response to *M. tuberculosis* is characterized by nodular collections of epithelioid macrophages with multinucleated giant cells, the tuberculoid granuloma, or *tubercle*. These can undergo central necrosis due to a cell-mediated hypersensitivity response. Sites of tuberculous infection may also show neutrophilic and eosinophilic exudates but these are generally not prominent features.

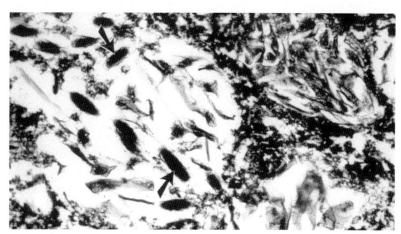

FIGURE 9.54 Ultrastructure demonstrates bacilli of *Tropheryma whippelii*.

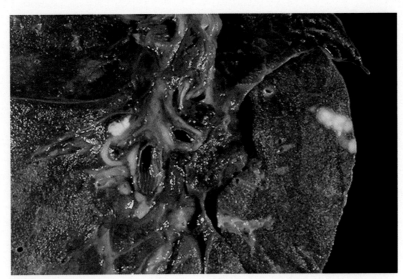

FIGURE 9.55 The *Ranke* complex includes the intrapulmonary *Ghon* focus of initial infection and the calcified hilar lymph node.

The mature tuberculoid granuloma is surrounded by a rim of T-lymphocytes and contained by an outer zone of fibrosis (Fig. 9.56). The presence of tuberculoid granulomas can be a distinguishing feature that differentiates tuberculosis from other forms of necrotizing granulomatous inflammation, e.g., Wegener's granulomatosis, in which tuberculoid granulomas are infrequent.

The term *caseous necrosis* properly refers to the "cheesy" appearance of the necrotic lesion on gross inspection. The histological correlate is destruction of lung tissue (Fig. 9.57) with loss of the underlying reticulin stroma. But as the latter finding is not apparent in H&E-stained sections, it is best, in practice, to term tuberculous lesions as either "necrotizing" or "nonnecrotizing," and to avoid the terms "caseating" versus "noncaseating."

As the response to mycobacterial infection reflects a component of immune hypersensitivity, even small numbers of organism can evoke substantial lung injury, and this complicates the task of identifying mycobacteria in situ. The organisms are best identified by their red color in acid-fast bacteria (AFB)-stained sections, i.e., with the Ziehl−Neelsen stain or its modifications, e.g., Fite−Ferraco. Mycobacteria can show substantial morphologic variability. They are curvilinear,

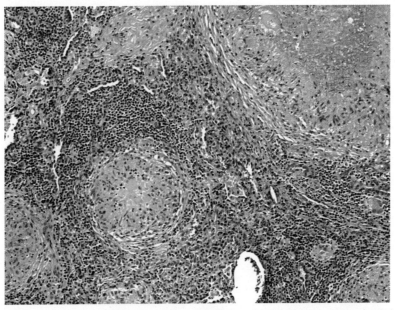

FIGURE 9.56 The tuberculoid granuloma is a collection of epithelioid histiocytes often with giant cells, surrounded by a lymphocytic infiltrate and fibrosis.

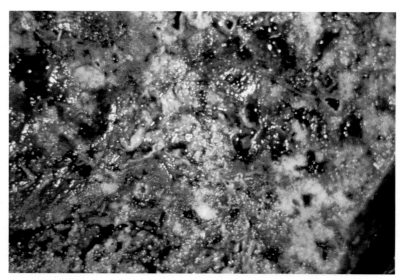

FIGURE 9.57 The cheesy gross appearance of a focus of caseous necrosis.

vary in length, and exhibit a characteristic "beaded" appearance attributable to nonhomogeneous uptake of the AFB stain (Fig. 9.58A).

However, there is no reliable way to distinguish *M. tuberculosis* from "atypical" mycobacteria by histochemical staining. As mycobacteria are weakly Gram positive, they can also be demonstrated in GMS-stained sections, a finding that is nonspecific, but can at times aid in the identification of the inconspicuous bacilli (Fig. 9.58B). As nonmycobacterial organisms can also be AFB positive, differential staining and culture results at times may be required to establish an accurate diagnosis (Table 9.5).

In practice the identification of *M. tuberculosis* by Ziehl–Neelsen staining is relatively insensitive, identifying mycobacteria in roughly 60% of culture-positive cases, so that the diagnosis may depend on isolating organisms in culture. However, examining multiple AFB-stained sections can improve the chance of identifying a pathogen.

Mycobacteria grow slowly in culture, and it can take weeks before they are eventually isolated. Consequently, ancillary methods have been developed with the aim of increasing the likelihood of establishing a diagnosis in a timely fashion. Sensitive fluorescent staining techniques, including the auramine–rhodamine stain, are used routinely in some laboratories in examining smears and tissues. Polymerase chain reaction (PCR) methods have been developed that can identify

(A)   **(B)**

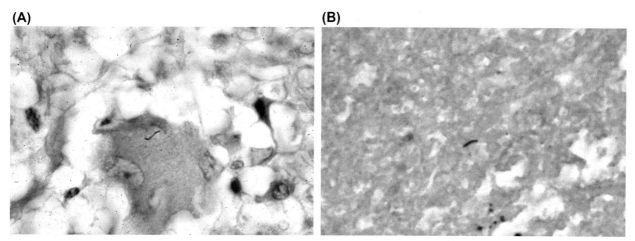

FIGURE 9.58 (A) Mycobacterial tuberculosis, the "red snapper," is a short beaded *bacillus* that can be equally well demonstrated with either the Ziehl–Neelsen stain or its weakly acid-fast modifications. (B) Mycobacteria are apparent in GMS-stained sections, but this finding is less specific than acid-fast bacteria staining.

TABLE 9.5 Acid-Fast Bacteria

Mycobacteria
Nocardia
Rhodococcus
Legionella micdadei

TABLE 9.6 Pulmonary Manifestations of Tuberculosis

Exposures with no disease (Ranke complex)
Caseating pneumonia
Acinar-no-dose pneumonia
Nodular disease (tuberculoma)
Cavitary pneumonia
Tracheobronchitis
Miliary disease
Lymphadenitis and calcification
Pleuritis
Fibrothorax (late)

M. tuberculosis and distinguish them from atypical mycobacteria, both in fresh tissues and in paraffin-embedded sections. The sensitivity and specificity of the latter approach are discussed elsewhere in this text.

Spectrum of Pulmonary Tuberculous Infection

Tuberculosis has protean manifestations in the lung (Table 9.6). Following most exposures, the primary infection is limited by the host's cellular immune system and the host remains asymptomatic, only showing evidence of previous limited infection via a positive tuberculin skin test. However, small numbers of mycobacteria can remain potentially viable and active infection may ensue if cell-mediated immunity is diminished by age, the use of corticosteroids, diabetes, ethanolism, or concomitant chronic infection. There is a marked increase in tuberculosis in patients with pulmonary silicosis and establishing the presence of mycobacterial infection in a patient with progressive massive fibrosis due to silicosis can be difficult.

As previously noted, histological examination of infection can provide an estimation of the adequacy of host immunity. The morphology of giant cells may be an indicator of whether host cell-mediated immunity is adequate in containing tuberculous infection. Prior to the development of effective cell-mediated immunity, many giant cells in the lesions show nuclei that have aggregated toward one pole of the multikaryon (Fig. 9.59A), whereas when infection is effectively contained, Langerhans' giant cells with peripheral nuclei or giant cells with centrally placed nuclei predominate (Fig. 9.59B). If cell-mediated immunity is profoundly diminished, granulomas may be poorly formed or absent.

Failure to contain the primary infection leads to its progression. Primary tuberculosis may show necrotizing pneumonia (Fig. 9.60), regionally involved lymph nodes, and a granulomatous pleuritis with lymphocytic effusion. Dissemination of organisms via the bloodstream can yield miliary disease, in which innumerable foci of active infection with poorly formed granulomas are seen (Fig. 9.61A−C).

Tuberculous acinar no-dose bronchopneumonia results from mycobacterial infection extending along the pulmonary acinus (Fig. 9.62). Tuberculomas are defined foci of nodular tuberculous infection (Fig. 9.63). When these cavitate into an adjacent bronchus, it can lead to discharge of numerous bacilli with cough or expectoration (Fig. 9.64A). Although bacillary counts in cavitary lesions are generally high, even large cavities at times will fail to show a single identifiable organism by histochemical staining, and diagnosis must be made presumptively based on histological appearance and

(A)   **(B)**

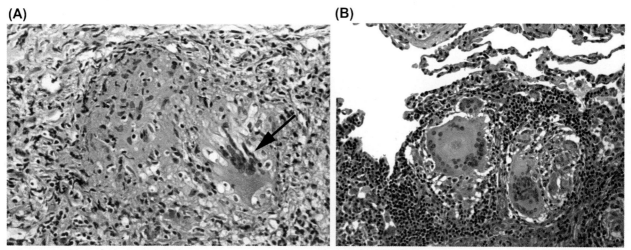

FIGURE 9.59   (A) Polarized giant cells in early infection with *Mycobacterium tuberculosis* (*arrow*). These cells may reflect host difficulties in containing the infection; (B) mature infection shows Langerhans' giant cells and giant cells with central nuclei.

empiric response to antituberculous medications. The extension of cavitary disease to involve an accompanying pulmonary artery may produce a Rasmussen's aneurysm and risk of vascular rupture leading to fatal hemoptysis (Fig. 9.64B).

Mycobacteria can spread along the mucosal surfaces of the airways to produce ulcerated lesions in the larynx and tracheobronchial tree that can mimic Wegener's granulomatosis (Fig. 9.65). However, as previously noted, the latter disorder rarely includes well-formed tuberculoid granulomas, and their presence favors the diagnosis of infection.

Reactivation Tuberculosis

In most clinically encountered cases of tuberculosis, previously asymptomatic individuals with positive cutaneous responses to purified protein derivative represent either reactivation or reinfection. Reactivation indicates an acquired defect in cell-mediated immunity. Tuberculosis tends to reactivate in the upper lobes of the lung where ventilation/perfusion ratios are high. Histologically, the lung shows necrotizing granulomas in areas of "scarring and traction bronchiectasis"

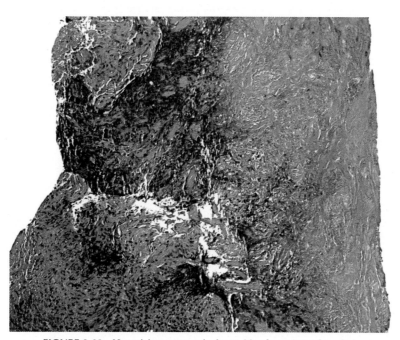

FIGURE 9.60   Necrotizing pneumonia due to *Mycobacterium tuberculosis.*

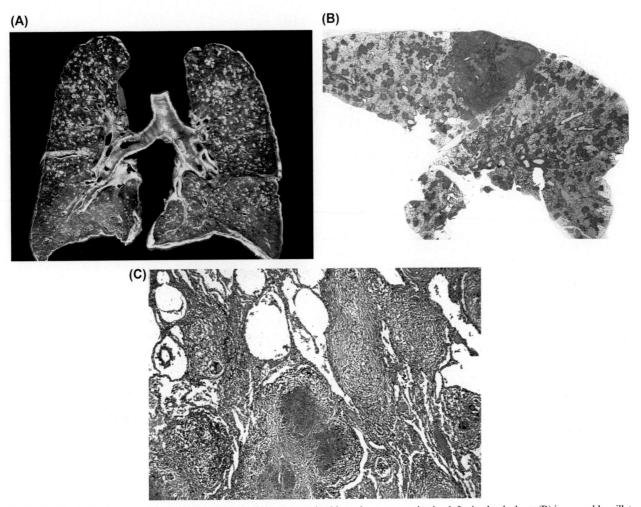

FIGURE 9.61 (A) Miliary tuberculosis in lung reflects the failure to contain either primary or reactivation infection by the host; (B) innumerable millet-sized nodules are seen in lung with focus of necrotizing pneumonia; (C) the response in miliary TB included poorly formed granulomas. With profound immunodeficiency, granulomatous changes may be absent.

(Fig. 9.66) due to the initial mycobacterial infection. But in addition to reactivation of tuberculosis, the differential diagnosis of necrotizing granulomatous inflammation in this setting includes fungal infection and atypical mycobacterial infection, as both have a predilection to develop in areas of old pulmonary apical scarring. The distinction may be difficult in the case of atypical mycobacterial infection, as special stains cannot distinguish these possibilities, and culture or ancillary diagnostic tests are necessary.

Atypical Mycobacteria

A number of mycobacteria that are genetically distinct from *M. tuberculosis* can produce pulmonary infection. These organisms vary in virulence and this is often reflected in the histologic appearance of the infection (Table 9.7). *Mycobacterium avium intracellulare* or *mycobacterium—avian complex* (MAC) can attack the lung in a variety of clinical settings. Patients immunosuppressed by HIV/AIDS can develop virulent infections with features that mimic tuberculosis. When adaptive T-cell-mediated immunity is severely compromised, the host response may be limited to foamy histiocytes that have ingested large numbers of mycobacteria. MAC can be demonstrated by both AFB and PAS stains, and clinical signs of both tuberculosis and MAC disease may only be recognized following treatment with antivirals, the so-called *immune reconstitution syndrome*.

In the immunocompetent host, MAC tends to affect elderly women (Lady Windermere's disease) and patients with bronchiectasis or bullous emphysema (Fig. 9.67). It is commonly seen in the right-middle lobe syndrome due to

FIGURE 9.62 Geographic infiltrates of necrotizing acinar-no-dose tuberculosis.

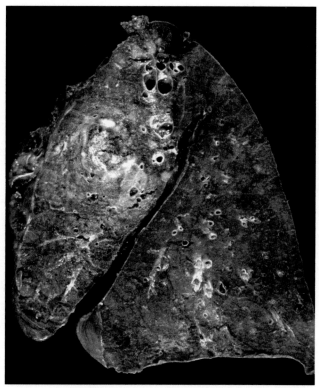

FIGURE 9.63 Localized tuberculoma.

(A)   **(B)**

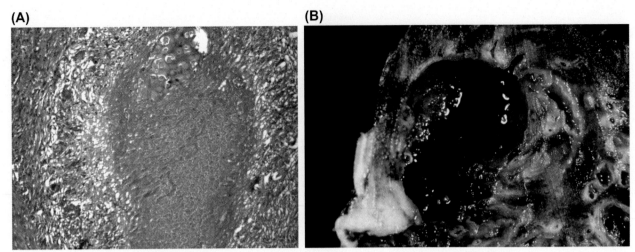

FIGURE 9.64   (A) Cavitary tuberculosis. (B) Erosion of a cavity into an adjacent pulmonary artery can produce a *Rasmussen's aneurysm* and lead to fatal pulmonary hemorrhage.

bronchiectasis and chronic atelectasis. The lesions of MAC infection are detected radiographically as "tree-in-bud" opacities reflecting terminal bronchiolar infection (Fig. 9.68A), together with nodules that may be either solid or cavitary. Histologically, the pathology often shows extensive areas of nonnecrotizing epithelioid histiocytes and is highly characteristic (Fig. 9.68B). The causative mycobacteria are indistinguishable from *M. tuberculosis* and only culture or PCR can accurately establish the diagnosis.

A recently described variant of MAC infection is "hot tub" lung. In these cases, the lung shows a microgranulomatous hypersensitivity pneumonitis that may be accompanied by necrotizing granulomatous inflammation. The pathology appears to represent primarily a cell-mediated hypersensitivity response to mycobacterial antigens (Fig. 9.69).

An unusual presentation of MAC infection in the immunosuppressed host is the *pseudosarcomatous* nodule. This can develop in the lung or in soft tissues and hematopoietic tissues. The nodules are composed of spindle cells, and may be mistaken for a low-grade spindle-cell neoplasm (Fig. 9.70A). However, examination reveals the foamy appearance of the spindle cells that prove to be CD68+ macrophages containing large numbers of ingested mycobacteria (Fig. 9.70B) and the absence of mitotic activity. Within the spectrum of unusual mesenchymal reactions seen in the immunocompromised one

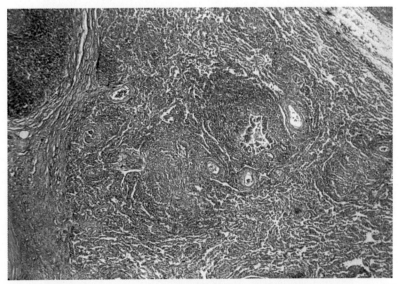

FIGURE 9.65   Tracheobronchial tuberculosis. These lesions can mimic Wegener's granulomatosis but the presence of tuberculoid granulomas is a distinguishing diagnostic feature, as they are rarely seen in Wegener's.

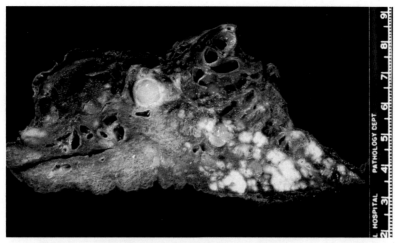

FIGURE 9.66 Bronchiectasis in tuberculosis.

must consider inflammatory myofibroblastic tumors that may include foamy histiocytes (Fig. 9.71A) associated with *human herpesvirus*-8 infection (Fig. 9.71B).

Other mycobacteria also cause pulmonary infection. *Mycobacterium kansasii* is a virulent species of atypical mycobacteria that produces necrotizing infection indistinguishable from tuberculosis. The organism often shows a prominent pattern of "cross-linking" on mycobacterial stains in situ that is characteristic but not diagnostic (Fig. 9.72A and B). Other rapid growing mycobacteria, including *Mycobacterium abscessus*, infect preexisting areas of active bronchiectasis, particularly in cystic fibrosis (Fig. 9.73). *M. fortuitum* produces limited infection in patients with diabetes, HIV/AIDS, and chronic upper gastrointestinal disease and all of the rapid growing nontuberculous mycobacteria, including *M. smegmatis*, can complicate pneumonia due to aspiration of lipid-based substances like nose drops.

TABLE 9.7 Classification of Mycobacteria

TB Complex

M. tuberculosis

M. bovis

Runyon Group I

M. kansasii

M. marinum

Runyon Group II

M. gordonae

M. scrofulaceum

Runyon Group III

M. intracellulare

M. avium

M. xenopi

Runyon Group IV

M. fortuitum

M. chelonei

M. abscessus

FIGURE 9.67   Necrotizing granulomas in area of old calcified focus of bronchiectasis were due to atypical mycobacterial infection.

Melioidosis

Rarely seen outside of southeast Asia where it is endemic, chronic infections due to *Pseudomonas pseudomallei* developed in veterans of the Vietnamese War many years after they had left the region. Acute melioidosis is a systemic infection that produces widespread coalescent microabscesses but it may resolve without being recognized only to recur many years later in lung, lymph nodes, and bone. The lung shows necrotizing granulomatous lesions surrounded by a zone of fibrosis (Fig. 9.74), and regional lymph nodes show stellate necrosis that can mimic "cat-scratch" disease due to *Bartonella henselae*. The offending organism is a Gram-negative motile *bacillus* that shows bipolar staining that can be difficult to demonstrate in situ.

Fungal Infection due to Yeasts

Fungi produce a spectrum of changes in the lung ranging from benign colonization of airways to malignant angioinvasive infections. Some fungi grow as yeast at body temperatures, whereas others are hyphate molds (Table 9.8). Most fungal

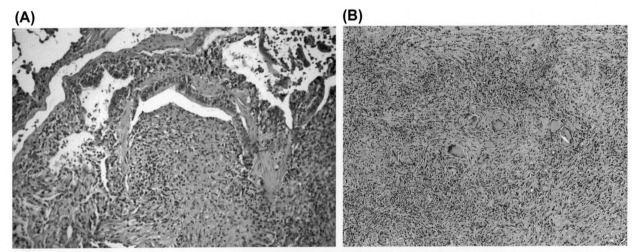

FIGURE 9.68   (A) Nonnecrotizing granulomatous inflammation ulcerates a small airway in a patient with MAC; (B) sheets of epithelioid histiocytes are characteristically seen in MAC infection in nonimmunosuppressed patients with chronic airways disease.

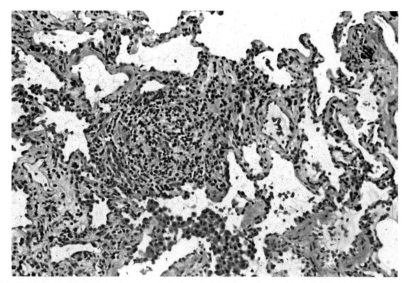

FIGURE 9.69   Micronodular granulomatous inflammation in "hot-tub lung," a hypersensitivity reaction to MAC.

yeasts are soil organisms that are topographically distributed in the United States. Whereas *Histoplasma capsulatum* may be encountered virtually anywhere that soil and water coexist, most cases in the United States are endemic to the mid-Western Mississippi and Ohio valleys. *Blastomyces dermatitidis* predominantly affects individuals living in the Great Lakes regions and in the southeast United States, whereas *Coccidioides immitis* is primarily encountered in the San Joaquin valley of the southwest. But despite their usual distribution, modern air travel and a highly mobile population has resulted in the possibility of these infections presenting virtually anywhere, and pathologists must be acquainted with their characteristic histological appearances. However, it is always prudent to enquire into possible travel prior to diagnosing an "exotic" fungal infection, as substantial overlap can exist in fungal morphologies.

Histoplasmosis

Primary *Histoplasma* infection produces a "viral-like" illness that generally resolves spontaneously. However, if there is a defect in cell-mediated immunity, or when the yeast burden is large, progression of infection may ensue. Chronic histoplasmosis tends to develop in the lung apices in areas of bullous emphysema or bronchiectasis. Although most yeast can

(A)   **(B)**

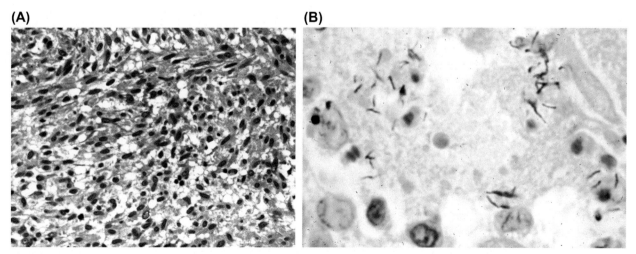

FIGURE 9.70   (A) A pseudosarcomatous nodule in a patient with HIV/AIDS due to MAC. (B) Large numbers of AFB/PAS+ mycobacteria are generally seen in this response. *AFB*, acid-fast bacteria.

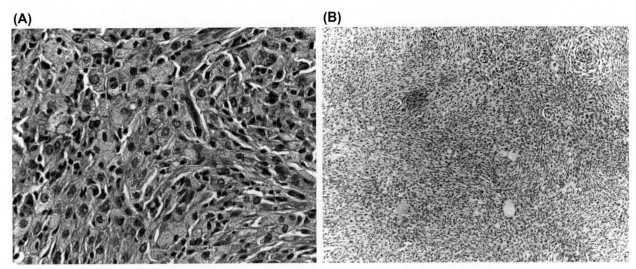

FIGURE 9.71   (A) Inflammatory myofibroblastic tumor with foamy histiocytes associated with; (B) *human herpesvirus-8* in patient with AIDS can mimic MAC infection.

be identified in H&E-stained sections, *Histoplasma* require special histochemical staining due to their small size. GMS is the stain of choice, as PAS at times fails to decorate the yeast.

Necrotizing granulomas due to *Histoplasma spp.* show central necrosis often surrounded by regions of mummefactive necrosis in which the "ghost outlines" of the underlying framework of the lung can still be distinguished in H&E and reticulin stained sections (Fig. 9.75A and B). The lining of the necrotic granuloma includes epithelioid histiocytes and giant cells but unlike tuberculosis does not show extensive tubercle granuloma formation, and the wall of the lesion characteristically exhibits paucicellular hyalinized basket weave fibrosis (Fig. 9.76).

Although they are facultative intracellular pathogens, *Histoplasma* cluster in areas of necrosis and are often seen outside of histiocytes. The 2−4 µM yeast show tear-drop shaped forms and reproduce by single narrow neck buds whose presence is diagnostic. Despite their name, no capsule is present (Fig. 9.77A). At times, pseudohyphae may be seen and this should not dissuade the pathologist from making the correct diagnosis (Fig. 9.77B), when all other diagnostic criteria are met. The differential diagnosis includes microforms of *Cryptococcus* or *Blastomyces, P. jirovecii*, and *Candida glabrata*. In the former, identifying the associated larger yeast forms eliminates the possibility of *H. capsulatum* infection. *P. jirovecii* shows irregularly shaped cysts with pericapsular accentuation on GMS stain and the organisms do not bud; *C. glabrata* can closely mimic the infection in GMS-stained sections, but unlike *Histoplasma*, they also stain amphophilic with H&E and are strongly Gram positive (Fig. 9.78A and B).

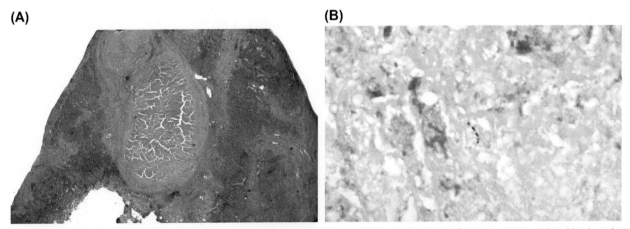

FIGURE 9.72   (A) *Mycobacterium kansasii* is a virulent organism that produces necrotizing granulomatous inflammation comparable to *M. tuberculosis*; (B) the elongate bacilli exhibit irregularity in their uptake of acid-fast bacteria stain producing a characteristic pattern of "cross-linking."

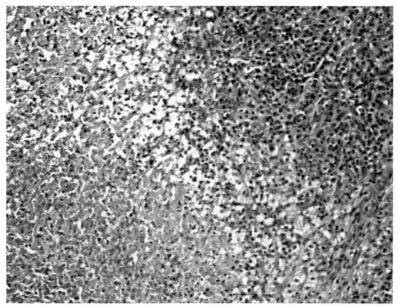

FIGURE 9.73 *Mycobacterium abscessus* complicates areas of bronchiectasis and may be exceedingly difficult to eradicate fully with antimycobacterial agents.

However, the most common and greatest difficulties in diagnosis can arise in distinguishing small regular micro-calcifications in GMS-stained sections as they can closely resemble degenerate yeast (Fig. 9.79). The presence of irregular calcifications is a clue to their actual nature but at times, ultrastructural examination may be required to exclude infection. In the immunosuppressed host, disseminated infection is primarily distributed within interstitial and alveolar macrophages (Fig. 9.80A and B).

Extension of infection from a peribronchial lymph node can produce *mediastinal granuloma*, a lesion characterized by dense paucicellular basket-weave hyaline fibrosis with aggregates of plasma cells (Fig. 9.81A). Organisms are rarely identified in the areas of paucicellular scarring and this form of the disease appears to be immunologically mediated. At times, necrotizing granulomatous inflammation is concomitantly present and confirms the diagnosis (Fig. 9.81B). The lesion can entrap the large vessels of the mediastinum leading to the superior vena caval syndrome and death. Surgical excision is required but not always be technically possible.

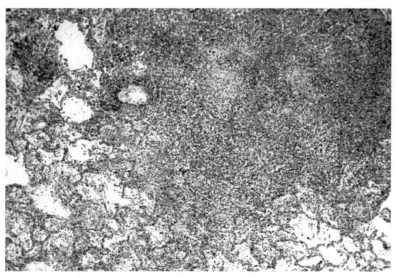

FIGURE 9.74 *Pseudomonas pseudomallei*, the cause of melioidosis, may reactivate many years following initial exposure to produce a necrotizing granulomatous pneumonia that resembles tuberculosis.

TABLE 9.8 Fungal Identification in Tissue

Organism	Size (Width, μM)	Defining Morphology
H. capsulatum	2–5	Narrow-neck bud
C. neoformans	5–20	Narrow-neck bud
B. dermatitidis	15–30	Broad-based bud
C. glabrata	3–5	Budding, no pseudohyphae
Candida spp.	2–3	Yeast, pseudohyphae, hyphae
Aspergillus spp.	3–5	Acute-angle branching, septate, conidial head
Zygomyces spp.	5–8	Right-angle branching, ribbons, pauciseptate
Pseudoallescheria spp.	3–4	Acute-angle branch, septate, terminal chlamydospore, pigmented conidia
Fusarium spp.	4–5	Acute and right-angle branch, septate, narrowed branch points
C. immitis	20–200	Endosporulation

Remotely infected calcified peribronchial lymph nodes due to histoplasmosis can erode into adjacent airways to be either expectorated or aspirated as broncholiths. Surprisingly, these may show persistently viable yeast forms and colonization by aspirated oropharyngeal bacteria (Fig. 9.82).

Old calcified granulomas are commonly encountered in surgical resection of lungs for neoplasia. Most of these are due to healed tuberculosis or histoplasmosis, depending on exposure. In the vast majority of cases no organism will be identified but occasionally nonviable degenerate histoplasma may be seen, although distinguishing them with confidence from microcalcifications can be difficult and a high threshold for the diagnosis of histoplasmosis should be maintained in this setting.

Blastomyces

B. dermatitidis shows a propensity to infect lung, skin, and bone. The spectrum of pulmonary presentations includes consolidative pneumonia, diffuse alveolar damage, and pulmonary nodules that radiographically may mimic pulmonary carcinoma (Fig. 9.83). The pulmonary lesion characteristically shows a granulohistiocytic response (Fig. 9.84A). The yeast

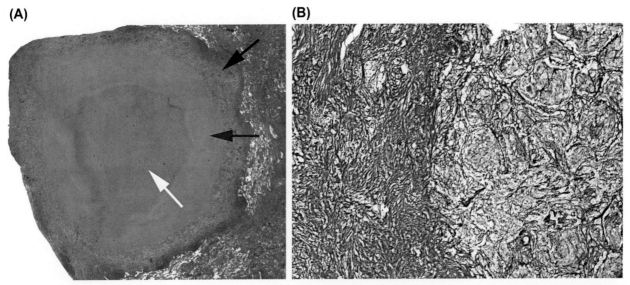

FIGURE 9.75 (A) Nodular pneumonia due to *H. capsulatum* often shows necrotizing granulomatous inflammation with three zones (*arrows*). The outer capsule encloses an area of mummefactive necrosis that in turn surrounds an area of caseating necrosis findings confirmed by (B) reticulin stains.

FIGURE 9.76   The wall of the necrotizing granuloma shows poorly formed granulomas and giant cells with a highly characteristic hyalinized basket-weave fibrosis.

are large (15–30 µM) and easily identified in H&E sections, where they are distinguished by their thick refractile cell wall (Fig. 9.84B). The organism is also multinucleate (Fig. 9.84C) and proliferates via single broad-based budding (Fig. 9.84D). Microforms may be present and should not be confused with coinfection by *H. capsulatum*. Giant yeast forms can also occur and may be confused with *C. immitis*.

Cryptococcus

Cryptococcus neoformans infects immunocompromised patients but can also be seen in apparently normal hosts. Meningoencephalitis is the most common clinical presentation and it represents a complication of subclinical pulmonary infection. *C. neoformans* can produce localized necrotizing cryptococcomas (Fig. 9.85A), confluent bronchopneumonia (Fig. 9.85B), granulomatous pneumonia (Fig. 9.85C and D), or a null response characterized by "yeast lakes" with minimal inflammation (Fig. 9.85E). Grossly, the infected lesions are glistening and "slimy."

(A) **(B)**

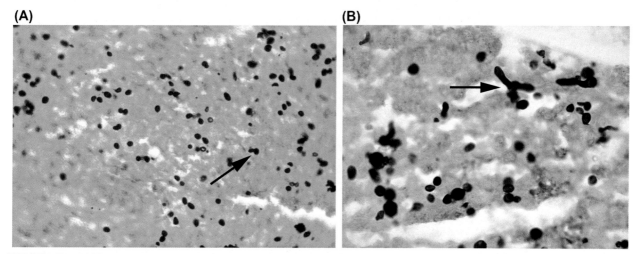

FIGURE 9.77   (A) Narrow necked budding 2–4 µM yeast of *H. capsulatum* in GMS stained section (*arrow*). PAS is not reliable for demonstrating this yeast (B) *H. capsulatum* infection with irregular yeast forms and pseudohyphae (*arrow*).

(A)

(B)

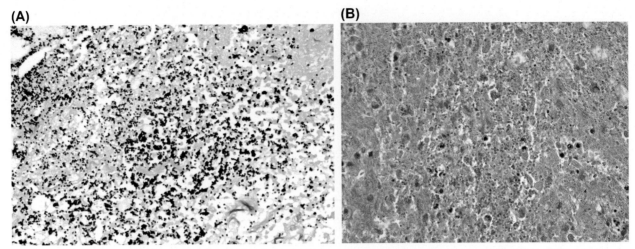

FIGURE 9.78 (A) *Candida glabrata* can easily be mistaken for *H. capsulatum* on GMS stain but (B) the yeast are amphophilic in H&E-stained sections and easily differentiated on this basis.

The organism shows substantial variability in size (2—15 µM) in size and shape (Fig. 9.86A) and innumerable microforms can occasionally be seen that must be distinguished from histoplasmosis (Fig. 9.86B). The yeast proliferates via single narrow-necked buds (secrete a capsule that is optimally visualized with mucicarmine in situ; Fig. 9.87A). The organisms in fluids stain with India ink (Fig. 9.87B). With GMS, the yeast body stains gray black and the capsule is not decorated (Fig. 9.88C). In capsular-deficient organisms, the Fontana—Masson stain reacts with a melanin precursor in the yeast wall, highlighting the organisms (Fig. 9.88A), although a careful examination of the mucicarmine stain will invariably reveal a poorly developed rim of capsular staining (Fig. 9.88B).

Coccidioides immitis

Generally affecting patients from the southwestern United States, this organism is distinct from other yeast by virtue of its size (20—200 µM) and its endosporulating mode of reproduction. *C. immitis* produces a spectrum of changes that includes fibrocaseous granulomas (Fig. 9.89), granulomatous pneumonia, and miliary disease, often accompanied by tissue

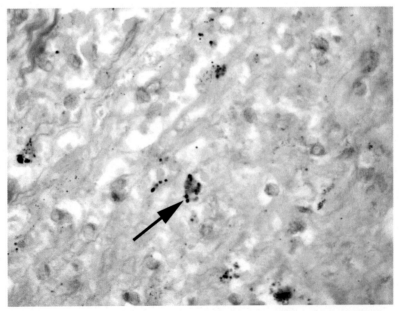

FIGURE 9.79 Microcalcifications (*arrow*) can closely resemble *H. capsulatum*. At times ultrastructural examination may be required in order to exclude infection.

(A)

(B)

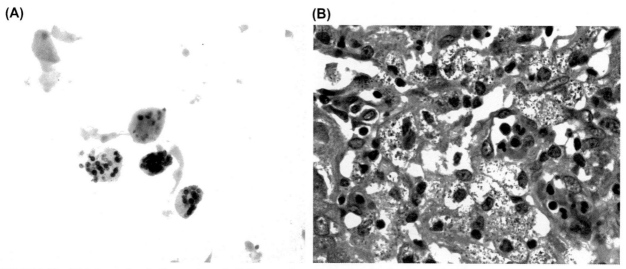

FIGURE 9.80 (A) Intracytoplasmic *H. capsulatum* in BAL macrophages in patient with disseminated infection and HIV/AIDS. (B) Debris within macrophage phagolysosomes can resemble intracellular yeast.

eosinophilia. The endospores are contained within a spherular capsule and both stain well with GMS (Fig. 9.90A), whereas spherules are variably stained by PAS. The cysts of *C. immitis* have a characteristic tendency to collapse after having discharged their endospores in situ (Fig. 9.90B).

In endemic areas, *C. immitis* can form fungus balls within preexisting pulmonary cavities (Fig. 9.91A). The organism is dimorphic and the presence of the infective hyphal arthroconidia should not be confused with a concomitant mold infection (Fig. 9.91B).

Paracoccidioides

Paracoccidioides braziliensis infection is endemic in South America, where it produces a range of pulmonary findings comparable to those of blastomycosis; but cases are rare in the United States. The organism is large (10−60 μM) and replicates by multiple narrow-necked buds that produce a "ships wheel" appearance that is pathognomonic (Fig. 9.92). However, when this feature is absent, the infection can be confused with other fungi.

(A)

(B)

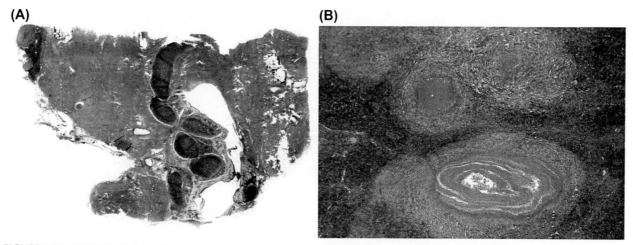

FIGURE 9.81 (A) Mediastinal granuloma is a serious complication of pulmonary histoplasmosis. The mediastinum shows paucicellular hyaline scarring and may include (B) foci of necrotizing granulomatous infection.

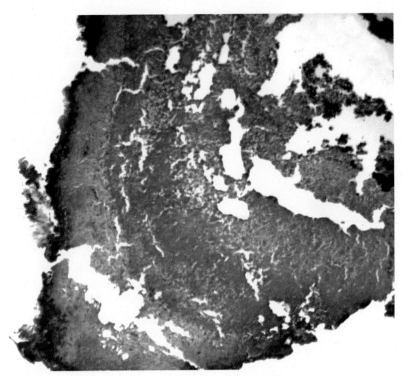

FIGURE 9.82 Expectorated broncholith due to histoplasmosis.

Candida spp.

Superficial colonization of the upper airways by *Candida spp.* is common in patients treated with inhaled corticosteroids or in chronically ill and diabetic patients (Fig. 9.93). Foci of aspiration pneumonia and abscess cavities may show colonization, but deep pulmonary infection is rarely seen in the absence of fungemia (Fig. 9.94A and B).

Candida can have a pleomorphic morphology that includes yeast (blastoconidia), pseudohyphae, and true hyphae, although in some cases, only yeast forms may be present. The organisms can be highlighted by either GMS or PAS and stain strongly Gram positive (Fig. 9.95A). At times, blastoconidia of *Candida spp.* can be large and mimic other fungal infections (Fig. 9.95B). *Candida glabrata* (*torula glabrata*) shows multiple 2–5 μM budding yeast that are amphophilic in H&E sections; they are distinct from other *Candida* species, as pseudohyphae and hyphae are never present.

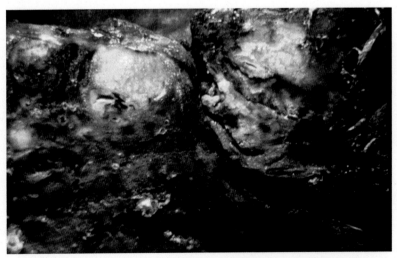

FIGURE 9.83 Pulmonary nodule excised as carcinoma due to blastomycosis.

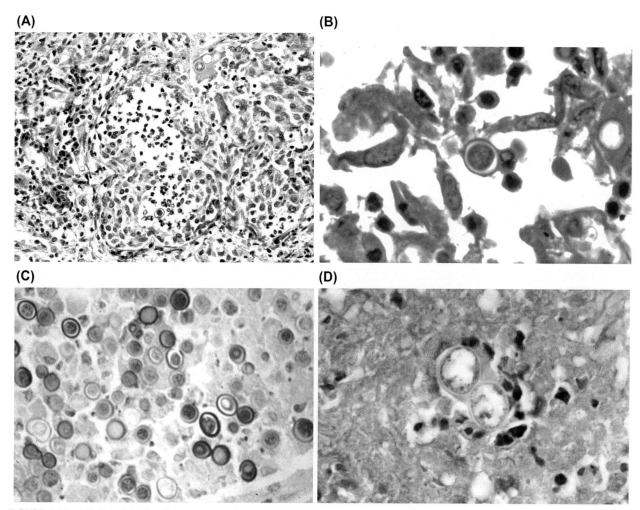

FIGURE 9.84 (A) Granulohistiocytic response to *B. dermatitidis*. (B) Yeast of *B. dermatitidis* with refractile cell wall. (C) *B. dermatitidis* with multinucleation. (D) *B. dermatitidis* showing broad-based budding.

Hyphate Fungi

Hyphate fungi or "molds" are responsible for a range of pulmonary disorders ranging from colonization of pulmonary airways to angioinvasive life-threatening infections. *Aspergillus spp.* account for the majority of pulmonary mold infections but other organisms including the *Zygomyces, Pseudoallescheria,* and *Fusarium* also produce pulmonary disease.

Aspergillus spp.

The hyphae of *Aspergillus spp.* range in diameter from 2.5 to 4.5 μM and show frequent septation. *Aspergillus spp.* branch progressively, primarily at acute angles of ∼45 degree, mimicking an arborizing tree branch (Fig. 9.96A) but when cut in cross-section may be mistaken for yeast, although the absence of budding suggests the correct diagnosis. In areas of mycelial growth, organisms become tangled, bulbous, and distorted and it may be impossible to confirm the diagnosis with accuracy based on morphology (Fig. 9.96B).

The *aspergil*, a ritual implement used in the Roman Catholic mass, resembles the fruiting body, and gives the fungus its name (Fig. 9.97A−C). Fruiting bodies develop from mycelia in areas of high oxygen tension, such as lung or sinus cavities but do not develop in tissues. They are composed of a vesicle with one or two layers of phialides that produce the infective conidial spores, and the morphology of the fruiting body allows for accurate speciated in situ. In general, the specific diagnosis of "aspergillosis" should be avoided unless the aspergil is identified, as other fungi can be morphologically virtually indistinguishable in tissue. Diagnoses are therefore optimally phrased as "acute-angle branching hyphae consistent with aspergillus."

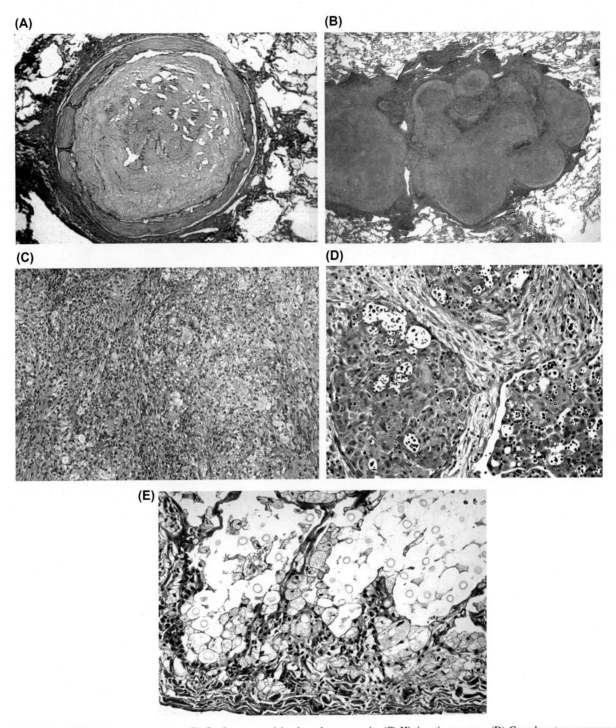

FIGURE 9.85 (A) Nodular cryptococcoma. (B) Confluent necrotizing bronchopneumonia. (C) Histiocytic response. (D) Granulomatous response to *Cryptococcus neoformans*. (E) Yeast lake.

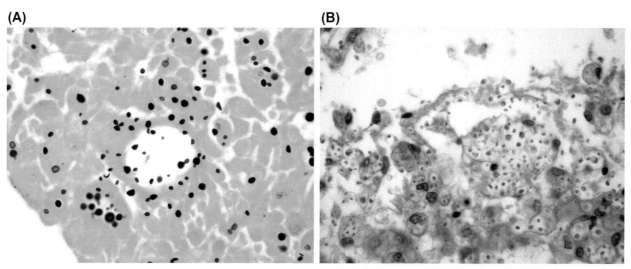

FIGURE 9.86 (A) Narrow necked budding of *Cryptococcus neoformans* showing variability in size and shapes; (B) microforms of *C. neoformans* within histiocytes must be distinguished from intracellular *H. capsulatum*.

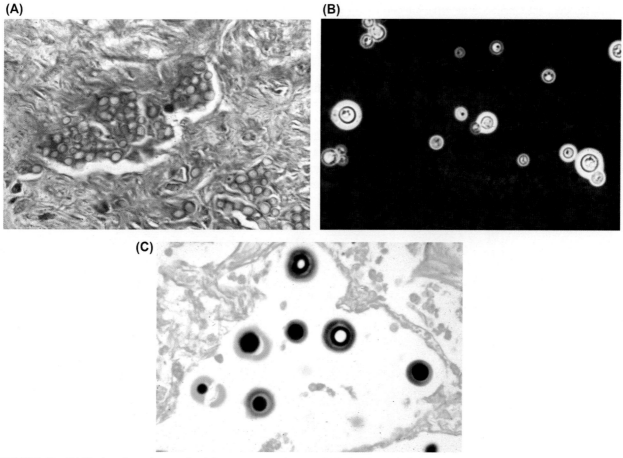

FIGURE 9.87 (A) Mucicarmine stains capsule of *Cryptococcus neoformans*. (B) India ink preparation shows capsule of yeast. (C) Both GMS (and PAS) stain body of yeast but not its capsule.

(A) **(B)**

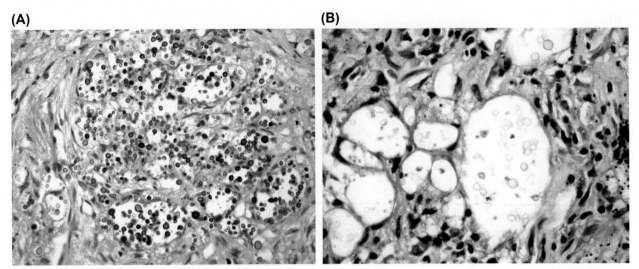

FIGURE 9.88 (A) "Capsular deficient" organisms stain with Fontana—Masson. (B) "capsular deficient" yeast invariably show a faint rim of muci-carminophilic staining.

Immune Disorders due to Aspergillus Infection

Aspergillus spp. give rise to a spectrum of disorders, some reflecting hypersensitivity responses to the organisms, whereas others are the consequence of invasive infection (Fig. 9.98). Distinguishing these is critical for the proper management of these disorders.

Allergic Bronchopulmonary Aspergillosis

Allergic bronchopulmonary aspergillosis shows a range of findings, including intractable asthma, proximal bronchiectasis, and both peripheral blood eosinophilia. It is not certain whether the fungus plays an opportunistic role in exacerbating atopic responses or is primary in its pathogenesis. Patients develop intractable bronchospasm with elevated serum IgE levels specific for *Aspergillus spp.* The pathology includes central cystic bronchiectasis with mucoid impaction (Fig. 9.99A). The impacted mucus is viscid and forms a cast of the airways, a disorder termed *plastic bronchitis*. Microscopically the mucus plugs show layers of degenerating eosinophils interspersed within the mucin (Fig. 9.99B), and the surrounding lung may show patchy eosinophilic pneumonia. The fragmented fungal hyphae can at times be difficult to

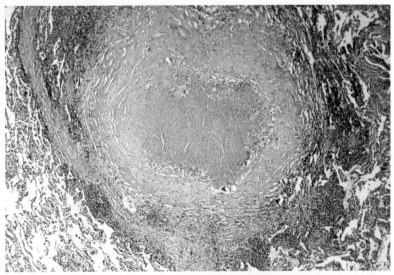

FIGURE 9.89 Fibrocaseous infection due to *C. immitis.*

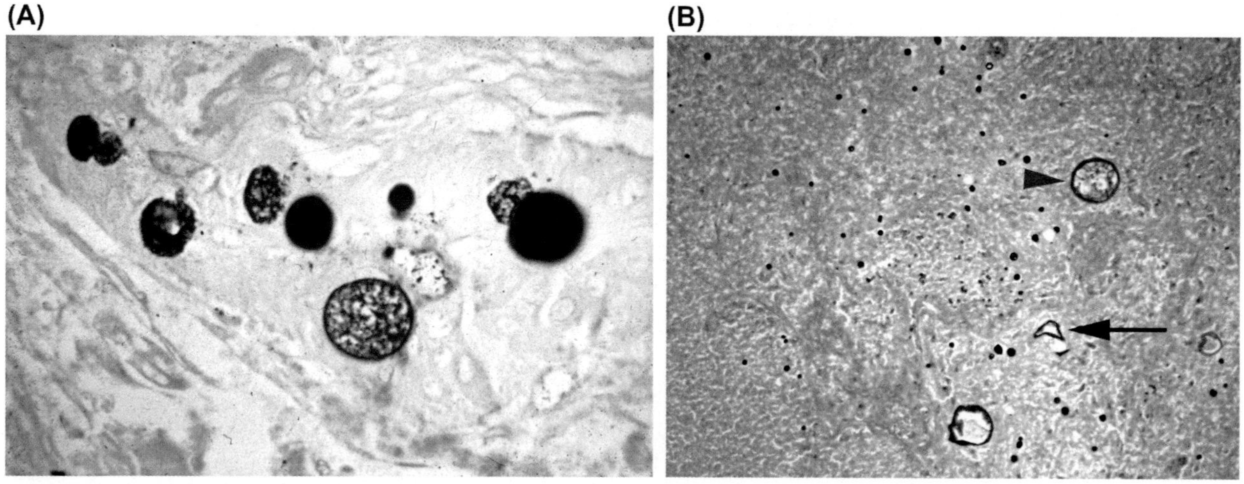

(A)

(B)

FIGURE 9.90   (A) Endospores are variably present within PAS+ cysts. (B) Cysts of *C. immitis* have a characteristic tendency to collapse after discharging their contents (*arrows*).

identify (Fig. 9.99C), so that silver stains should be applied routinely to the evaluation of allergic mucus plugs. Although the clinical and histologic features of this disorder are most frequently caused by hypersensitivity to *Aspergillus spp.*, other fungi, e.g., *Candida spp.*, can yield a comparable syndrome. A subset of patients with cystic fibrosis develops concomitant ABPA, and establishing the diagnosis in this setting requires evidence of elevated IgE in serum or BAL that is especially reactive with aspergillus antigens.

Cases of ABPA vary in severity but can be inordinately difficult to manage. Corticosteroids remain the mainstay of treatment for the asthmatic component of the disease. Flutter valves instruments (Acapella) can assist in loosening the areas of mucoid impaction. Finally, azole (itraconazole or voriconazole) treatment may reduce fungal colonization in the airways.

Bronchocentric Granulomatosis

Bronchocentric granulomatosis (BCG) appears to reflect an abnormal cell-mediated response to *Aspergillus spp.*, in which small caliber airways develop circumferential granulomatous inflammation, loss of the normal lining respiratory epithelium, and impaction of the airway lumen by granular basophilic mucin admixed with cellular debris (Fig. 9.100A−C). The disorder may be first noted radiographically as isolated or multiple airway centered nodules. BCG may be seen as part of the spectrum of findings in ABPA, or as an isolated disorder. As in ABPA, fragmented hyphae may be difficult to identify

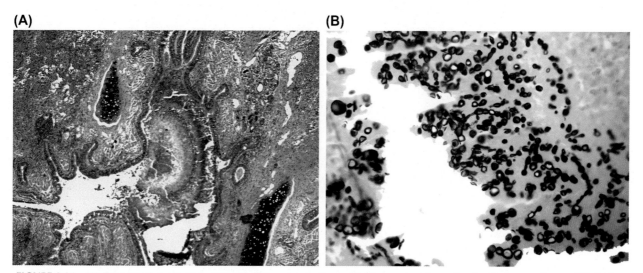

FIGURE 9.91   (A) Fungus ball due to *C. immitis* with (B) yeast and arthroconidia that must not be confused with a concomitant mold infection.

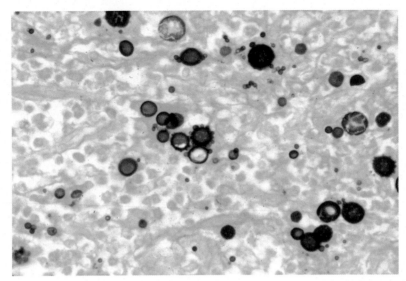

FIGURE 9.92 *Paracoccidioides brazilienses* shows multiple narrow buds that resemble a ship's wheel but this diagnostic feature is not always present.

and surrounding areas of eosinophilic pneumonitis are common. When the disease is suspected clinically, it can be treated noninvasively with corticosteroids; however, definitive resection may be undertaken to exclude neoplasia.

Hypersensitivity Pneumonitis

Hypersensitivity pneumonitis reflects a combined abnormality of humoral and cell-mediated immunological responses to organic antigens. Most cases of HP are caused by thermophilic actinomycetes, but hypersensitivity to *Aspergillus spp.* is well documented. Upper lobe predominance is the rule and this can be a helpful feature in establishing the diagnosis.

The diagnosis of HP is based primarily on establishing a historical link between antigen exposures and the clinical findings but lung biopsies can establish the diagnosis. Microscopically, the lung shows bronchiolocentric lymphohistiocytic interstitial infiltrates with poorly formed microgranulomas (Fig. 9.101) and giant cells that may contain birefringent crystals and cholesterol crystals. Although CD8+ lymphocytes characteristically predominate in BAL fluid specimens,

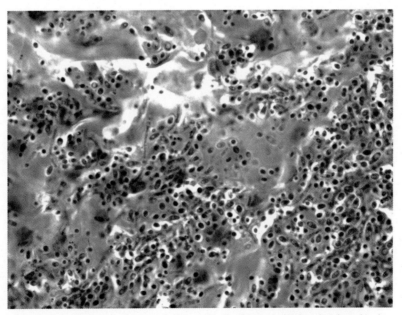

FIGURE 9.93 Dense colonization of airway with *C. albicans* in intubated patient showing.

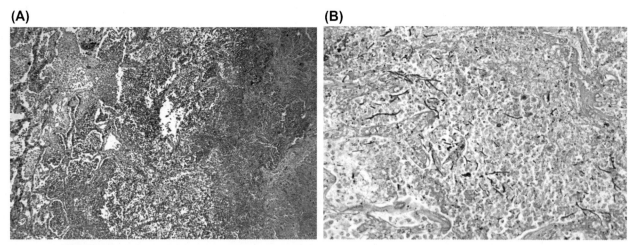

FIGURE 9.94 (A) Necrotizing pneumonia due to *C. albicans* in fungemic patient. (B) PAS stain highlights the organism.

immunostains will reveal dominance of either CD4+ or CD8+ lymphocytes in situ. In addition, other histopathologies, including, nonspecific cellular interstitial pneumonitis, organizing pneumonia, lymphoid interstitial pneumonitis, and nonnecrotizing granulomatous inflammation resembling sarcoidosis, can be caused by HP. The presence of interstitial and alveolar eosinophils is characteristically seen in HP due to *Aspergillus* antigens and is rare in the response to other antigens.

Aspergillus bronchitis and Chronic Necrotizing Aspergillosis

The presence of primary airway infection by *Aspergillus spp.* is a poorly recognized entity. It is generally seen in the setting of modest immunosuppression accompanying disorders like diabetes mellitus or the use of aerosolized steroids in asthmatics. In some cases, it is a precursor lesion for invasive disease. Fungal hyphae can be seen filling the lumen of airways (Fig. 9.103A), without evidence of frank tissue invasion. Elastic stains are helpful in determining whether organisms have begun to transgress normal tissue barriers (Fig. 9.102B). Centrally necrotic lesions (Fig. 9.103A) of chronic necrotizing aspergillosis can resemble those due to mycobacteria and other fungi (Fig. 9.103B). Treatment with voriconazole is indicated and if disease is localized surgical resection may be required.

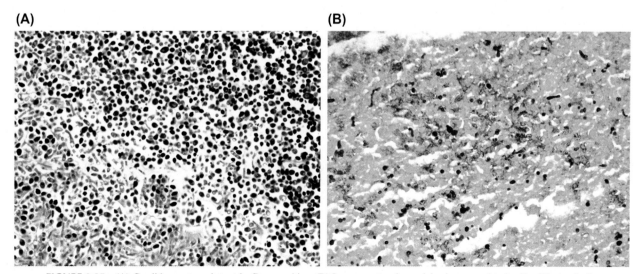

FIGURE 9.95 (A) Candida yeast are intensely Gram positive. (B) Large yeast and pseudohyphae proved to be *Candida tropicalis*.

(A)   **(B)**

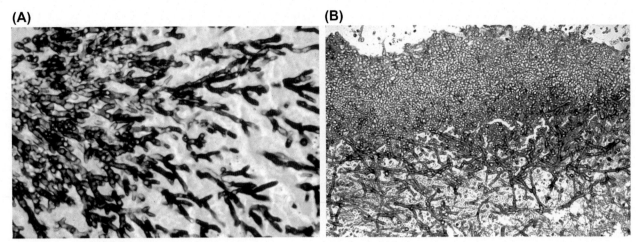

FIGURE 9.96 *Aspergillus* spp. characteristically branch dichotomously and progressively at acute angles. (B) Tangled and distorted mycelial growth. It is impossible to speciate fungi based on this morphology.

Fungus Balls

The colonization of old fibrocavitary disease, e.g., areas of bronchiectasis due to healed tuberculosis and sarcoidosis, or emphysematous bullae, by *Aspergillus spp.*, is the most common cause of a pulmonary fungus ball (Fig. 9.104A and B). The term mycetoma should not be applied to intracavitary fungal mycelial growth as it accurately applies only to soft tissue infections. Patients may be asymptomatic, or alternatively present with episodes of hemoptysis, at times massive and requiring emergent bronchial arterial embolization or definitive resection.

The morphology of the hyphae in a fungus ball is frequently distorted and it can be impossible to identify diagnostic septate acute angle branching forms. *Aspergillus* fungus balls show heterogeneous staining intensity (Fig. 9.104C), giving the impression of alternating zones of growth. The wall of the fungus ball frequently shows increased numbers of tissue eosinophils (Fig. 9.104D). The *Splendore−Hoeppli* phenomenon is invariably present and is reassuring evidence that angioinvasion is unlikely, unless there has been a recent supervening cause of immunosuppression or neutropenia. The walls of the cavity are lined with granulation tissue, granulomatous inflammation, or metaplastic squamous epithelium, depending on the activity of the disease. The occasional presence of germinative fruiting bodies of *Aspergillus spp.* with characteristic phialides and conidial forms allows definitive speciation.

The definitive treatment of fungus balls is surgical extirpation. However, the degree of underlying lung disease and adhesions to the chest wall make this form of resection technically difficult. If patients are actively hemoptysizing, coiling of a feeding bronchial artery may suffice as treatment. However, recurrent bleeding or constitutional symptoms should lead to reconsideration for surgical intervention. There is evidence that voriconazole may assist as adjuvant therapy but the thick-walled nature of these cavities limits its efficacy.

One may rarely see rapid expansion of a cavity due to vascular thrombosis induced by calcium oxalate crystal deposition, a disorder termed *chronic pulmonary oxalosis*. Oxalic acid is produced by a variety of *Aspergillus spp.* but is most commonly a feature of *Aspergillus niger* infection (Fig. 9.105A). Diffusion of oxalate into the surrounding blood vessels is prothrombotic and can lead to extensive ischemic necrosis (Fig. 9.105B and C). Oxalate crystal deposition in the renal tubules may also be seen. Emergency resection of the fungus ball is the only effective treatment.

Angioinvasive Aspergillosis

This life-threatening infection is seen in patients who have been chronically immunosuppressed and/or neutropenic. It is a complication of bone marrow and solid organ transplantation, as well as of antileukemic chemotherapies. Angioinvasive aspergillosis is an uncommon complication of HIV/AIDS, despite the associated profound immune deficiency. Grossly, the lung shows "targetoid" lesions showing central thrombosed vessels secondary to angioinvasion, surrounded by a rim of consolidated lung, confluent bronchopneumonia, or lobar consolidation (Fig. 9.106A) and microscopically a necrotizing pneumonia (Fig. 9.106B) at times featuring giant cells (Fig. 9.106C) may be present. Angioinvasion is identified microscopically and may be enhanced with silver and elastic stains (Fig. 9.107A and B). Rarely, foci of infarcted lung can produce an infected nonviable pulmonary sequestrum (Fig. 9.108).

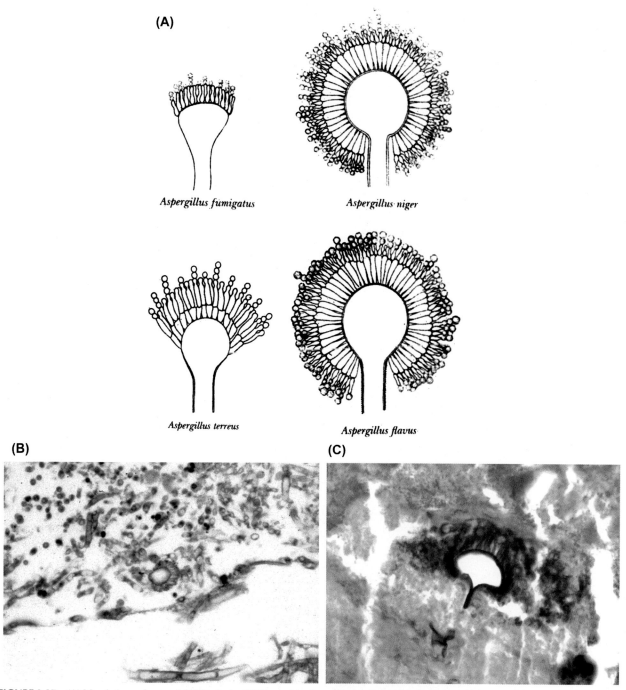

FIGURE 9.97 (A) Morphology of aspergil fruiting body; (B) Gram stain shows fruiting body of A. *fumigatus* with innumerable Gram-positive conidia; (B) A. *niger* with innumerable pigmented conidia.

The fungal hyphae tend to invade blood vessels and to "metastasize" to other organs. The presence of sunburst vasculocentric hyphal growth is diagnostic of a "metastatic" focus of infection (Fig. 9.109) and virtually any organ may be secondarily involved. Circulating *Aspergillus spp.* can also seed abnormal cardiac valves to produce endocarditis with fatal embolic hemorrhagic infarctions to brain and other vital organs.

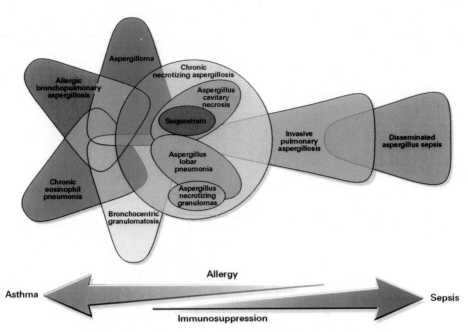

FIGURE 9.98 Spectrum of disease due to *Aspergillus spp.*

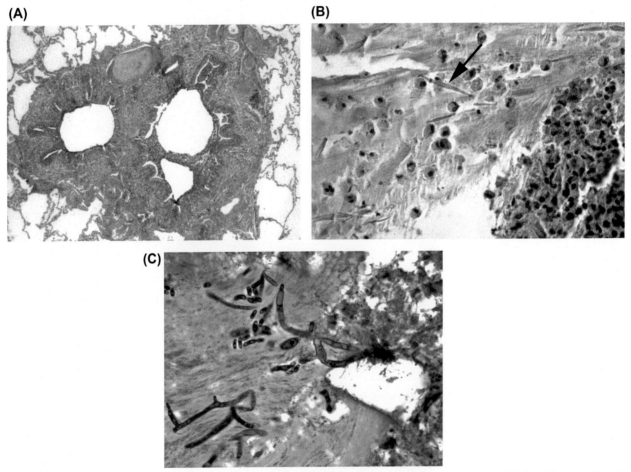

FIGURE 9.99 (A) Central bronchiectasis with dense peribronchiolar scarring in patient with ABPA. (B) Expectorated "allergic" mucus plug with eosinophils and Charcot–Leyden crystals (*arrow*). (C) Aspergillus hyphae in mucus plug.

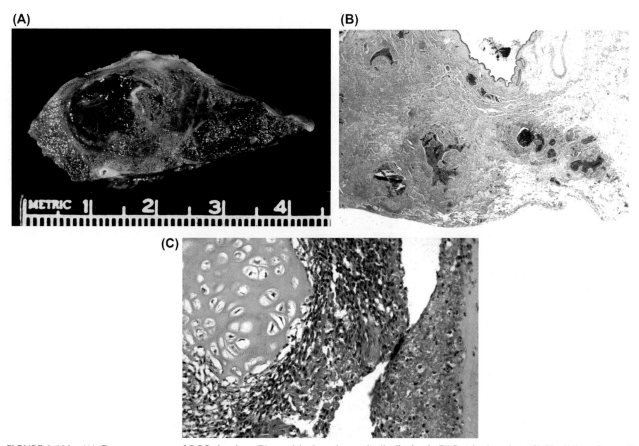

FIGURE 9.100 (A) Gross appearance of BCG showing; (B) exquisite bronchocentric distribution in PAS-stained section. (C) The lining of a small airway is replaced by granulomatous inflammation and the lumen is filled with a granular basophilic exudate.

Other *Aspergillus* Species

Aspergillus terreus is an opportunistic fungus that infects patients with chronic granulomatous disease, as well as other immunocompromised hosts. Its hyphae are pyriform with orthogonal branches (Fig. 9.110A and B). Fungus balls due to *Aspergillus nidulans* have a propensity to produce pale staining Hulle cells with Maltese cross birefringence when examined under polarized light (Fig. 9.111A and B).

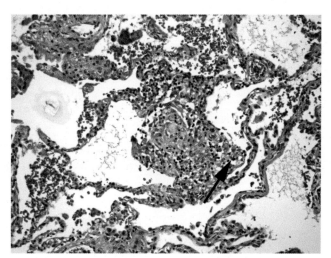

FIGURE 9.101 The microgranulomatous hypersensitivity response to aspergillus often includes tissue eosinophils.

(A)

(B)

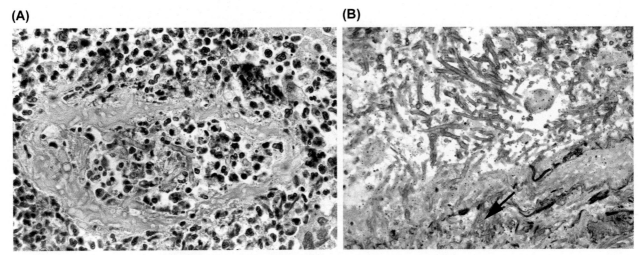

FIGURE 9.102   (A) *Aspergillus bronchitis*/bronchiolitis is a precursor to invasive disease; (B) elastic stains can determine whether tissue invasion has occurred (*arrow*).

Other Hyphate Fungi

Although the majority of acute-angle branching septate hyphae encountered in medical practice prove to be *Aspergillus spp.*, exceptions do occur and can be difficult to distinguish in situ. Organisms that closely mimic *Aspergillus* include the *Zygogomyces, Pseudallescheria boydii (Scedosporium)*, and *Fusarium spp*. Pulmonary infection by *Zygomyces* is due to organisms of the order *Mucorales* (Mucor), including *Mucor, Rhizopus, Absidia, Rhizomucor,* and *Apophysomyces*. Infection generally occurs in immunosuppressed hosts and in patients with diabetic ketoacidosis or disorders of iron metabolism. As treatment invariably includes surgical resection, establishing an accurate diagnosis is critical.

The histologic responses to infection are comparable to those caused by angioinvasive aspergillosis. The hyphae of the *Zygomyces* are broad (5–25 μM), pauciseptate, tending to branch at right angles (Fig. 9.112A), although acute angle branching does occur and its presence should not dissuade one from the correct diagnosis. The hyphae of the *Zygomyces* are well stained by hematoxylin and they have a proclivity to produce ribbon-like structures (Fig. 9.112B) but caution in diagnosis is required as treated *Aspergillus* can assume this appearance. The fungus rapidly invades vessels, perineural lymphatics, cartilage, and tends not to respect tissue boundaries. *Zygomyces*, as well as all other hyphate fungi can rarely form noninvasive fungus balls in the lung.

(A)

(B)

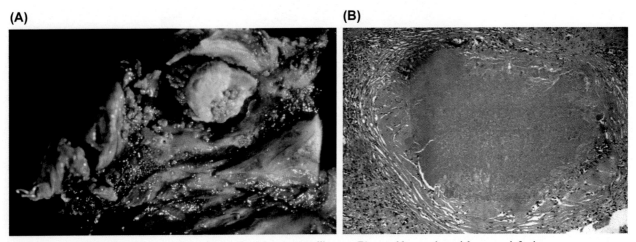

FIGURE 9.103   (A) Necrotizing lesions due to aspergillus can (B) resemble mycobacterial or yeast infection.

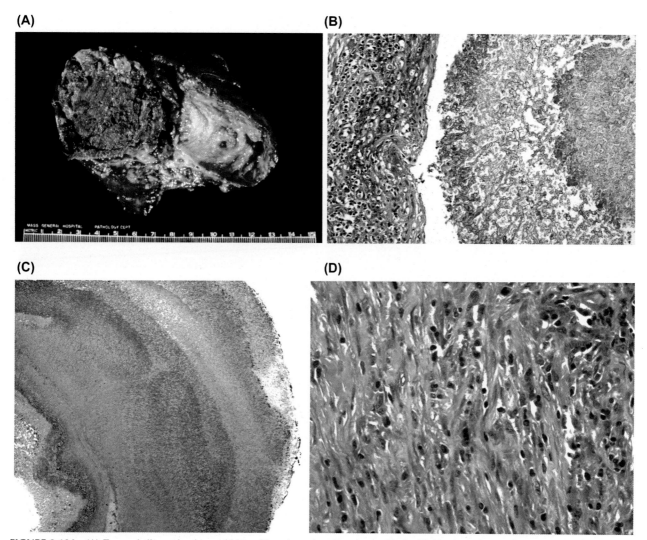

FIGURE 9.104 (A) Fungus ball growing in an old bronchiectatic cavitary due to sarcoidosis; (B) histology shows tangled mass of hyphae and the Splendore—Hoeppli phenomenon. (C) Fungus balls often show alternating "zones" of staining intensity with H&E (and GMS). (D) Fungus balls frequently contain eosinophils in their wall.

Pseudoallescheria (*Scedosporium*)

The hyphae of *Pseudallescheria* are smaller (2—4 µM) than those of *Aspergillus spp.*, the fungus branches predominantly at acute angles, but unlike *Aspergillus*, tends to branch haphazardly rather than progressively, and terminal chlamydospores mimicking a "tennis racket" may be apparent (Fig. 9.113). Although it is a hyaline fungus, the conidia produced by *Pseudoallescheria* in fungus balls are ovoid and pigmented.

Fusarium

Fusarium is an opportunistic infection of lung that is seen in patients who are severely immunosuppressed. The septate hyphae of *Fusarium spp.* branch irregularly, and at right angles, showing constrictions at branch points (Fig. 9.114).

Differential Diagnosis

Although distinguishing hyphate fungal pathogens from one another in tissue can be exceedingly difficult, it is important, as the efficacy of available fungal antibiotics depends on the diagnosis. For example, although both *Aspergillus spp.* and *Pseudallescheria* are sensitive to voriconazole, *Pseudallescheria* are resistant to amphotericin. The sensitivity of *Fusarium spp.* to most antifungal antibiotics appears to be both unpredictable and limited. In all cases, culture remains the

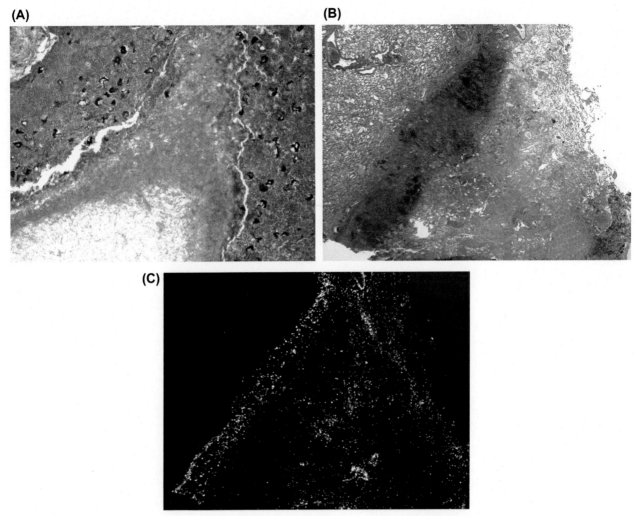

FIGURE 9.105   (A) Fungus ball due to *A. niger*; (B) extensive local ischemic infarction of lung; and (C) deposition of calcium oxalate crystals are seen with polarized light.

gold standard for diagnosis. Immunohistochemical reagents that can distinguish between hyphate fungi exist but are not commercially widely available and all reagents must be carefully tested with appropriate controls. The PCR assay is presently not an effective diagnostic approach.

Serological testing can at times assist in the distinction between invasive "mucormycosis" and aspergillus infections. 2,3-Beta-glucan is generally not produced by the zygomycetes but the test is also nonspecific and positive results can be seen in other conditions and circumstances that are noninfectious. The galactomannan assay was developed as a specific marker of aspergillus infection; however, its sensitivity in part depends on the fungal burden and on whether there is tissue invasion. Results from the BAL are more sensitive but may not distinguish noninvasive colonization from angioinvasive disease. For these reasons, these tests should be used prudently based on the context of clinical and radiographic presentation.

Dematiaceous (Pigmented) Fungi

Pigmented fungi can be divided into those forming yeast (chromoblastomycosis) and those due to hyphate fungi (phaeohyphomycosis). The specific organisms producing disease cannot be diagnosed morphologically and isolation in culture is required. The Fontana—Masson stain can be applied when questions exist as to whether a fungus seen in H&E sections is pigmented. *Bipolaris, Curvularia, Alternaria, and others,* are pigmented hyphate fungi that cause allergic pulmonary disorders mimicking those due to *Aspergillus spp.* (Fig. 9.115). The correct diagnosis is suggested by the presence of pigmented hyphae within allergic mucus or within fungus balls. *Chromoblastomycosis* is appreciated by the

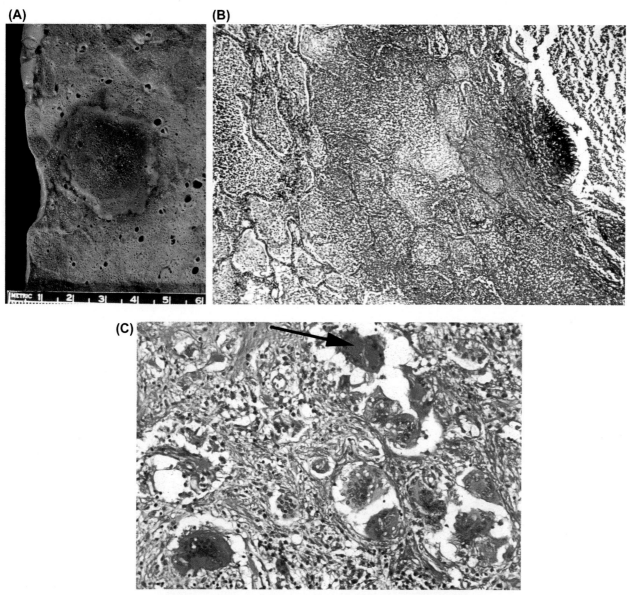

FIGURE 9.106   (A) Angioinvasive aspergillosis with targetoid lesion due to vascular thrombosis, ischemic infarction and fungal invasion into surrounding tissues; (B) necrotizing pneumonia due to angioinvasive aspergillosis with mycelial fungal growth developing at surface of cavitary lesion. (C) Giant cells in pneumonia due to *A. fumigatus.*

presence of sclerotic (Medlar) bodies. These are clusters of pigmented yeast that show characteristic multiaxial septa (Fig. 9.116).

Other Fungi

Pulmonary infection due to *Penicillium marneffei* is rarely seen outside of patients from southeast Asia with HIV/AIDS, but it is a significant cause of necrotizing bronchopneumonia in this region. The organism is a small (2−5 μM) dimorphic fungus with an elongate sausage shape showing a characteristic septum. The organisms can be identified free within areas of necrosis or within foamy histiocytes (Fig. 9.117). The infection may also disseminate, commonly to involve the skin. Other fungi, including *Geotrichum* and *Sporothrix*, are rare causes of pulmonary infection.

(A)   **(B)**

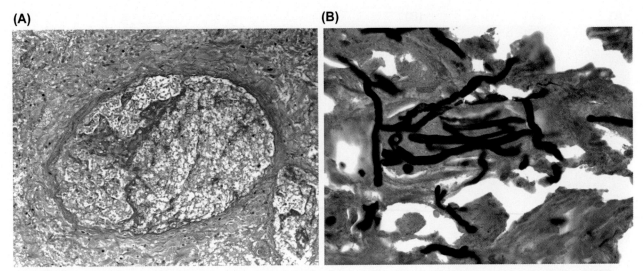

FIGURE 9.107   (A) Blood vessel invasion may be seen with H&E and to better advantage in (B) section costained with GMS and H&E.

Parasites

The lung can be host to a variety of parasitic infections. Although many are primarily seen primarily in tropical climates, others occur in temperate climes, or as the result of immunosuppression. As previously noted, the current ease of global travel has greatly increased encounters with tropical diseases.

PROTOZOA

Several species of protozoa can produce pulmonary infection. *Entamoeba histolytica* affects the lung secondary to the transdiaphragmatic extension of a hepatic amoebic abscess, or less commonly as the result of blood-borne spread from an intestinal source. A pulmonary amoebic abscess may extend to the pleura or rupture into an airway leading to

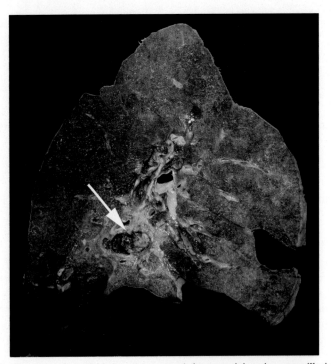

FIGURE 9.108   Area of nonviable lung (sequestrum) due to angioinvasive aspergillosis (*arrow*).

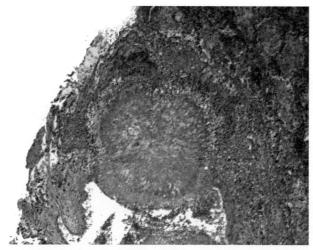

FIGURE 9.109 Focus of "metastatic" fungal infection showing sunburst pattern of growth out of blood vessel.

intrapulmonic dissemination. The amoebic abscess shows liquefactive necrosis that grossly resembles "anchovy paste" (Fig. 9.118). The histological response in areas of necrosis primarily includes neutrophils but the margins of the lesion will show a polymorphous infiltrate of macrophages, lymphocytes, and plasma cells and tissue eosinophilia is uncommon (Fig. 9.119A). Amoeba can be distinguished from macrophages by their larger size, amphophilic bubbly cytoplasm, a sharply defined nuclear karyosome (Fig. 9.119B), and in the case of *E. histolytica*, by the ingestion of erythrocytes.

Free living amoebae, including *Acanthamoeba, Balamuthia*, and *Naegleria*, rarely affect the lung and primarily cause meningoencephalitis. However, in immunosuppressed patients, infection may spread beyond the confines of the nervous system to include the lung.

Toxoplasma

Toxoplasma gondii produces pneumonia in patients who are immunosuppressed most often with HIV/AIDS. Although *Toxoplasma* is a common infection of the central nervous system and retina in HIV/AIDS, pulmonary infection is rare. When present, it includes a fibrinopurulent pneumonitis with areas of necrosis (Fig. 9.120A). The organisms are obligate intracellular parasites and are identified either as engorged pseudocysts containing crescent-shaped tachyzoites and/or GMS+ and PAS+ true cysts containing bradyzoites (Fig. 9.120B).

(A)　　　　　　　　　　　　　　　　　　**(B)**

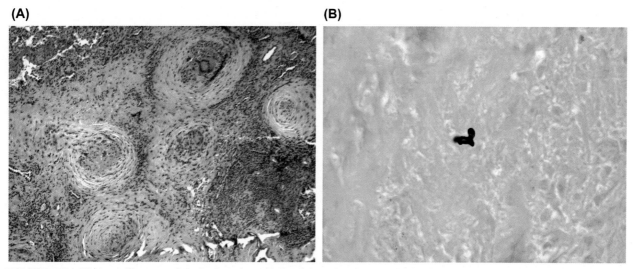

FIGURE 9.110 (A) *Aspergillus terreus* infection in patient with chronic granulomatous disease showing (B) fragmented pyriform hyphae branching at right angle.

(A)

(B)

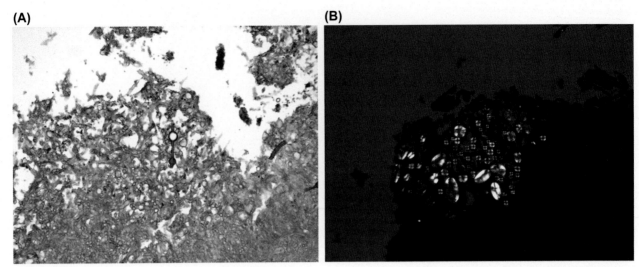

FIGURE 9.111 (A) Fungus ball due to *Aspergillus nidulans* with *Hulle* cells that (B) show Maltese cross configuration when examined with polarized light.

Cryptosporidium

Cryptosporidium parvum is water-borne opportunistic infection that affects patients with HIV/AIDS but has also been seen as outbreaks among children in day care centers. The disease primarily affects the small bowel leading to cholera-like watery diarrhea. In patients with HIV/AIDS, the infection can spread to the hepatobiliary tree as well as to the pulmonary airways.

The diagnosis is made by identifying the amphophilic spores (3−5 μM) of *Cryptosporidia* along the surface of infected respiratory epithelium (Fig. 9.121). Although apparently extracellular by light microscopic determination, ultrastructural analysis demonstrates that the cysts are actually intracytoplasmic and invested by apical cytoplasm. The underling mucosa shows a mild lymphocytic infiltrate. Although well seen with H&E, the cysts are also stained by GMS, modified AFB, and Giemsa. The diagnosis is obvious once considered, but can easily go unnoticed unless considered in the differential diagnosis.

(A)

(B)

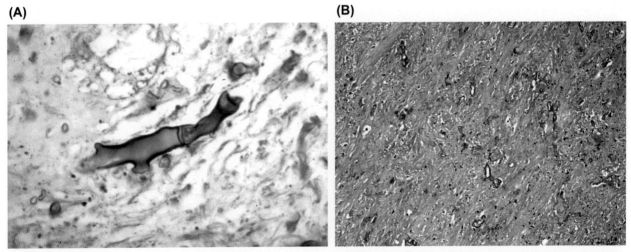

FIGURE 9.112 (A) *Zygomyces* spp. hyphae are broad, pauciseptate, and tend to branch at right angles; (B) area of necrotizing angioinvasive disease with ribbon-like hyphae that stain well with hematoxylin.

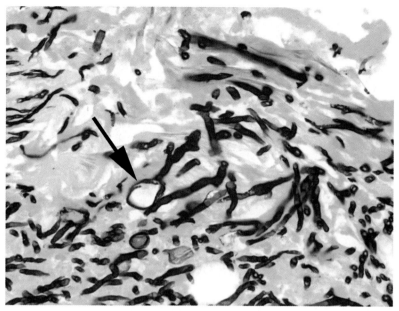

FIGURE 9.113 *Pseudeallescheria boydii* are slightly smaller than *Aspergillus* spp. tend to grow randomly with prominent terminal chlamydospores (*arrow*). Definitive diagnosis is generally not possible based solely on morphology in tissue sections.

Microsporidia

Microsporidia are spore-forming protozoa that like *Cryptosporidium* infect immunosuppressed patients to produce a watery diarrheal illness. Extraintestinal disease, including bronchial involvement, is uncommon but has been observed. The organisms are intracytoplasmic and generally located within the apical cytoplasm. Because of their small size (1−2 μM) they may be exceedingly difficult to detect, especially when the organism load is low. Special stains, including tissue Gram stain, trichrome, Warthin−Starry, and GMS can assist in their identification (Fig. 9.122). Further speciation requires ultrastructural examination.

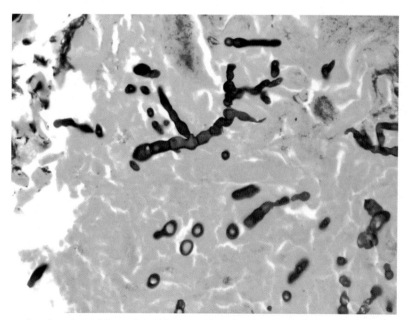

FIGURE 9.114 *Fusarium* tend to branch at right angles with narrow branch points but definitive classification is generally not achieved in tissue sections.

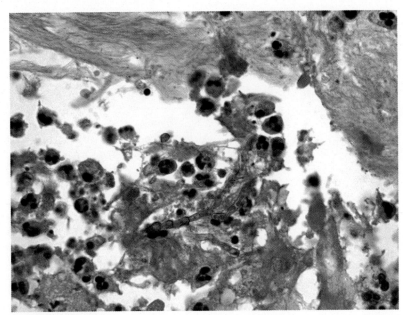

FIGURE 9.115 Phaeohyphomycosis show pigmented fungal hyphae. Subclassification requires microbiological culture.

NEMATODES (ROUND WORMS)

Several nematodes (round worms) have a larval developmental phase in the lung (Table 9.9). The larvae of *Ascaris, Necator, Ancylostoma,* and *Strongyloides* migrate through the airways toward the mouth where they are swallowed or expectorated. This process can evoke wheezing, migratory pneumonia, and blood eosinophilia, a complex termed Loeffler's syndrome. The presence of track-like necrotizing granulomatous bronchitis and bronchopneumonia with prominent eosinophilic infiltrates should alert the pathologist to the presence of a migratory parasitic pulmonary infection (Fig. 9.123).

Strongyloides stercoralis is most often seen in the tropics, however, cases are endemic to the southeastern United States. Patients receiving high doses of corticosteroids are susceptible to infection and what is termed "hyperinfection." In hyperinfection, filariform organisms exit the gut and migrate to the lungs. There, they cause hemorrhagic and eosinophilic pneumonia. When normal pathways of maturation are inhibited, the filariform larvae mature to egg-laying adults that give rise to rhabditiform larvae and the presence of expectorated ova indicates hyperinfection (Fig. 9.124A). Parasitemia

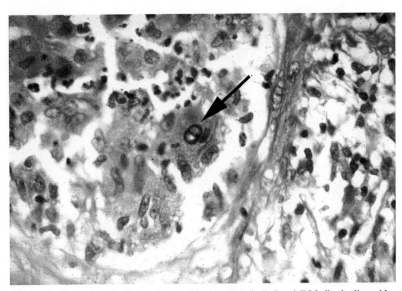

FIGURE 9.116 Chromblastomycosis shows pigmented yeast forms with characteristic "sclerotic" Medlar bodies with multiaxial septation (*arrow*).

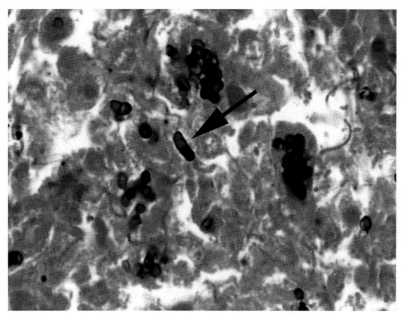

FIGURE 9.117 *Penicillium marneffei* is a dimorphic fungus that is often seen within histiocytes and has a "sausage" shape with characteristic septation (*arrow*).

predisposes to Gram-negative sepsis that can lead to concomitant DAD (Fig. 9.124B). Due to the risk of activating the hyperinfection syndrome via immunosuppression, patients should either be tested prior to immunosuppressive treatment or treated empirically and prophylactically with ivermectin.

Dirofilaria

In temperate climates, *Dirofilaria immitis* is a zoonosis that causes canine "heartworm." Mosquitoes transfer microfilariae into subcutaneous tissues where they mature silently and enter the systemic venous circulation. From there, they travel to the right heart, but in humans they do not mature further. Pulmonary disease reflects embolism of a nonviable helminth from the heart into the pulmonary circulation where it generally lodges in a small muscular artery to produce an area of localized rounded infarction with a chronic immunological reaction to the dead worm. The embolic event may be clinically

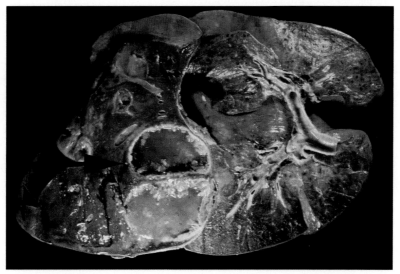

FIGURE 9.118 Amebic abscess with "anchovy paste" gross appearance.

(A)

(B)

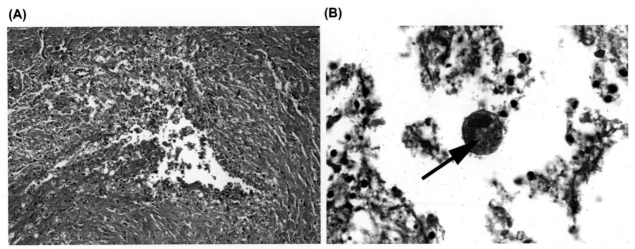

FIGURE 9.119   (A) Liquefactive necrosis in lung abscess due to (B) *E. histolytica* showing ingestion of erythrocyte (*arrow*).

silent or associated with chest pain, fever, chills, hemoptysis, and peripheral blood eosinophilia. The disease is often first recognized as a solitary pulmonary nodule that is resected to exclude neoplasia.

Microscopically one sees a rounded area of pulmonary infarction (Fig. 9.125A). The diagnosis is facilitated by identifying the coiled nematode within an occluded muscular pulmonary artery. The area of surrounding necrosis is surrounded by a zone of granulomatous inflammation with lymphocytes, plasma cells, and variable degrees of tissue eosinophilia, all contained within a dense fibrous capsule. Trichrome, elastic, and reticulin stains highlight the features of the helminth and help to localize it in a vessel (Fig. 9.125B and C). *Dirofilaria* are distinguished from other nematodes by their prominent muscular lateral cords and a striated cuticle that bulges inward in the region of the lateral cords, yielding a bat wings appearance.

Trematodes (Flukes)

A number of trematodes produce pulmonary disease that can present as solitary pulmonary nodules, pulmonary infarctions, necrotizing granulomas, or eosinophilic pneumonia. The most frequent fluke infection in the United States is *Schistosomiasis* due to its wide distribution in both the eastern and western hemispheres.

Patients who have traveled to endemic areas may develop Katayama fever 6–10 weeks following their return. This disorder consists of fever, pulmonary infiltrates, and eosinophil. It results from the antigenic response to the egg laying of the female schistosome in the bloodstream. Patients should be treated with praziquantel.

(A)

(B)

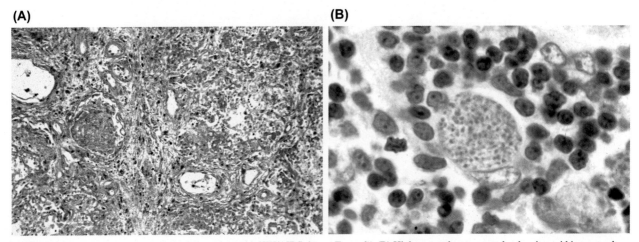

FIGURE 9.120   (A) Necrotizing pneumonia in patient with HIV/AIDS due to *T. gondii*. (B) High power demonstrates bradyzoites within macrophage.

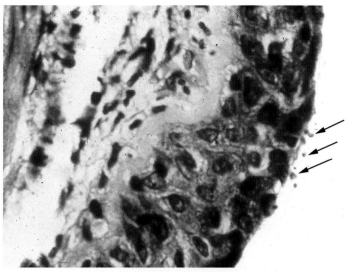

FIGURE 9.121 *Cryptosporidium parvum* seen along the surfaces of ciliated pulmonary epithelium (*arrows*). The cysts are actually invested by apical cytoplasm.

TABLE 9.9 Nematodes with Pulmonary Larval Phase

Ascaris lumbricoides

Strongyloides stercoralis

Necator americanus

Ancylostoma duodenale

Subsequent pulmonary involvement reflects aberrant migration of adult worms into the lung, where they are found within pulmonary vessels (Fig. 9.126A). The helminth evokes an intense granulomatous and eosinophilic response and extravascular refractile ova may be present and like the helminth can evoke granulomatous and eosinophilic reactions.

The ova are large (70–170 μM) depending on the species. The ova of *Schistosomiasis mansoni*, which produces most cases of pulmonary disease, show a prominent lateral spine that can be seen with H&E and that is highlighted by modified

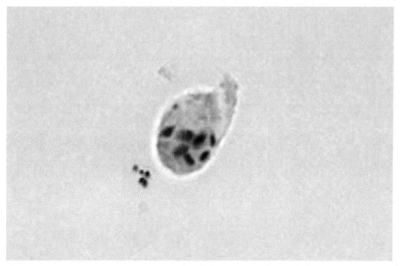

FIGURE 9.122 BAL macrophage contains *Microsporidium* spp. demonstrated by Brown–Hopps stain.

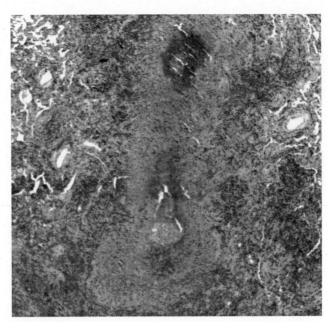

FIGURE 9.123 Necrotizing granulomatous and eosinophilic inflammation due to *S. mansoni.*

acid-fast stains, although the latter finding is inconstant (Fig. 9.126B). Other species, including *Schistosomiasis hematobium* and *Schistosomiasis japonicum,* rarely cause pulmonary disease and their ova show either a prominent or inconspicuous terminal spine, respectively.

Schistosomiasis can also produce a granulomatous pulmonary hypertensive arteriopathy due to the presence of either ova or migrating schistosomules within small caliber pulmonary arteries and arterioles (Fig. 9.127). The thickened vascular walls show epithelioid histiocytes and giant cells with adventitial fibrosis and eosinophils. This disorder usually arises from infection by *S. mansoni* or less commonly *S. japonicum* that has already produced pipestem hepatic fibrosis and

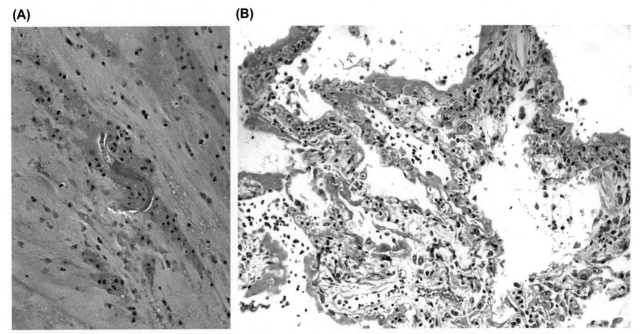

FIGURE 9.124 (A) Rhadbitiform larval form of *Strongyloides stercoralis* in patient with hyperinfection. (B) Patient died from DAD due to superimposed Gram-negative sepsis.

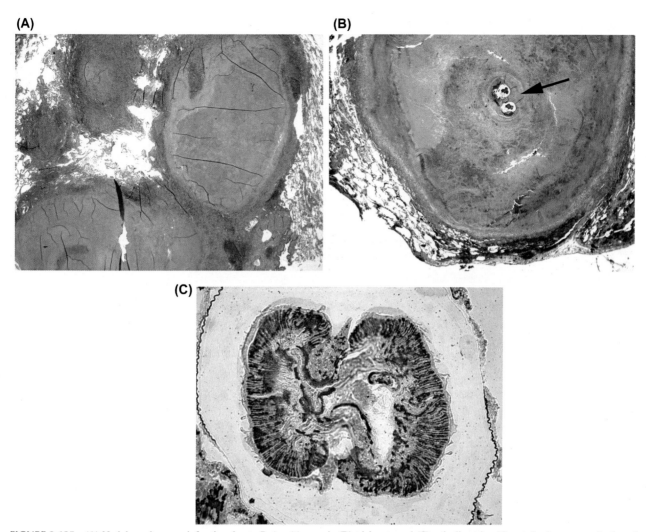

FIGURE 9.125 (A) Nodular pulmonary infarction due to *D. immitis* seen in (B) trichrome and (C) reticulin stains. Nematode shows muscular lateral chords with invagination of overlying cuticle.

presinusoidal portal hypertension, leading to shunting of portal blood into the systemic venous circulation, with ova swept into the pulmonary arterial bed where they evoke the pathological response.

Paragonimiasis

Paragonimus, often referred to as the "lung fluke," is a globally distributed trematode with human disease limited to endemic regions. Most cases in Asia are due to *Paragonimus westermani,* but other species are responsible for disease seen in Africa, Central and South America. Zoonotic forms of human infection rarely occur in North America secondary to *Paragonimus kellicotti* that affects rodents and cats.

The organism is a fresh water species that infects crabs and crayfish as intermediate hosts; the disease is transmitted to man via ingestion of these crustaceans, although cases may be seen that are due to ingestion of contaminated seaweed or watercress. The larvae mature in the bowel and migrate through the diaphragm to infect the lung. Patients present with pulmonary infiltrates, fever, weight loss, and malaise. Hemoptysis is common. Pleural involvement is present in roughly half of the patients. Peripherally located cystic lesions develop in the lung, together with areas of pneumonic consolidation, and concomitant bacterial and mycobacterial infections may be present.

Grossly, *Paragonimus* is a large 10-mm reddish fluke. Its wall consists of a tegument with prominent spiny projections best seen on high-power examination of trichrome stained sections (Fig. 9.128A). It exhibits prominent oral and ventral suckers and a loose internal stroma. The worm evokes a cystic necrotizing granulomatous and fibrotic response with tissue

(A)   **(B)**

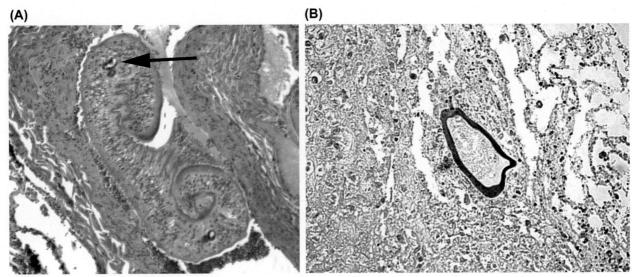

FIGURE 9.126   (A) Coiled tuberculated schistosome within pulmonary artery (*arrow*). (B) Large lateral spine and cortex of *S. mansoni* ovum is decorated by modified acid-fast bacteria stain.

eosinophilia. The ova of *Paragonimus* are ~80 μM and show a flattened operculum (Fig. 9.128B). They are also intensely birefringent under polarized light, and this is an important distinguishing factor with respect to schistosome ova (Fig. 9.128C).

Cestodes (Tapeworms)

The most common cestode pathogen in the lung is *Echinococcus*, although other tapeworm larvae, including *Cystercerca* and *Sparganum* can rarely infect the lung (Fig. 9.129). The adult tapeworm lives attached to the wall of the small intestine of carnivorous canids. Intermediate hosts are the grazing ungulates, including sheep, goats, deer, and bison. In the continental United States, the disease is endemic to midwestern and western states.

Pulmonary disease is generally the result of preexisting hepatic involvement by *Echinococcus granulosus*, although other species including *Echinococcus vogeli* and *Echinococcus multilocularis* are also pathogenic. The appearance of the

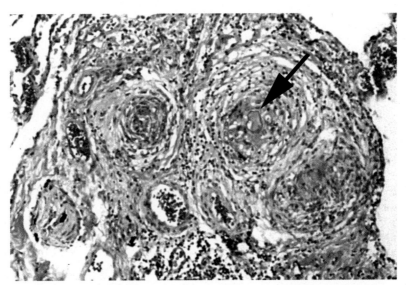

FIGURE 9.127   Granulomatous pulmonary arteritis due to *S. mansoni*. Both eggs (*arrow*) and schistosomules can be found in vessels although their demonstration may require the examination of multiple sections.

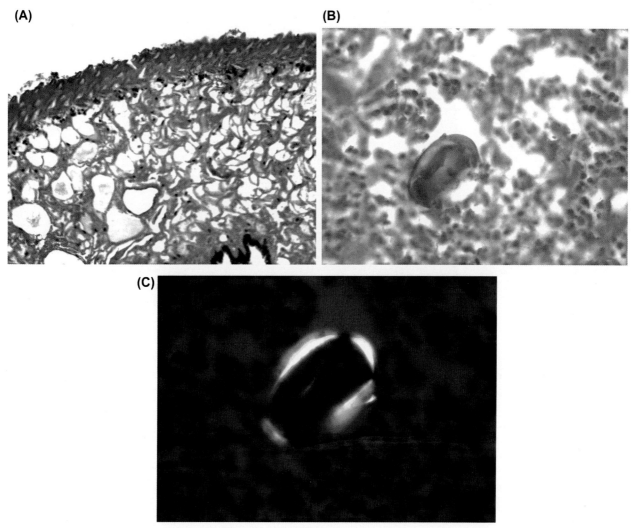

(A)

(B)

(C)

FIGURE 9.128   (A) *Paragonimus* spp. are trematodes that show characteristic surface tegumental spikes. (B) The ova of *Paragonimus* spp. are operculated, refractile, and show diagnostic birefringence with polarized light.

disease is characteristic on chest radiographs that show single or multiple large fluid filled cysts (Fig. 9.130). These cysts can be asymptomatic or cause symptoms due to compression of the surrounding airways and lung. Rupture into an airway can lead to suppurative pneumonia, the proliferation of new cysts, and generate fatal anaphylactic reactions.

The cyst wall shows contributions from the cestode as well as from the host. The inner germinative layer of the lamellar cyst wall is the matrix for protoscolices that detach to form secondary brood capsules (Fig. 9.131A). The rostellum of the attachment apparatus of the protoscolex contains rows of hooklets that are refractile and stain with modified AFB (Fig. 9.131B). The cestode component is acellular chitinous lamellar wall that is apparent with H&E and further highlighted with GMS and within the cyst wall, detritus and hooklets give rise to so-called "hydatid sand" (Fig. 9.132).

If the disease is isolated to a single cyst, percutaneous aspiration, injection of 20% hypertonic saline and 80% ethanol, followed by reaspiration in 5–30 min (PAIR procedure) may be effective therapy especially if coupled with.

Albendazole therapy. Complications of this therapy, which may be coupled with subsequent resection, include rupture of the cyst wall with anaphylaxis.

MICROBES ASSOCIATED WITH BIOTERRORISM

Recent world events have prompted interest in biological agents that can potentially be used as weapons of mass destruction. Several of these produce pneumonia and depend on dissemination via aerosolized secretions to achieve their

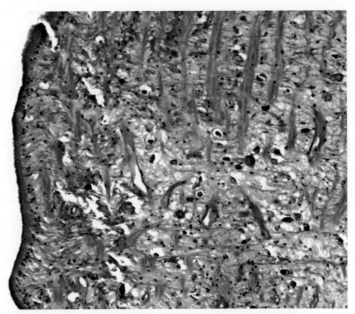

FIGURE 9.129   A cestode larval *sparganum* removed from lung.

ignominious goal. Although most of these infections are naturally virulent, they tend to occur under situations that are no longer commonly encountered in modern societies. Their adoption by bioterrorists may include the need to bioengineer the organisms in order to promote their attack rate.

Anthrax

Bacillus anthracis, a toxin producing Gram-positive *bacillus*, was first isolated by Robert Koch as the cause of anthrax, a disease that has primarily involved sheep and other farm animals. However, transmission to man from infected animals has been recognized from antiquity as a complication of wool sorting (wool sorter's disease). An epidemic of inhalational anthrax occurred in 1979 at a biofacility in Sverdlovsk, in the former U.S.S.R., and in

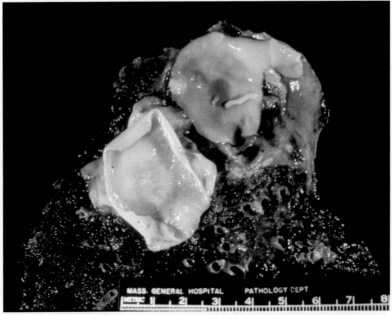

FIGURE 9.130   Pulmonary cyst due to *Echinococcus granulosus*.

(A) **(B)**

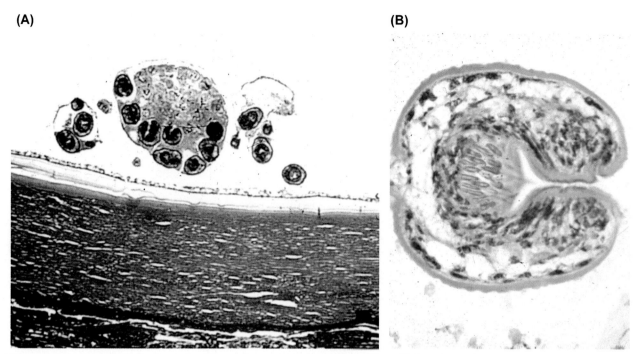

FIGURE 9.131 (A) Germinative layer of echinococcal cyst giving rise to protoscolices. (B) Rostellum and hooklets of a protoscolex stain with modified acid-fast bacteria.

2001, there was a limited epidemic of anthrax due to contaminated letters sent through the US mail service by an unidentified terrorist.

Cutaneous penetration by the *bacillus* produces a necrotic eschar with rapid extension to blood vessels resulting in bacteremia, sepsis, and death. However, the most deadly form of infection is pulmonary. In these cases, the bacilli are inhaled which evokes a localized hemorrhagic pneumonia and associated pleural effusion. Organisms proliferate rapidly in the lung where they produce a localized hemorrhagic pneumonia (Fig. 9.133) and spread via the pulmonary lymphatics to the regional lymph nodes to produce hemorrhagic mediastinitis, followed by bacteremia, toxic shock, and death in a high percentage of cases. The diagnosis must be suspected and treated early in

FIGURE 9.132 Chitinous cyst wall (left panel) and refractile hooklet of *Echinococcus granulosus* (right panel).

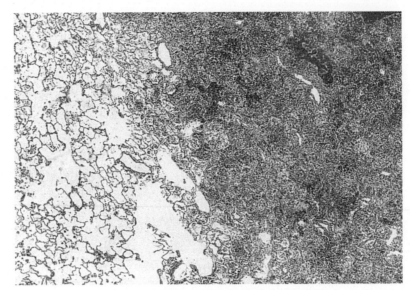

FIGURE 9.133 Hemorrhagic pneumonia due to anthrax.

order to be curable. However, it has most often first been recognized at autopsy, where tissues prove to be teeming with bacteria.

Yersinia pestis (Plague Pneumonia)

Plague has played an important role in world history. The causative agent is *Yersinia pestis*, a Gram-negative rod that is carried by animal fleas. Infection occurs as the result of contact with infected animals via aerosol or direct contact with infected secretions. Throughout history, the black rat, *Rattus rattus*, has been most responsible worldwide for the persistence and spread of plague in urban epidemics, but any rodent can mechanically transmit infected fleas. Although *Y. pestis* has not yet been seen as a bioterrorist agent, it has received attention as a potentially weapon of mass destruction, because as few as 1–10 bacilli are sufficient to cause infection when introduced via the oral, intradermal, subcutaneous, or intravenous routes.

Yersinia produces a necrotizing hemorrhagic pneumonia and large numbers of extra cellular organisms that can be seen with H&E (Fig. 9.134). Pulmonary infection leads rapidly to bacteremia and to death by sepsis.

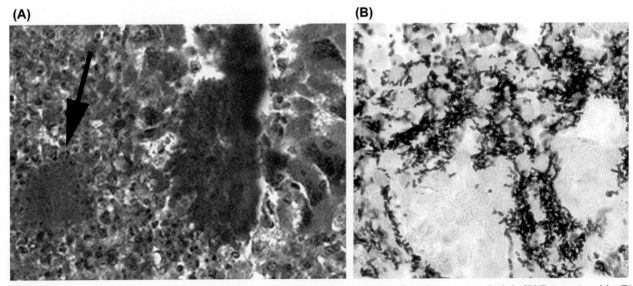

FIGURE 9.134 (A) Necrotizing hemorrhagic pneumonia due to plague (*Yersinia pestis*). Organisms are seen both in H&E (*arrow*) and in (B) silver-stained sections.

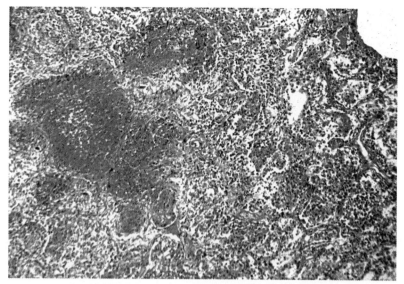

FIGURE 9.135   Granulomatous pneumonia due to tularemia.

FIGURE 9.136   Organizing fibrinous pleuritis due to pneumococcal pneumonia. Patient had parapneumonic effusion.

(A)   **(B)**

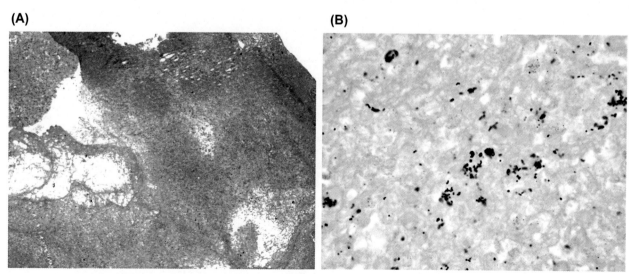

FIGURE 9.137   (A) Empyema secondary to staphylococcal pneumonia. (B) Clusters of cocci are stained by GMS.

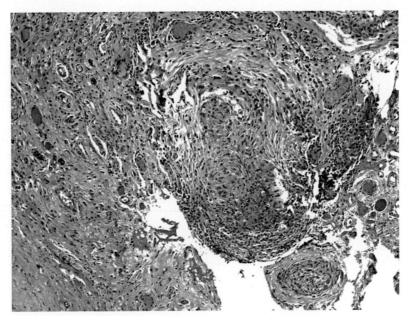

FIGURE 9.138 Granulomatous pleuritis in tuberculosis.

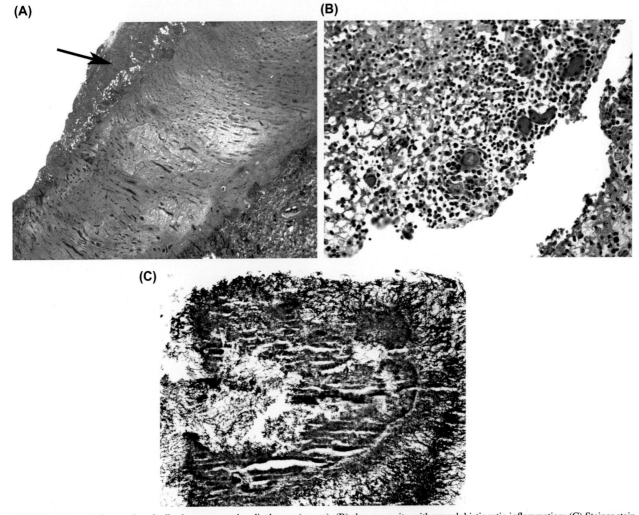

FIGURE 9.139 (A) Dense pleural adhesion transgressing diaphragm (*arrow*); (B) abscess cavity with granulohistiocytic inflammation; (C) Steiner stain shows filamentous bacteria in sulfur granule. Organisms were also positive with tissue Gram stain and GMS.

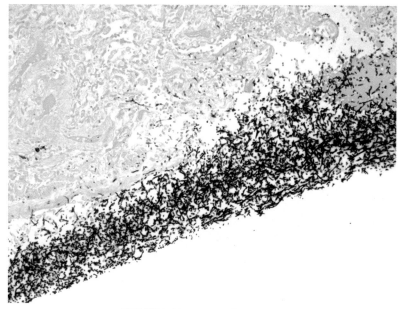

FIGURE 9.140   *Aspergillus pleuritis.*

Francisella tularensis (Tularemia Pneumonia)

Tularemia causes a necrotizing bronchopneumonia that leads to sepsis and death. However, the Gram-negative *bacillus* is less virulent than either anthrax or plague, and it produces a relatively slow progression of disease, a fact that limits its potential role as an agent of bioterrorism. The histological response in the lung is polymorphic and includes an early hemorrhagic granulohistiocytic response with microabscess formation (Fig. 9.135), followed by granulomatous inflammation. When these coexist, the appearance of this infection is characteristic. The short coccobacillary forms require silver impregnation to be visualized in tissues.

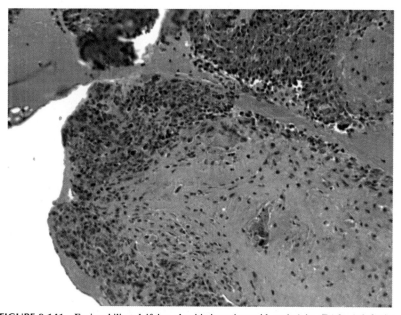

FIGURE 9.141   Eosinophilic calcifying pleuritis in patient with underlying *Trichuris* infection.

Pleural Infection

Parapneumonic effusions can complicate bacterial pneumonias but they are rarely biopsied unless they lead to a restrictive rind around the lung requiring decortication (Fig. 9.136A). Both Gram-positive and Gram-negative bacteria can produce empyema, i.e., abscess in the pleural space (Fig. 9.137A and B). Mycobacteria (Fig. 9.138), actinomyces (Fig. 9.139A and B), fungi (Fig. 9.140), and parasites can all yield exudative effusions, and the presence of necrotizing granulomatous inflammation will substantially assist in narrowing the differential diagnosis, even when organisms cannot be identified. Pleural eosinophilia can be a clue to the presence of an underlying parasitic infection (Fig. 9.141) but can also be seen in fungal and mycobacterial infections, in response to pleural metastases, as well as following pneumothorax.

FURTHER READING

Pound, M.W., Drew, R.H., Perfect, J.R., 2002. Recent advances in the epidemiology, prevention, diagnosis, and treatment of fungal pneumonia. Curr Opin Infect Dis 15 (2), 183—194.
An important topic in the management of an increasingly immunosuppressed patient population.
Ruuskanen, O., Lahti, E., Jennings, L.C., et al., 2011. Viral pneumonia. Lancet 377, 1264—1275.
A review of a broad and ever-changing topic.
Saunders, B.M., Cooper, A.M., 2000. Restraining mycobacteria: role of granulomas in mycobacterial infections. Immunol Cell Biol 78 (4), 334—341.
An examination of granuloma formation in the host response to mycobacterial infection.

Chapter 10

Lung Cancer

Despite inroads into understanding its molecular pathogenesis and therapeutic advances, cancer of the lung continues to be the most lethal human malignancy. Smoking remains the greatest risk factor for developing lung cancer. Although far fewer people smoke cigarettes in the United States today, than in the 20th century, smoking remains a major medical and socioeconomic problem among women and teenagers.

Pathologists have traditionally refined the histological classification of lung cancer in the hope of determining prognostic features (Table 10.1). A recent modification in the pathological classification of lung *adenocarcinomas* (ACA) includes the abandonment of the term "bronchioloalveolar carcinoma," and the adoption of the phrase *lepidic* growth to describe well-differentiated ACA whose growth is limited to the alveolar scaffolding of the normal lung (Fig. 10.1). Tumors with lepidic growth and no evidence of invasion into the lung parenchyma are currently classified as ACA in situ pTis and ideally should have 100% long-term survival following complete resection. Tumors with minimal invasion and those pT1a carcinomas that are ≤02.0 cm also have an excellent prognosis, but mortality then increases with tumor size, pleural invasion, lymph node involvement, and nonlymphoid tissue metastases (Table 10.2).

There are four common histological types of lung carcinoma. Large-cell undifferentiated (LCC) carcinoma is diagnosed less frequently due to the application of immunostains that have demonstrated that most LCCs are in fact poorly differentiated ACAs. All of the major types of lung carcinoma have been associated with cigarette smoking, but squamous cell and small-cell carcinomas are rarely seen except in heavy smokers. Second-hand smoke has also been estimated to increase the risk of developing lung cancer.

TABLE 10.1 Lung Cancer

Non-small-cell carcinoma	A term that is best used to refer to small biopsies before immunoprofiling is available. These cases in the vast majority of instances will be subsequently reclassified as squamous, ACA, or LCCs.
SCC	Almost always in smokers. Most common type of cancer complication is IPF and dermatomyositis. Central (85%), keratin, intercellular bridges, necrosis, hypercalcemia. Immunostains positive for p40, p63, CK5/6. Basaloid carcinoma, a tumor that mimics the appearance of cutaneous basal cell cancers with prominent central necrosis is a variant of squamous cell cancer with a slightly worse prognosis for equivalent stage
ACA	Most common type of lung cancer. Mostly occurs in lung periphery but may arise from proximal airway glands. Smoking related but substantial sunset never smoke. May arise in focal scars. Growth types include acinar (gland forming), solid, and lepidic. Tumors that are 100% lepidic are recognized as in situ carcinoma.
	Immunostain for TTF-1, Napsin A, and CK7 (latter can also be seen in squamous cancers). Mucin-secreting tumors may immunostain only with CK7 and CK20 but there is much variation in this subset. The term bronchioloalveolar carcinoma should no longer be used. ACAs may also immunostains for p63 and rarely p40 but are still classified as ACAs if TTF and Napsin positive. Adenosquamous carcinoma is unusual and the diagnosis requires clear presentation of both glandular and distinct squamous areas of growth
LCC	Most of these prove to be ACA by immunostaining or ultrastructural analysis. However, ~10% of lung carcinomas express neither TTF-1 nor p40 and these are properly termed LCC
Pleomorphic carcinomas	A term used to refer to a poorly differentiated non-small-cell carcinoma that may show either ACA or SCC immunostaining or may not. It is a histological variant based on the presence of spindle cell (sarcomatoid) elements or tumor giant cells

Continued

Understanding Pulmonary Pathology. http://dx.doi.org/10.1016/B978-0-12-801304-5.00010-1

TABLE 10.1 Lung Cancer—cont'd

Small-cell carcinoma	Almost always seen in heavy cigarette smokers. A tumor of basal cell origin that expresses neuroendocrine antigens. Tumor cells are ~2× the size of a lymphocyte, i.e., ~30 µM and may show rosette formation, extensive necrosis, and extracellular precipitation of DNA (Azzopardi effect). Cells contain neuroendocrine granules and immunostains for synaptophysin, chromogranin, cytokeratins, CD56, and TTF-1 in some combination. It is important to note that some ACAs may also express neuroendocrine antigens in relative small numbers of cells and this has no known prognostic implication
Large-cell neuroendocrine carcinoma	A large-cell "variant" of small-cell carcinoma. Cells may show spindle cell configuration or show prominent nucleoli and organoid growth. Immunostains are the same as in small-cell carcinoma and the treatment approach is largely the same.
Carcinoid tumor	A low-grade neuroendocrine carcinoma. Can secrete humoral factors but these rarely cause the carcinoid syndrome. May produce adrenocorticotrophic hormone and glucagon with systemic effects. Tumors tend to be central but ~15% are peripheral with spindle morphology. Immunostains strongly for synaptophysin and chromogranin but weakly if at all for TTF-1
Atypical carcinoid	Intermediate between carcinoid and small-cell carcinoma both histologically and prognostically. Mitotically active (2—4) per hpf. Prognosis is guarded

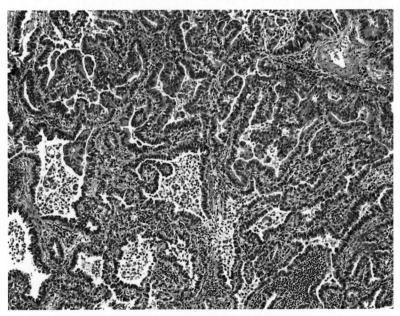

FIGURE 10.1   Lepidic growth of ACA in situ.

TABLE 10.2 WHO Pathology Staging Schema

Tumor Stage	
pTis	In situ carcinoma
pT1a	Tumor less that 2 cm
pT1b	Tumor≥ 2—3 cm
pT2a	Tumor ≥3—5 cm or visceral pleural invasion
pT2b	Tumor ≥5—7 cm
pT3	Tumor >7 cm or two tumors in same lobe
pT4	Invasion into chest wall, pericardium, great vessels, or different ipsilateral lobes

Continued

TABLE 10.2 WHO Pathology Staging Schema—cont'd

Lymph Node Stage

N1	Positive peribronchial or hilar lymph nodes
N2	Positive mediastinal or subcarinal lymph nodes
N3	Positive paratracheal, scalene, or contralateral lymph node groups
Metastasis	
M1	Metastatic disease

TABLE 10.3 Immunostains Used to Diagnose Lung Tumors and Benign Proliferations

Immunostain	Diagnosis
p40	Squamous carcinoma and benign squamous proliferations
p63	Same as above but somewhat less specific, mucoepidermoid carcinoma chief cells
TTF 1	ACAs, small-cell carcinomas, benign alveolar epithelium
Napsin A	ACAs and benign alveolar epithelium
Keratin 7	Tumors of foregut origin, both ACAs and SCCs. A subset of malignant mesotheliomas
CD56	Nonspecific marker but often strongly expressed by neuroendocrine malignancies
Calretinin	Often expressed by malignant mesothelioma, most frequently those showing epithelioid growth and a small percentage of lung ACAs
D2-40 (podoplanin)	Malignant mesothelioma and benign lymphovascular structures
CK5/6	SCC, a large percentage of malignant mesotheliomas, benign squamous epithelium
Wilms' tumor (WT)-1	Malignant mesothelioma, normal blood vessels
S-100	Langerhans' cell histiocytosis, neural tumors
CD1a	Langerhans' cell histiocytosis
HMB-45	Proliferating cells in lymphangioleiomyomatosis, peri-endothelial cell tumors
CD5	Thymic tumors

Currently, immunohistochemical staining plays a critical role in the subclassification of lung cancer and has become a standard technique in pathology. An immunohistochemical staining panel used to distinguish the common types of lung carcinoma is shown in Table 10.3. Despite this diagnostic approach, a small percentage of cases do not fall neatly into any category, and these include LCC and pleomorphic carcinomas.

Molecular phenotyping of lung carcinomas has also become a routine practice in pathology, and oncologists are likely to insist on it to determine whether a targeted molecular therapy is available for patients with unresectable disease. But despite the current enthusiasm surrounding this novel and effective therapeutic approach, targeted molecular therapies rarely result in cure, and most patients eventually die from their disease. Nevertheless, sustained tumor responses may be observed.

SQUAMOUS CELL CARCINOMA

Squamous cell carcinoma (SCC) generally arises in the trachea and proximal airways although ~15% of cases occur in the lung periphery. As in other squamous-lined areas of the body, including the skin, esophagus, and cervix, squamous neoplasia progresses through a series of stages from mild to severe dysplasia, and from carcinoma in situ to invasive carcinoma (Fig. 10.2A). SCC recapitulates the epidermis of the skin in its expression of cytokeratins (CKs) (Fig. 10.2B). It is almost always seen in heavy smokers, and it is common to have a field effect with multiple sites of squamous dysplasia

(A)

(B)

(C)

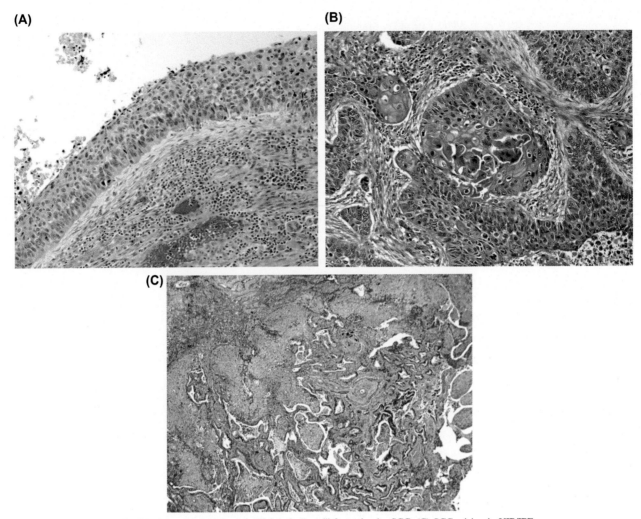

FIGURE 10.2   (A) SCC in situ, (B) keratin "pearl" formation by SCC, (C) SCC arising in UIP/IPF.

or carcinoma in situ present in the airways of patients with SCC. The risk of developing SCC falls after smoking cessation but never fully reaches that of a nonsmoking population, particularly in heavy smokers.

A significant subset of the peripheral SCCs is associated with diffuse interstitial fibrosis, often due to IPF or dermatomyositis (Fig. 10.2C). Emphysema appears to be an independent factor of risk for carcinoma in smokers and whether smoking-related fibrosis is also an independent risk factor has not yet been adequately evaluated.

Patients with SCC may be asymptomatic or alternatively may present with cough, dyspnea, hemoptysis, or chest pain. The latter suggests extension of the tumor into the chest wall. Tumors arising in the proximal airways tend to invade lymphatics early or can spread by contiguity to peribronchial lymph nodes. SCC has an unusual propensity to develop central necrosis and to cavitate, so that radiographic evidence of a cavitary mass in a smoker should raise the possibility of SCC.

Basaloid carcinoma is a pathological variant of poorly differentiated SCC that may have a slightly worse prognosis. The tumor cells in this variant recapitulate the appearance of a basal cell carcinoma of the skin, do not show obvious keratin formation, and have a peculiar propensity to undergo early central comedonecrosis (Fig. 10.3).

The presence of hypercalcemia in the absence of bony metastases is also seen in SCC, where it reflects the secretion of a parathormone-like substance by the tumor.

Unlike ACA, finding targetable mutations for SCC has been challenging. Immune interventions that target the programmed cell death-1 subset of T lymphocytes and its epithelial cell ligands appear to reduce local immunosuppression and have shown antitumor responses in early clinical trials. It is noteworthy that therapies that target tumor angiogenic factors, e.g., bevacizumab, have resulted in fatal pulmonary hemorrhage in patients with SCC and are contraindicated in treatment.

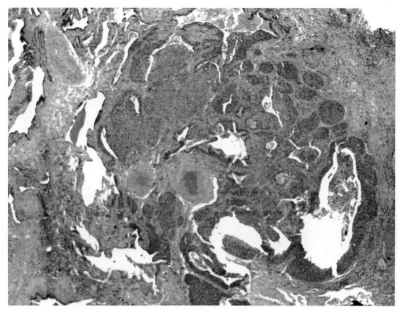

FIGURE 10.3   Basaloid carcinoma with central "comedonecrosis."

Squamous cell malignancies in the lung in patients with known SCC at other sites can be a diagnostic challenge. Examining whether the histology of the previous tumor is comparable or distinct from the lung primary can clarify whether the lesion in the lung is metastatic, but this is not always the case. An area of in situ SCC growth in the biopsy is *prima facie* evidence that the lesion is primary in the lung. Factors favoring a metastasis include peripheral location, multiple and bilateral tumors, and hemolymphatic invasion of tumor in a biopsy.

As the head and neck accounts for many of the extrapulmonary SCCs, evidence of human papilloma virus (HPV) infection in the tumor or p16 expression, which is a surrogate marker for HPV infection, favors a metastasis from the head and neck region. A comparable statement applies as well to less common metastases from SCCs of the uterine cervix.

SMALL-CELL AND LARGE-CELL (NEUROENDOCRINE) CARCINOMA

Small-cell carcinoma is virtually always seen in heavy cigarette smokers, and in the absence of a smoking history the pathologic diagnosis should be approached with caution, and potential mimics in small biopsies, including carcinoid tumors should be excluded. Another mimic is the rare primitive neuroectodermal tumor (Ewing's sarcoma).

Similar to SCC, small-cell carcinoma generally arises in the proximal airways, although small-cell carcinoma in situ has never been identified. Bulky malignant lymphadenopathy and distant metastases are common. Cases of small-cell carcinoma limited to the lung without lymph node or distant metastasis have a better prognosis and should be staged like other lung carcinomas. Oncologists approach small-cell carcinoma as either limited or extensive disease, and patients with the latter receive prophylactic brain irradiation as the tumor tends to relapse in the central nervous system following chemotherapy. Paraneoplastic syndromes in a patient with a lung mass, including Cushing's syndrome, inappropriate secretion of antidiuretic hormone, and myasthenia-like disease (Eaton−Lambert) syndrome, are virtually always a complication of small-cell carcinoma.

The tumor cells of small-cell carcinoma are ~ 30 μM, i.e., about twice the size of a normal lymphocyte. They exhibit an organoid growth pattern and rosette formation. Their nuclei show stippled chromatin, often referred to as a "salt and pepper" appearance, and molding of cells may be present on high power examination (Fig. 10.4A). Mitotic activity is brisk and necrosis is apparent. There is also a peculiar tendency of these cells to show a characteristic crush artifact, which although nondiagnostic is a helpful feature in diagnosis (Fig. 10.4B). DNA released by dying tumor cells is intensely basophilic and streams out of the cells to produce the *Azzopardi effect* (Fig. 10.4C).

Large-cell neuroendocrine cell (NEC) can show a histologically distinct appearance, with prominent central nucleoli (Fig. 10.5), but at times the distinction between large versus small-cell NEC is subjective, and it is not uncommon to see cases that show both small- and large-cell tumor elements.

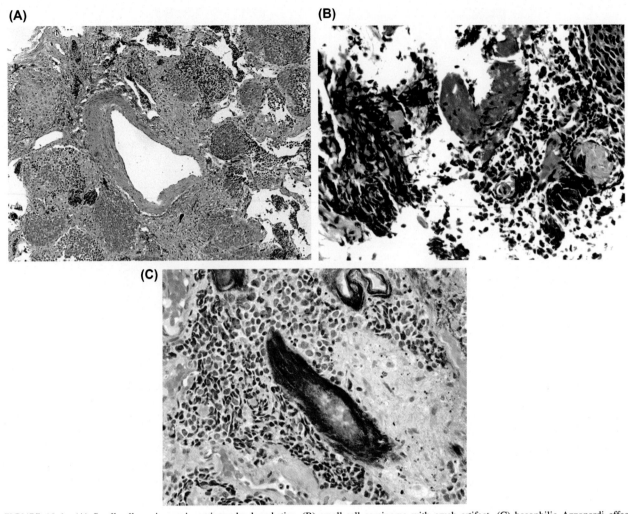

FIGURE 10.4 (A) Small-cell carcinoma in perivascular lymphatics, (B) small-cell carcinoma with crush artifact, (C) basophilic Azzopardi effect surrounding vessel.

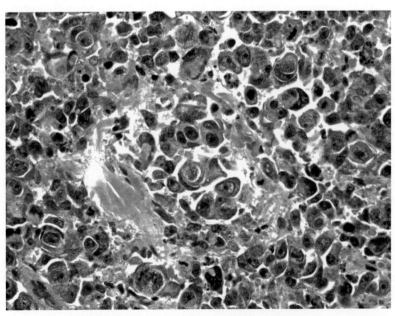

FIGURE 10.5 Large-cell neuroendocrine carcinoma.

There is some evidence to suggest that large-cell neuroendocrine carcinoma responds more like non-small-cell carcinomas than small-cell carcinoma to therapy. It is noteworthy that a subset of ACA also shows evidence of neuroendocrine differentiation based on their immunostaining profile. The diagnosis of large-cell neuroendocrine carcinoma should be limited to cases where the dominant malignant cell population shows diffuse histological and immunohistochemical features of neuroendocrine differentiation. To complicate matters, some carcinomas may show areas of glandular differentiation together with small-cell NEC. In such cases, the tumor should be treated as a small-cell carcinoma. It is also well recognized that following treatment for small-cell carcinoma, the rebiopsy of tumor may show non-small-cell carcinoma, presumably reflecting treatment-induced differentiation.

ADENOCARCINOMA

ACA is the most common type of lung cancer and currently the most well characterized. The reason for its apparent increased incidence in recent years is multifactorial and includes improved diagnostic modalities for distinguishing ACA from SCC, as well as yet undetermined environmental factors. Although linked to cigarette smoking, ACAs also occur in nonsmokers, and in such cases, other potential environmental carcinogens, e.g., asbestos, radon, etc., should be excluded.

ACAs can arise either in the proximal airways where they are likely caused by mutations in the minor salivary glands of the airways, but more often they arise in the lung periphery as part of the terminal acinar unit (Fig. 10.6A). Patients with preexisting pulmonary scars (Fig. 10.6B) and those who have received previous radiotherapy show an increased incidence of ACA that appears to arise in scars. Atypical adenomatous hyperplasia is generally held to be a precursor lesion for at least some ACAs. These lesions are commonly seen in surgical specimens and at autopsy, show a lepidic architecture, are limited in size (<5–10 mm), and display low grade cytological abnormalities (Fig. 10.7A). Marked cytological atypicality even when the lesions are small suggests that the lesion is best characterized as in situ ACA (Fig. 10.7B).

Substantial efforts have been directed at the histological subclassification of ACAs. Papillary and micropapillary growth patterns are recognized as poor prognostic features (Fig. 10.8). Mucin-secreting ACAs (Fig. 10.9A) can show lepidic growth (Fig. 10.9B); however, they are considered to be invasive and there is currently no agreed upon in situ status for these lesions. They may also be multicentric in the lungs due to airway spread of tumor. A small percentage of these patients present with bronchorrhea and show hypoxemia with physiological shunt. In such cases, the lung often shows abundant macrophages that have ingested mucin (muciphages). A high percentage of tumors showing mucinous signet-ring cells (Fig. 10.10) has been associated with *ALK* fusion products (Fig. 10.10B and C).

Differentiating mucin-secreting primary ACAs from extrathoracic primaries in the gut or pancreato-biliary tree may be challenging. Most mucinous lung ACAs express either nuclear thyroid transcription factor (TTF-1) or cytoplasmic Napsin A but a significant number do not, and these may immunostain comparable to an extrathoracic gastrointestinal primary with expression of CK-20 and CDX-2. In such cases, clinical correlation is required and molecular analysis may assist in making the distinction.

(A)   **(B)**

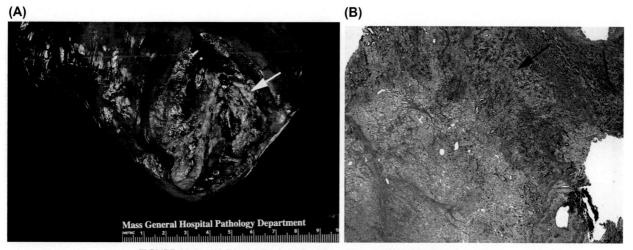

FIGURE 10.6   (A) Distal ACA in unfixed lung (*arrow*), (B) ACA arising in a scar.

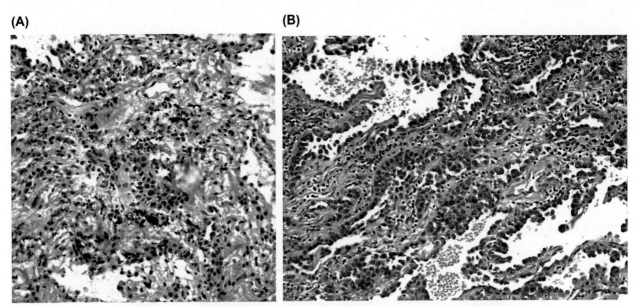

FIGURE 10.7 (A) Atypical adenomatous hyperplasia, (B) highly atypical cells in small lepidic ACA.

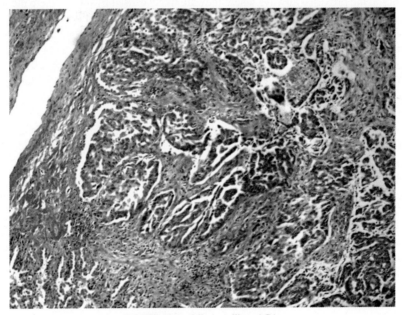

FIGURE 10.8 Micropapillary ACA.

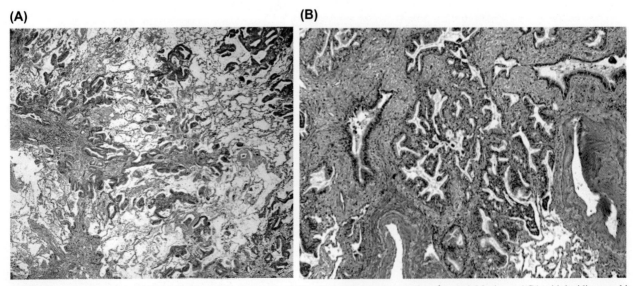

FIGURE 10.9 (A) Mucin-secreting ACA, (B) multifocal mucinous ACA with lepidic airway growth and spread. Mucinous ACA with lepidic spread is considered invasive.

(A)

(C)

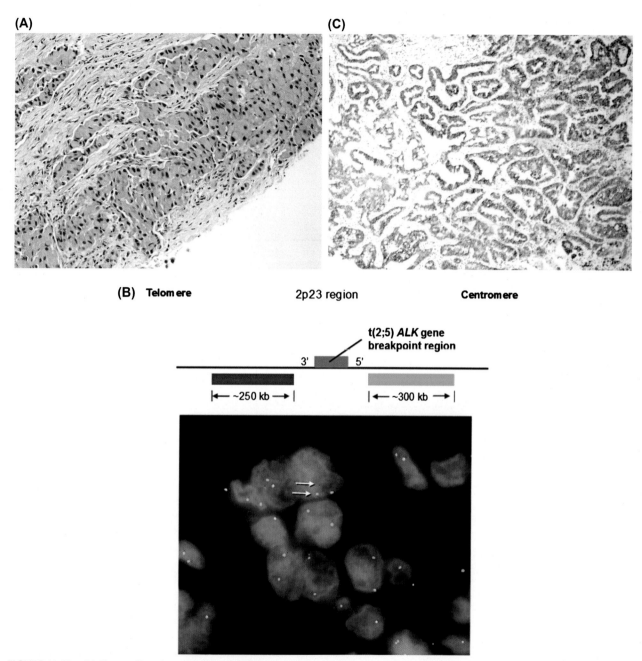

(B) **Telomere**   **2p23 region**   **Centromere**

t(2;5) *ALK* gene breakpoint region

3'   5'

|←── ~250 kb ──→| |←── ~300 kb ──→|

FIGURE 10.10   (A) Signet cell carcinoma with mucin-filled cells and eccentric nuclei in patient with ALK 1 mutation, (B) fusion product in Alk-1 seen by in situ hybridization, (C) Alk-1 immunostain of carcinoma.

MULTIPLE TUMORS

ACAs can arise in the lung but may also represent metastases from other organs. It is important to provide information about previously resected cancers so that the pathologist can do due diligence in ascertaining whether an ACA in the lung is primary or metastatic. Bilateral lung involvement generally suggests metastatic disease but can be seen from primary lung carcinomas that metastasize to lung or spread via shedding malignant cells along the airways.

A common scenario is the presence of more than one ACA in a lobe of the lung or in more than one lobe. If the histological appearance of the lesions is distinct, they are likely the result of different synchronous primary ACAs. On the other hand, if they share a similar appearance, they may be clonally related. These cases merit both immunohistochemical and molecular analyses. This can help to distinguish, e.g., two distinct ACAs from stage IV disease, as treatment options and prognosis are vastly different in such scenarios. Currently, accepted pathological staging classifies multiple tumors in a single lobe as pT3 disease, mostly to avoid having to determine clonality of the lesions histologically.

MOLECULAR ANALYSIS

The analysis of proto-oncogenes has yielded important information concerning the molecular pathogenesis of non-small-cell lung cancer (NSCLC) and opened the door to new therapeutic options (Fig. 10.11). Currently, virtually all new cases of ACA of the lung are reflex tested for evidence of a targetable mutation, and immunotherapeutic approaches to NSCLC have also begun to show promise (Fig. 10.12).

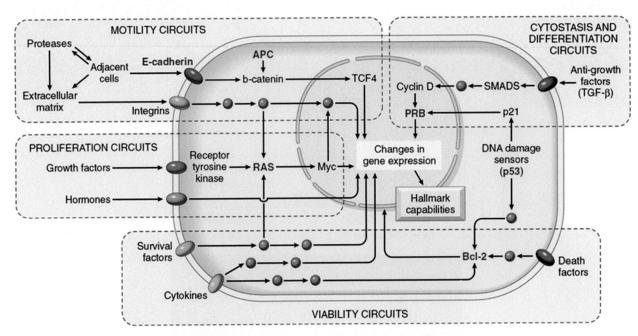

FIGURE 10.11 Molecular pathways that drive growth in lung cancer.

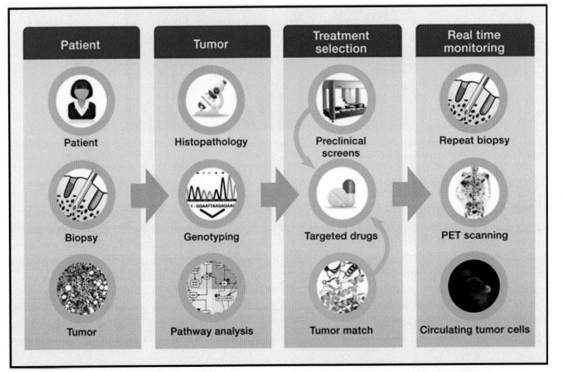

FIGURE 10.12 New paradigm for treatment of patients with lung cancer. *Courtesy of J. Iafrate.*

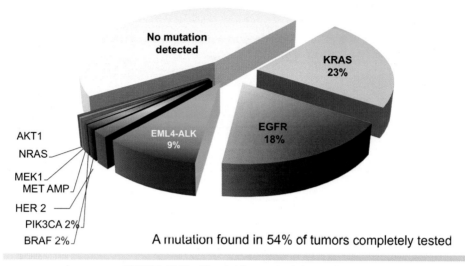

FIGURE 10.13   Mutation frequencies in lung ACA.

Unfortunately, the most common proto-oncogene mutation, *K-RAS*, which occurs in 20% of ACA and 5% of SCC, has not yet been successfully targeted by pharmacological agents (Fig. 10.13). The first reports of a targetable epidermal growth factor receptor (*EGFR*) mutation were from a subset of patients with ACA of the lung. *EGFR* mutations occur in 10–15% of patients with lung cancer. *EGFR* exon 19 deletions and exon 21 point mutations are the most common, and they confer tumor susceptibility to tyrosine kinase inhibitors (Fig. 10.14). Exon 20 mutations are generally resistant to tyrosine kinase inhibitors. These *EGFR* mutations have been seen in nonsmokers, young women, and those of Asian descent. Preliminary evidence suggests that nonsmoking asbestos-exposed patients may show *EGFR* mutations, raising the question as to whether asbestos may be a cause of these mutations.

Due to the downstream pathway of molecular carcinogenesis a *KRAS* mutation effectively excludes a targetable *EGFR* mutation. Other potentially targetable driver mutations in ACA include *BRAF, HER2, PTEN, AKT,* and *PIK3CA*. In SCC, potentially actionable mutations include *DDR2, PIK3CA, PTEN, AKT, KEAP1,* and *NFE2L2*.

Certain gene rearrangements can drive tumor growth in NSCLC. *ALK-1* rearrangements occur in 3–5% of ACA. These may be detected by immunohistochemistry or by in situ hybridization. These patients respond to crizotinib therapy that inhibits the ALK-1 enzyme. *ROS-1* and *RET* fusions are seen in 1–2% of patients with ACA. Gene amplification due to *MET* is present in 1% of ACA, and *FGFR1* amplification is seen in 20% of patients with SCC. Various techniques ranging from immunohistochemistry (*EGFR, ALK-1,* and *ROS-1*), in situ hybridization and gene sequencing are currently used in the diagnostic laboratories to establish the presence of these mutations.

IMMUNOHISTOLOGY

At the histological level, a substantial number of ACAs will show squamous differentiation and it can be impossible to determine whether one is dealing with a squamoid ACA or a poorly differentiated SCC without the aid of immunohistochemistry. By convention, immunohistochemical evidence of glandular differentiation, as judged by positive immunostaining for TTF-1 and Napsin A, trumps expression of either p63 or less commonly p40, by ACAs, and these malignancies are by convention diagnosed as ACAs. The diagnosis of adenosquamous carcinomas should be limited to cases where there is a distinct component of both ACA and SCC or to those cases where glandular units share a distinct squamous component as is commonly seen in the lower esophagus.

LARGE-CELL UNDIFFERENTIATED CARCINOMA

Large-cell undifferentiated carcinoma (LC) is a diagnosis that is made less commonly than in the past. The reason is largely based on how tumor lineage is defined, i.e., based on histological features versus immunohistochemical and molecular

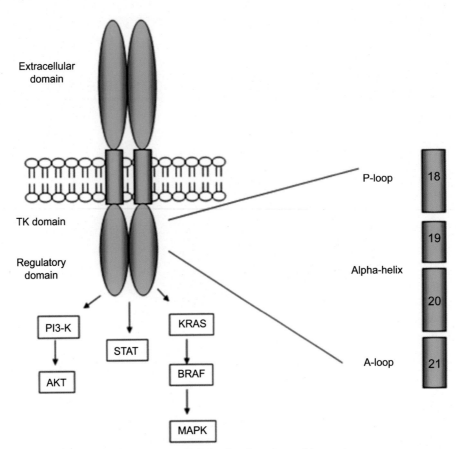

FIGURE 10.14   EGFR showing sites of targetable mutation.

features. It is recognized that most cases of LCs are ACAs as judged by immunohistochemistry (TTF-1, Napsin A) and ultrastructural analysis.

However, there are cases that show no clear evidence of differentiation and lack positive immunostaining for p40 (SCC), TTF-1, or Napsin A (ACA). Most of these are CK7 positive; others may be only CK positive without further evidence of differentiation. LC makes up a large percentage of Pancoast or "superior sulcus" tumors, although the term "superior sulcus" is a misnomer as no such anatomic entity exists. Rather, these are carcinomas that arise in the apex of the upper lobes and grow into the adjacent chest wall and brachial plexus. Patients may develop weakness of the upper extremity and *Horner's syndrome* due to involvement of the preganglionic sympathetic chain leading to meiosis of the ipsilateral pupil of the eye, ptosis, and unilateral anhidrosis.

PLEOMORPHIC CARCINOMAS

Pleomorphic carcinomas are poorly differentiated carcinomas that may show immunohistochemical features of either SCC or ACA. What distinguishes them is a spectrum of histological changes including large numbers of tumor giant cells (giant cell carcinoma) (Fig. 10.15) or spindle cell features (sarcomatoid carcinoma). These carcinomas tend to be clinically aggressive, inoperable, and have a bad prognosis despite treatment. Rare tumors termed *pulmonary blastomas* may include an epithelial component and heterologous sarcomatoid elements (chondrosarcoma, rhabdomyosarcoma, and osteosarcoma). These tend to be large when detected and have a poor prognosis.

MALIGNANT MELANOMA

Malignant melanoma rarely arises in the airways, but is one of the tumors, along with breast and renal cell cancer, that commonly metastasize to the airway mucosa, where they can mimic a lung primary. When a primary cutaneous melanoma

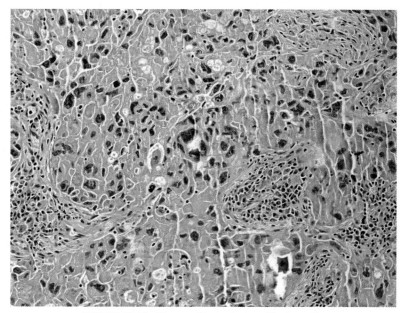

FIGURE 10.15   Pleomorphic carcinoma with giant cells.

has been diagnosed, or when the tumor cells produce melanin, the diagnosis is generally straightforward. However, poorly differentiated amelanotic melanomas can mimic primary NSCLC (Fig. 10.16). The diagnosis is suggested by failure of the cells to immunostain for CKs. Malignant melanomas immunostain positive for one or more melanocytic antigens including S-100, melan A, and MART-1. These tumors are great mimics and can present as spindle cell sarcomas and one must then delve beyond the absence of keratin immunostaining—a feature shared by most sarcomas and melanomas to demonstrate melanogenesis via the immunostains noted above.

Renal cell carcinoma (RCC) may metastasize to the bronchi or produce cannonball lesions within the lung (Fig. 10.16B). It immunostains with Pax 8, RCC-1, and CD10.

(A)   **(B)**

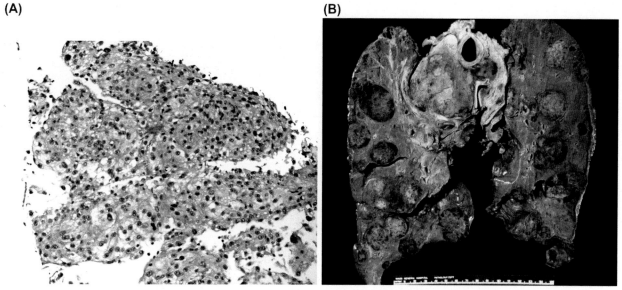

FIGURE 10.16   (A) Metastatic melanoma to bronchus, (B) meta renal cell.

METASTATIC CARCINOMA

The lung is a common site for metastatic lesions from virtually all sites. In most cases, a primary site is recognized. Surgical resection has been shown to improve survival for melanoma, colorectal carcinoma, and RCC when there is limited oligometastatic disease.

It is not uncommon to be faced with the dilemma of having to distinguish a metastatic malignancy from a primary in the lung. To further complicate matters, occasionally primary malignancies in the lung closely mimic those from other sites. In practice the lesions that are most problematic are metastatic SCCs from the head and neck, esophageal carcinomas, breast cancer (Fig. 10.17A), transitional cell carcinomas (Fig. 10.17B), colorectal carcinomas (Fig. 10.17C), pancreatobiliary carcinomas, hepatocellular carcinomas (Fig. 10.17D), and thyroid carcinomas (Fig. 10.17E). In general, metastases are multiple, often bilateral, and peripherally located.

In the case of SCCs from the head and neck and esophagus, distinctions can be difficult as most of these patients are also smokers and therefore at risk for squamous neoplasia at both sites. As already noted the presence of multiple peripheral tumors favors metastasis. Expression of p16, a surrogate marker for papillomavirus, favors a metastasis from the head and neck area.

The time-honored way to distinguish primary from secondary malignancies in pathology is to compare the histologic appearance of a known primary tumor with that of the lung tumor in question. If this is indeterminate, or if sections from the original tumor are not available, immunohistochemical stains and genetic molecular profiling can help to distinguish primary from metastatic tumors. In general, it is wise to exercise caution before diagnosing a new primary lung cancer when there is a known malignancy at another site, especially if they potentially may share common histological appearances. It is, therefore, incumbent upon the clinician to advise the pathologist of a known primary outside of the lung and not to assume that this will be obvious to the pathologist.

Papillary thyroid carcinomas can be indolent and present as multiple small nodules that grow slowly in the lung. Radioiodine scans may detect the lesions noninvasively. Thyroid tumors will immunostain for TTF-1 like papillary lung carcinomas but the former can be distinguished by their immunostaining for PAX-8 and thyroglobulin. Follicular carcinomas of the thyroid are unusual in that they tend to spread via the blood circulation rather than via lymphatics.

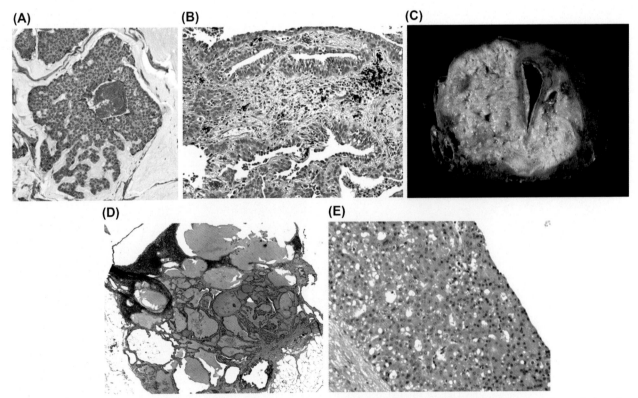

FIGURE 10.17   Lungs with (A) metastatic breast cancer, (B) glistening and necrotic appearance of metastatic colorectal carcinoma to bronchus, (C) metastatic transitional cell carcinoma of bladder, (D) papillary thyroid carcinoma, (E) hepatocellular carcinoma.

SARCOMAS OF THE LUNG

Sarcomas differ from carcinomas in their histological appearance, differentiation, and mode of spread. Unlike carcinomas that spread via lymphatics, sarcomas spread via the circulating peripheral blood. Most primary sarcomas arise in the proximal portions of the lung and can grow to large size before being detected. Leiomyosarcomas (desmin-immunopositive) (Fig. 10.18) and malignant fibrous tumors (CD34 and STAT 6 immunopositive) are the most commonly encountered primary pulmonary sarcomas (Fig. 10.19). However, virtually any mesenchymal element, cartilage, skeletal muscle, etc., can potentially undergo malignant transformation.

Tumors arising from endothelial cells (angiosarcomas or epithelioid hemangioendothelioma) exhibit distinct histological appearances. The latter tumor was originally described by Liebow in the lung as an "intravascular bronchio-loalveolar tumor." This was an error based on the tumor's epithelioid growth pattern and unusual myxoid stroma

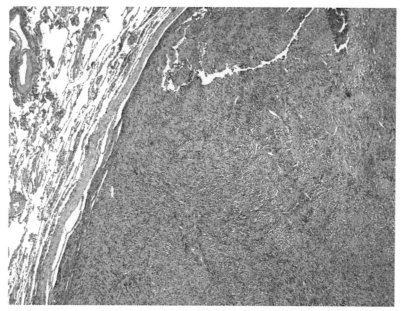

FIGURE 10.18   Pulmonary leiomyosarcoma.

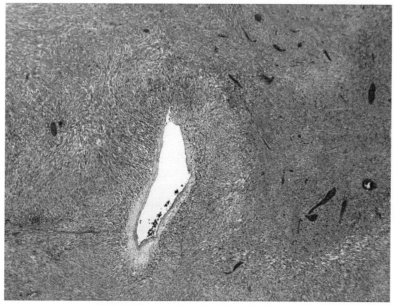

FIGURE 10.19   Malignant solitary fibrous tumor.

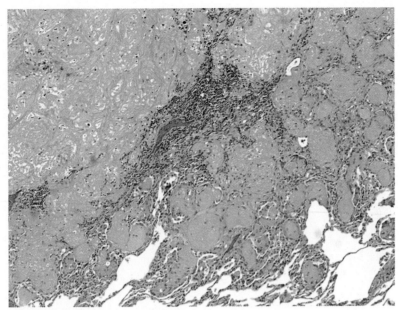

FIGURE 10.20   Hemangioendothelioma shows characteristic dense matrix deposition.

(A)   **(B)**

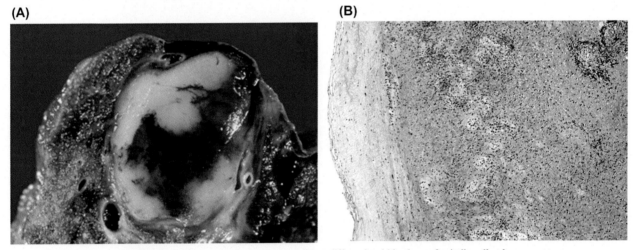

FIGURE 10.21   (A) Pulmonary artery sarcoma in large artery, (B) undifferentiated histology of spindle cell pulmonary artery sarcoma.

(Fig. 10.20). The tumor immunostains positive for vascular antigens and ultrastructurally shows Weibel–Palade bodies seen in normal endothelial cells.

Sarcomas can arise in the intima of the pulmonary artery (pulmonary artery intimal sarcomas) and may mimic thromboembolic disease, resulting in dyspnea and pulmonary hypertension (Fig. 10.21A and B). Computed tomography (CT) imaging and magnetic resonance imaging are most helpful in suggesting the diagnosis, which ultimately requires endovascular pulmonary artery biopsy for confirmation.

Metastatic sarcomas from peripheral sites frequently involve the lung. Current practice includes the surgical extirpation of small numbers of these tumors (as a life extending but noncurative procedure in conjunction with chemotherapies). Similar approaches are effective with oligometastatic colorectal and RCCs.

Kaposi's sarcoma can involve the lung and is seen in patients with acquired immunodeficiency syndrome and CD4+ counts <200/ml. In the lung, the lesions tend to have a plaque-like configuration along bronchovascular septa and the lesions may be seen at bronchoscopy as violaceous plaques. The microscopic pathology shows a bland spindle cell proliferation with slit-like spaces, PAS-positive inclusions, and extravascular red blood cells (Fig. 10.22). The proliferating cells are immunopositive for HHV-8 as display vascular antigens, e.g., CD31 and CD34.

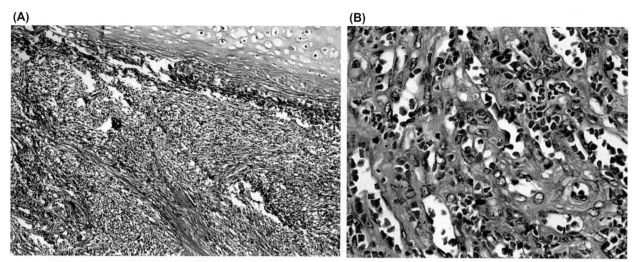

FIGURE 10.22   (A) Kaposi sarcoma growing in a bronchus, (B) histology showing vascular spaces and diapedesis of red blood cells.

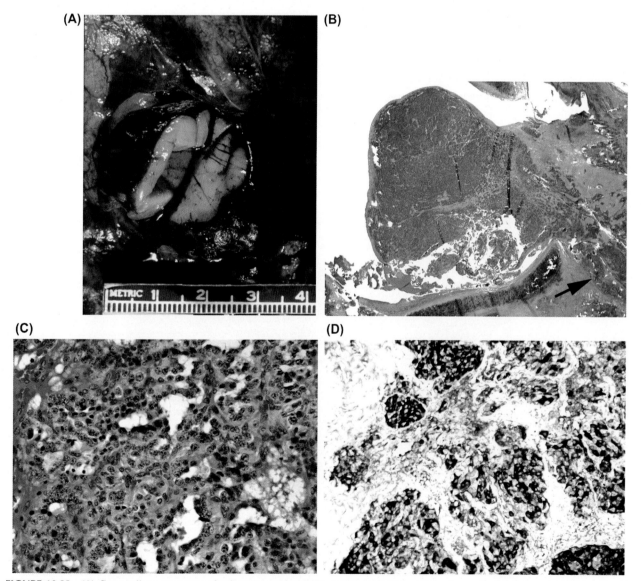

FIGURE 10.23   (A) Gross yellow appearance of pulmonary carcinoid, (B) typical carcinoid in a large airway showing iceberg type of growth with invasion through cartilaginous plates, (C) salt and pepper nuclear chromatin seen in carcinoid tumors, (D) and adrenocorticotrophic hormone secreting carcinoid in patient with Cushing's syndrome.

TUMORS OF INTERMEDIATE MALIGNANT POTENTIAL

A number of low-grade malignant tumors occur in the lung. Carcinoid tumor is the most common, and it accounts for ~1% of pulmonary malignancies. Roughly 10% of these are associated with multiple endocrine neoplasias. Although carcinoids are usually limited to the lung, they may metastasize to hilar lymph nodes. Despite this, carcinoid tumors rarely lead to death.

The tumors arise primarily in the trachea and mainstem bronchi (Fig. 10.23A). Their growth shows an "iceberg" effect, so that only the most superficial part of the tumor is seen through the bronchoscope, whereas the bulk of tumor may invade deep into the airway wall (Fig. 10.23B).

Many patients with carcinoid tumors are asymptomatic but they may obstruct the airway causing stridor; other patients be erroneously diagnosed with asthma. The distinction can often be made by examination of a flow volume loop that shows diminished peak flow rates with flattening of the expiratory loop and is subsequently confirmed by CT scans or endoscopy. The tumors are highly vascular, and biopsy can rarely lead to brisk bleeding, but is safe in most instances. The tumors should never be "shelled" out due to their invasive nature, and the proper approach is either lobectomy or a "sleeve" resection.

Microscopically, carcinoids have a characteristic appearance. They are composed of monotonous polygonal cells that show bland nuclear features with neuroendocrine "salt and pepper" nuclear chromatin (Fig. 10.23C). By immunohisto-chemistry they are virtually always strongly immunopositive for chromogranin and synaptophysin and negative for TTF-1. This pattern of immunostaining can be helpful when trying to distinguish a carcinoid from either atypical carcinoid or small-cell carcinoma in a small biopsy where the distinction by light microscopy alone can be difficult. Along the spectrum of neuroendocrine neoplasia, the higher the degree of malignancy the greater the immunoexpression of TTF-1 and the lower the expression of both chromogranin and synaptophysin. As noted, the absence of a significant smoking history should be considered before making a diagnosis of small-cell carcinoma.

Some carcinoids in the lung are hormonally active producing neuroendocrine factors, e.g., adrenocorticotrophic hormone leading to Cushing's syndrome. A serotonin (5-hydroxytyptamine) secreting carcinoid in the lung may cause the carcinoid syndrome with facial flushing, wheezing, diarrhea, and tricuspid valvular fibrosis leading to insufficiency (Fig. 10.23D).

Atypical carcinoids differ from typical carcinoids by virtue of their increased mitotic activity (two to five mitoses per 10 hpf) and the presence of necrosis (Fig. 10.24). Other features, including increased cellularity and cytological atypia,

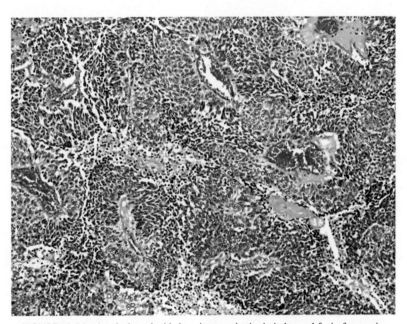

FIGURE 10.24 Atypical carcinoid show increased mitotic index and foci of necrosis.

(A)   **(B)**

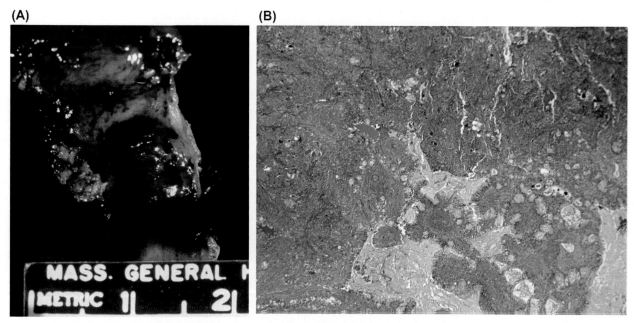

FIGURE 10.25   (A) Mucoepidermoid carcinoma in major bronchus, (B) histology with mucus secreting and squamoid cells.

suggest malignancy but are not considered diagnostic criteria. TTF-1 positivity may be present together with diminished immunostaining intensity for synaptophysin and chromogranin.

SALIVARY GLAND TUMORS

The airways are invested with seromucous glands that are morphologically comparable to those seen in the minor salivary glands. These cells may undergo neoplastic transformation leading to tumors that are generally low-grade malignancies but can, based on their degree of differentiation, behave aggressively locally and metastasize.

MUCOEPIDERMOID TUMORS

Mucoepidermoid tumors are unusual and generally occur in the major bronchi where they may present as obstructing lesions (Fig. 10.25A). Patients who present with wheezing or stridorous sounds should undergo a chest radiograph before being labeled as having asthma. Pulmonary function tests may show evidence of obstruction. These tumors are graded as "low" or "high grade" dependent on a number of intermediate cells with squamoid differentiation (Fig. 10.25B). High-grade tumors tend to be aggressive and careful follow-up is required following excision. It may be difficult histologically to distinguish high-grade mucoepidermoid carcinomas from SCC, and the absence of a smoking history favors the former.

ADENOID CYSTIC CARCINOMA

Adenoid cystic tumors mostly involve the trachea (Fig. 10.26A). Although morphologically they are low-grade neoplasms, they have a propensity to be locally invasive and to recur, eventually leading to respiratory disability and death, although this may occur after many years and repeated local excisions.

The appearance of these tumors is characteristic; they produce a cribriform or "swiss cheese" pattern of growth when observed at low power (Fig. 10.26B). They secrete basement membrane material with pseudolumens. Adenoid cystic carcinoma has a peculiar propensity to invade the perineural lymphatics and this may contribute to its malignant behavior.

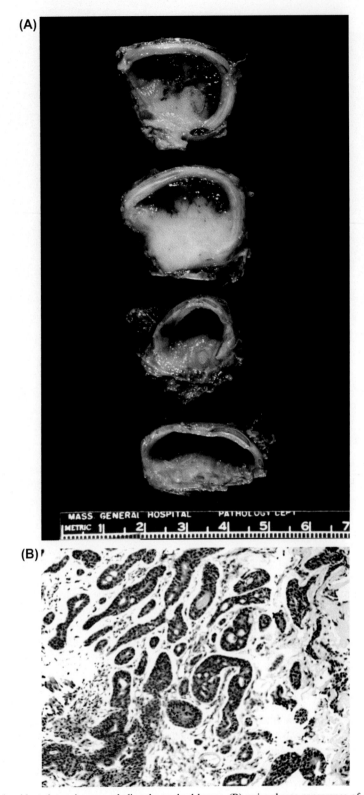

FIGURE 10.26 (A) Adenoid cystic carcinoma occluding the tracheal lumen, (B) swiss-cheese appearance of adenoid cystic carcinoma.

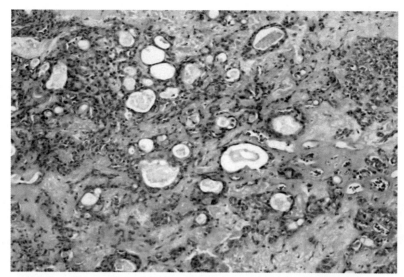

FIGURE 10.27   Pleomorphic adenoma arising from minor salivary glands of a major bronchus.

MIXED TUMORS

These are very unusual but can involve the major airways (Fig. 10.27). They are morphologically identical to the lesions seen in the salivary glands. Some tumors may be aggressive with local recurrence and metastasis following excision.

BENIGN TUMORS

Hamartoma

This term is likely a misnomer as these lesions are benign neoplasms rather than malformations. They characteristically occur in the lung as circumscribed bosselated lesions that tend to calcify to produce the radiographic appearance of "popcorn" calcifications (Fig. 10.28A). The lesions contain combinations of hyaline cartilage and smooth muscle with entrapment of type II pneumocytes and bronchiolar epithelium (Fig. 10.28B). When there is a history of cigarette smoking they should be removed out of caution, otherwise radiographic follow-up to ensure that stability is indicated.

SCLEROSING HEMANGIOMA

Most commonly seen in women in the lung periphery of the lower lobes, these tumors are benign and well demarcated. Their etiology is uncertain. They generally show proliferation of alveolar type II cells and a vascularized fibrous stroma with old hemorrhage with hemosiderin deposition (Fig. 10.29). Some variants of a prominent epithelial component can be mistaken for a malignant ACA, especially on rapidly frozen sections. The tumor appears to be of epithelial origin and the proliferating epithelial cells express TTF-1 and CKs.

BENIGN CLEAR CELL "SUGAR" TUMOR (PERIVASCULAR CELL TUMOR)

This lesion is also benign and well demarcated and occurs in the lung periphery. The cells show a characteristically clear cell appearance that is strongly positive for glycogen on PAS histochemical staining. The cell of origin appears to be a perivascular cell and the tumors immunostain strongly for HMB-45 and S-100. Ultrastructurally, they show intra-cytoplasmic melanosomes and abundant glycogen granules (Fig. 10.30).

GLOMUS TUMORS

These are rarely seen in the major airways or lung periphery and can be confused with typical carcinoids, although they appear to be of smooth muscle origin and immunostain for smooth muscle actin and are negative for S-100 and vascular markers (Fig. 10.31).

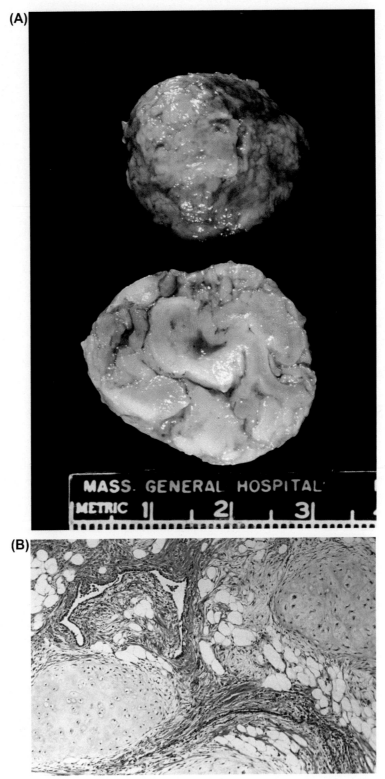

FIGURE 10.28 (A) Nodular hamartoma with calcifications seen as "popcorn lesion" radiographically, (B) cartilage and respiratory epithelium in hamartoma.

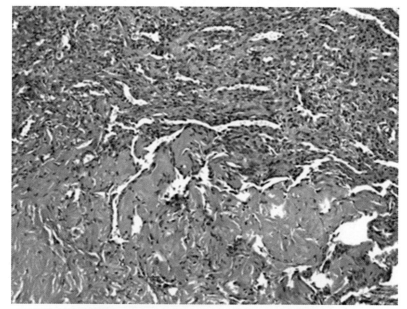

FIGURE 10.29   Sclerosing hemangioma with dense matrix and slit-like vascular spaces.

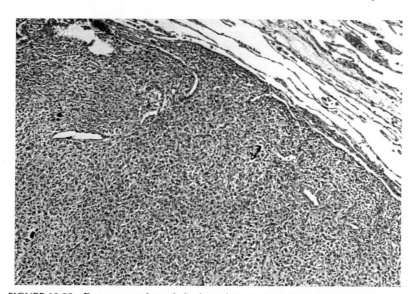

FIGURE 10.30   Foamy macrophages in benign pulmonary "sugar tumor" (pericellular tumor).

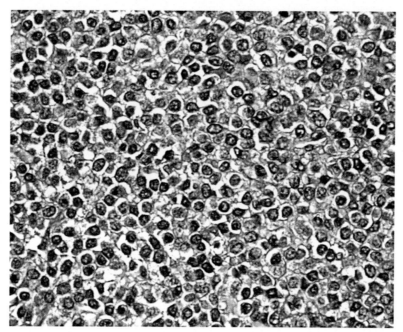

FIGURE 10.31   Glomus tumor with small cells that mimic carcinoid. *Courtesy of P. Cagle.*

SQUAMOUS PAPILLOMAS

These tumors occur in the tracheobronchial tree and are caused by HPV infection (Fig. 10.32A). They may be isolated or numerous and lead to respiratory failure. The cells will immunostain for p16, a surrogate marker for human papilloma virus. In some long-standing cases, the proliferating cells can develop severe dysplasia and even develop frank malignant transformation (Fig. 10.32B and C). The prognosis varies with the extent of airway involvement and can be fatal due to progressive suffocation.

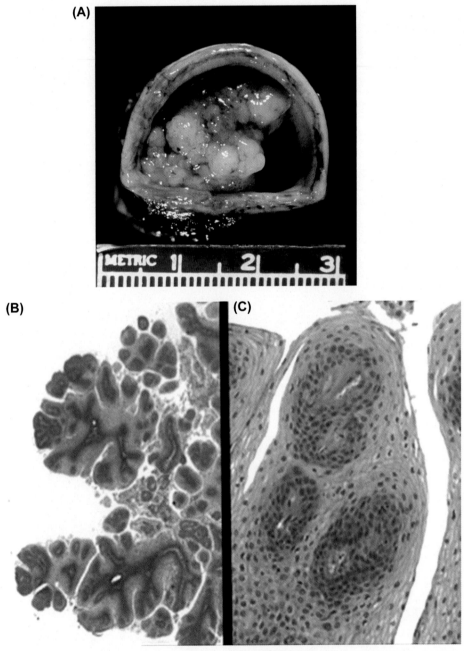

FIGURE 10.32 (A) Papillomatosis in airway, (B) papillary squamous epithelium protruding into airway lumen, (C) nuclei with viral induced koilocytotic change.

ALVEOLAR AND PAPILLARY ADENOMA

A rare benign peripheral tumor, the alveolar adenoma is microcystic and lined by benign alveolar type II cells (Fig. 10.33). Papillary adenomas show papillary growth of alveolar type II cells (Fig. 10.34). However, they can vary with respect to their level of differentiation so that the diagnosis should be made only when there is no question of malignancy. Otherwise they should be considered lesions of uncertain malignant potential.

Benign soft tissue lesions in the lung include leiomyoma, chondroma, neural tumors, and solitary fibrous tumors. The pathologist must use the standard criteria of atypical cytological features and mitotic rate to differentiate these from their sarcomatous counterparts.

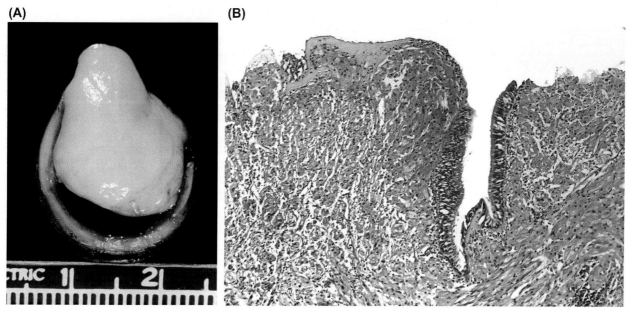

FIGURE 10.33   (A) Granular cell tumor in trachea, (B) cells showing eosinophilic granular staining beneath hyperplastic lining epithelium.

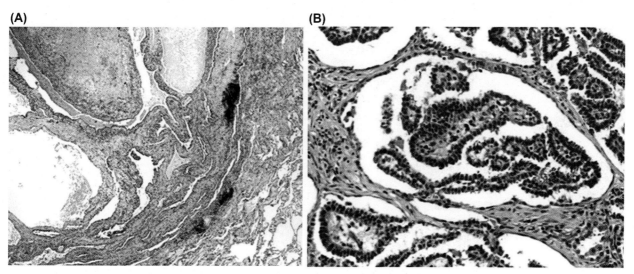

FIGURE 10.34   (A) Benign pneumocyte adenoma, (B) benign papillary adenoma.

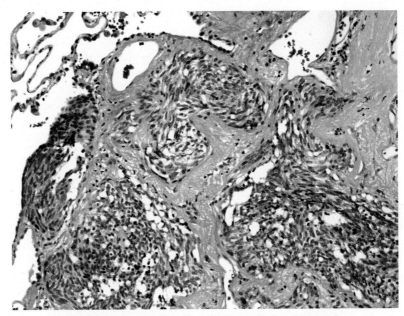

FIGURE 10.35 Spindle cell carcinoid tumorlet invading vessels.

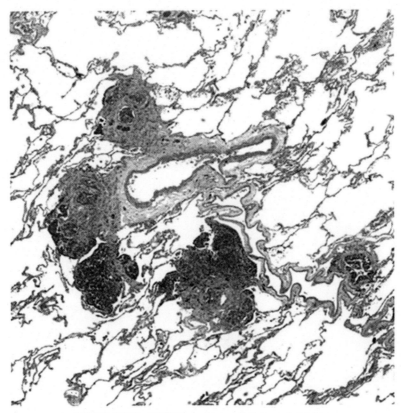

FIGURE 10.36 DIPNEC causing small airways obstruction.

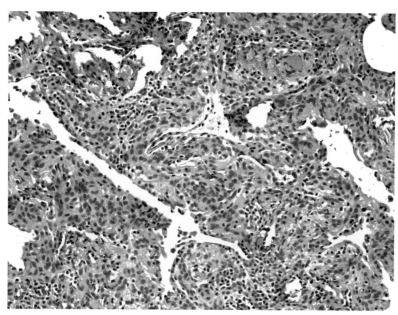

FIGURE 10.37   Benign meningothelial tumorlet.

BENIGN TUMORLETS

These are common pathological curiosities in resections of lung tissue for carcinoma or other disorders. They rarely come to attention radiographically and except in extraordinary cases do not produce symptoms.

There are two variants. Carcinoid tumorlets are essentially small carcinoids located adjacent to intralobular septa (Fig. 10.35). They are rarely larger than 1.0 cm and often show focal intravascular invasion that appears to have no clinical implication. They are frequently multiple and are seen in lungs with underlying scarring. When identified, the pathologist should exclude the possibility of diffuse neuroendocrine pulmonary endocrine cell hyperplasia (DIPNECH), which can be seen as a cause for small airways obstruction (Fig. 10.36), as well as the possibility of an underlying multiple endocrine neoplasia type 1 genetic abnormality. However, in most cases the changes are idiopathic and no further work-up or treatment is indicated. Rarely, patients with large numbers of tumorlets become clinically symptomatic.

Meningothelial tumorlets are the most common variant and their biological significance is comparable to those of carcinoid tumorlets (Fig. 10.37), although the possibility of a diffuse abnormality need not be considered. Rarely, a patient with a history of cranial meningioma who has undergone surgical extirpations can present with metastatic meningioma to the lung that mimics a tumorlet. Otherwise, the finding is best ignored.

FURTHER READING

Sequist, L.V., Yang, J.C., Yamamoto, N., et al., 2013. Phase III study of afatinib or cisplatin plus pemetrexed in patients with metastatic lung adeno-carcinoma with EGFR mutations. J Clin Oncol 31 (27), 3327–3334.
The study that demonstrated the greater efficacy of targeted molecular therapy above combination chemotherapy.
Travis, W.D., Brambilla, E., Noguchi, M., et al., 2011. International association for the study of lung Cancer/American thoracic Society/European respiratory Society International Multidisciplinary classification of lung adenocarcinoma. J Thorac Oncol 6 (2), 244–285.
The current detailed pathology subclassification of pulmonary adenocarcinoma.
Garcia-Yuste, M., Matilla, J.M., Cueto, A., et al., 2007. Typical and atypical carcinoid tumours: analysis of the experience of the Spanish Multi-centric study of neuroendocrine tumours of the lung. Eur J Cardiothorac Surg 31 (2), 192–197.
A review of a low grade malignancy with unpredictable clinical behavior.

Chapter 11

Lymphoid Lesions

BENIGN INTRAPULMONARY LYMPH NODES

The human lung lacks well-organized bronchial-associated lymphoid tissue, although one can identify follicular lymphoid tissue irregularly distributed throughout the airways. Encapsulated lymph nodes are found in the parenchyma of ~18% of excised lungs. Intrapulmonary lymph nodes are generally small but may enlarge as a result of anthracosilicotic dust deposition (Fig. 11.1). They may be seen on chest computed tomography (CT) scans often in the peripheral lung adjacent to interlobular septa. They have a pyramidal appearance and can generally be distinguished radiographically from malignancy.

NODULAR LYMPHOID HYPERPLASIA

Some patients with immunodeficiency or autoimmune disorders show prominent nodules of lymphoid hyperplasia that can be detected on chest radiographs and are a concern for neoplasia (Fig. 11.2). Nodular lymphoid hyperplasia by definition does not meet the criteria for monoclonal lymphoid expansion required to diagnose malignant lymphoma. The lesions show lymphoid tissue with germinal center formation and may be associated with a background of organizing pneumonia and fibrosis. Many of these lesions were termed "pseudolymphoma" in the past. Although the lesions do not meet the criteria for malignancy, they are associated with an increased risk of lymphoma, although the frequency is uncertain. In general, once the diagnosis has been established, these patients merit noninvasive follow-up, unless there is a striking change in the appearance of the lesions and the onset of clinical symptoms, in which case a repeat biopsy is indicated.

LYMPHOID INTERSTITIAL PNEUMONITIS

As previously discussed, lymphoid interstitial pneumonitis (LIP) is an unusual disorder that must be distinguished from malignant lymphoma. The lesion involves the lower lobes predominantly and often is associated with dysgammaglobu-linemia, Sjögren's syndrome, or other autoimmune disorders. In children, LIP is seen in patients with congenital HIV, which should be excluded. Radiographically, there may be cyst-like lesions that may reflect air trapping behind small airways encroached upon by micronodules of lymphoid tissue.

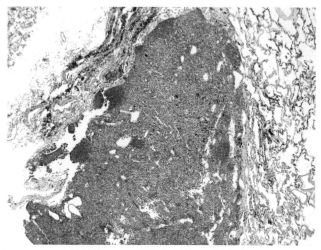

FIGURE 11.1   Intraparenchymal lymph node.

Understanding Pulmonary Pathology. http://dx.doi.org/10.1016/B978-0-12-801304-5.00011-3

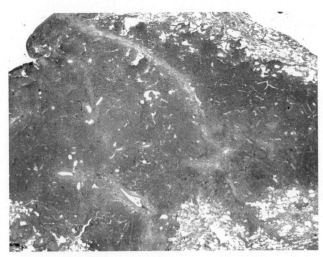

FIGURE 11.2   Nodular lymphoid hyperplasia.

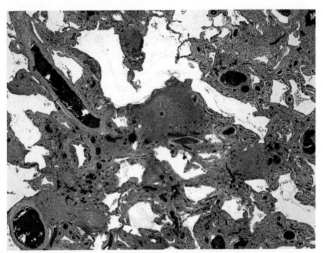

FIGURE 11.3   Lymphoid interstitial pneumonia showing expansion of interstitium.

The low-power appearance of the lesion suggests a "snowball" effect and the lymphoid cells are exclusively interstitial. Invasion of airways or vessels is a feature of malignancy and not LIP (Fig. 11.3). Sarcoidal granulomas are seen in 20% of cases. At times LIP may be difficult to distinguish from sarcoidosis or hypersensitivity pneumonitis and clinical correlation is required. Treatment of the disorder generally includes corticosteroids, but anti—B-cell interventions, e.g., rituximab, may be helpful.

LYMPHOMATOID GRANULOMATOSIS

Lymphomatoid granulomatosis (LYG) as currently defined is an angiocentric B-cell lymphoma associated with EBV infection. The lesion may present with single or multiple nodules in the lung associated with cough, hemoptysis, or constitutional symptoms. Extrapulmonary involvement of the kidney, skin, liver, nervous system, and other organs is seen.

LYG often shows areas of ischemic infarction attributable to angioinvasion by lymphoid cells (Figs. 11.4A,B). The lymphoid infiltrates range from polymorphic and "granulomatous" to monomorphic. They are angiodestructive and bronchodestructive but do not yield fibrinoid necrosis of the invaded structures and as such technically the disease does not qualify as a vasculitis. Clonal analysis shows the presence of monotypic CD20 + cells some of which may be CD30+. The tumor cells by definition are EBV positive as judged by in situ EBV RNA expression analysis, and grading of the tumor is based on the number of EBV + cells.

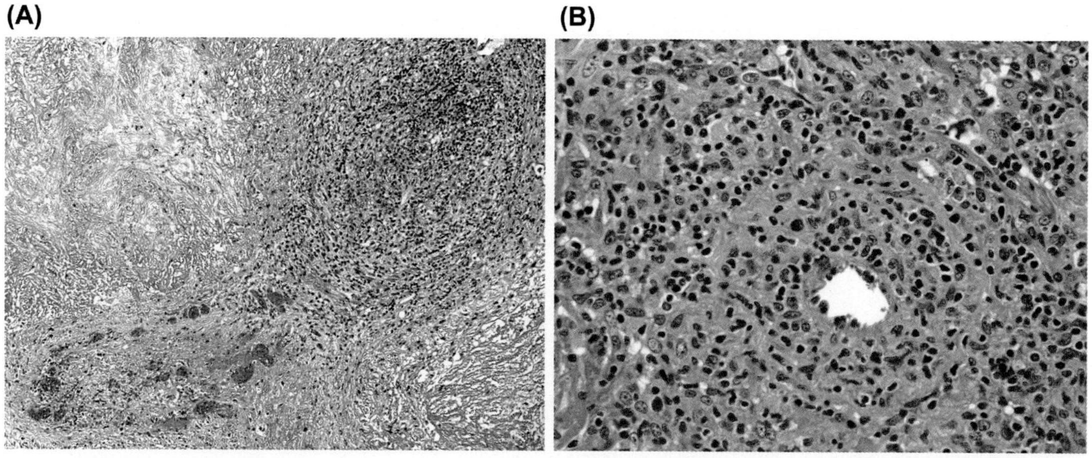

(A)   **(B)**

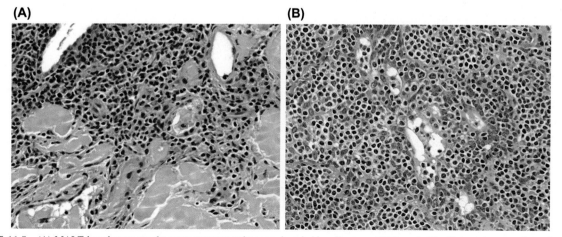

FIGURE 11.4   (A) Low-grade lymphomatoid granulomatosis with angioinvasion by lymphoid cells. (B) Microinfarction due to angiodestructive ischemic change. (Courtesy of J. Ferry.)

There are pulmonary lesions that appear to meet the histological criteria for LYG but lack evidence of EBV expression. Some prove to be T-cell lymphomas but others are unclassified and referred to as "lymphocytic angiitis, not otherwise specified." There is no consensus concerning the long-term prognosis of such lesions.

EXTRANODAL MARGINAL ZONE B-CELL LYMPHOMA (MALT LYMPHOMA)

This is the most common type of lymphoma to affect the lung primarily. The symptoms are nonspecific and may include cough, dyspnea, weight loss, and night sweats. The disease usually presents as single or multiple nodules. There is an increased incidence in autoimmune disease, especially Sjogren's disease. The tumor may show a lymphangitic distribution but hilar nodes are generally negative. There may be plasmacytic differentiation with associated amyloid production (Fig. 11.5A). The tumor is often polymorphic with intermediate sized lymphoid cell with pale cytoplasm. Invasion of the airways epithelium (lymphoepithelial lesions) is common (Fig. 11.5B). The malignant marginal zone lymphocytes are CD20 + CD79a + and PAX-5 + and may coexpress CD43. A characteristic translocation t(1; 18) can be identified by fluorescence in situ hybridization or PCR. The prognosis of this tumor is relatively good when limited to the lung. It tends to respond to either surgical resection or chemotherapy and at times it may be wise to observe the progression of the disease without intervention, as some lesions will spontaneously regress.

(A)   **(B)**

FIGURE 11.5   (A) MALT lymphomas are the most common pulmonary lymphoid malignancy. (B) Lymphoid cells have a propensity to invade this airway epithelium to produce "lymphoepithelial lesions."

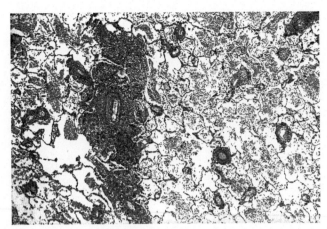

FIGURE 11.6   Secondary involvement of the lung by lymphoma showing lymphangitic growth around vessels.

PRIMARY B-CELL LYMPHOMA

This is the second most common lymphoid malignancy in the lung. Patients present with dyspnea, cough, and hemoptysis, but also have constitutional symptoms of fever, night sweats, and weight loss. By definition, the tumor does not involve the lymph nodes primarily. These tumors are classified like other B-cell lymphomas and range from well differentiated to large cell and anaplastic. Growth tends to progress along lymphatic pathways (Fig. 11.6). Unlike LIP, the tumor cells often invade the airways, vessels, and pleura, indicating their malignant potential. Prognosis is dependent on the cell type.

HODGKIN'S LYMPHOMA

Hodgkin's lymphoma can involve the lung and at times may be primary there. The criteria for diagnosis are the same as in the lymph nodes. However, it is worth noting that some cases of Hodgkin's disease can induce a prominent granulomatous reaction that closely mimics that of mycobacterial infection, and cases have been presumptively and erroneously been treated for tuberculosis for months to years with no response before the correct diagnosis was made (Fig. 11.7). As Hodgkin's disease can skip adjacent nodes or involve them only partially, establishing the diagnosis may require generous sampling of lymph nodes in order to identify diagnostic Reed—Sternberg cells (Fig. 11.8A). Immunohistochemistry may be helpful as Reed—Sternberg cells are CD30+ (Fig. 11.8B) and usually CD15+ as well. They are negative for CD45, CD79a, and CD43.

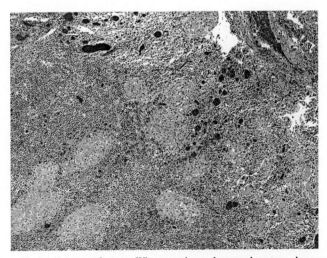

FIGURE 11.7   Hodgkin's disease with granulomatous features. When prominent, the granulomatous changes can obscure the actual diagnosis.

(A)   **(B)**

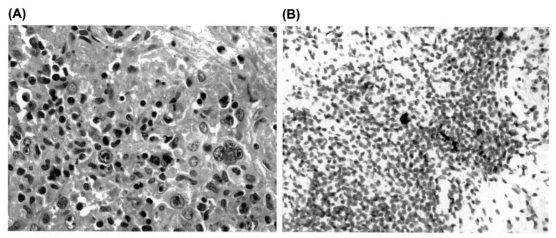

FIGURE 11.8   (A) Above case with area of diagnostic Reed—Sternberg cells that are (B) immunopositive for CD30.

INTRAVASCULAR LARGE B-CELL LYMPHOMA

This is a rare disease that can be inordinately difficult to diagnose premortem. Patients present with a variety of chest-related complaints and radiographic appearances including ILD, organizing pneumonia, and DAD. The diagnosis should be considered when the biopsy shows an increase in the number of large lymphocytes within the alveolar wall microvasculature, and this may easily be misinterpreted as chronic inflammation. A CD20 immunostain shows numerous B cells in small vessels, which establish the correct diagnosis (Figs. 11.9A,B). The prognosis of this disease overall is poor.

T-CELL LYMPHOMA

Other lymphoid malignancies involve the lung and should always be considered when a diagnosis is preexisting. These include T-cell lymphomas (either mycosis fungoides or peripheral T-cell lymphoma) (Figs. 11.10A,B). This may be associated with lymph node involvement but ∼20% do not involve lymph nodes. Involvement of skin, lung, and central nervous system are most common. The diagnosis requires molecular confirmation of T-cell receptor clonality by PCR.

Treatment efficacy is limited. It is worth noting that this and other lymphomas, e.g., IV B-cell lymphoma can evoke other prominent histologic responses in the lung, e.g., organizing pneumonia, so that one must be careful not to overlook the malignant component of the change. Lymphoplasmacytic lymphomas and multiple myeloma may be associated with AL amyloid deposition as nodules or around vessels.

(A)   **(B)**

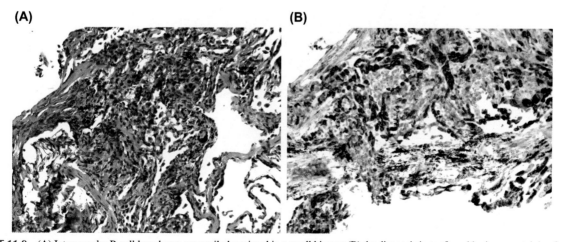

FIGURE 11.9   (A) Intravascular B-cell lymphoma can easily be missed in a small biopsy; (B) the diagnosis is confirmed by immunostaining for CD20.

(A)

(B)

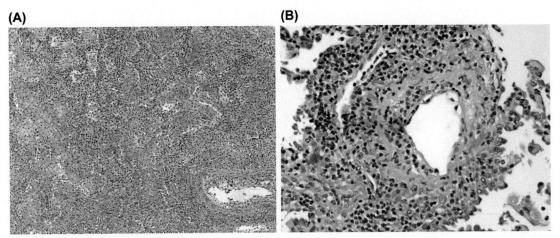

FIGURE 11.10   T-cell lymphomas can produce a variety of lung reactions including (A) fibrinous organizing pneumonia, (B) diagnosis is suggested by the invasion of blood vessel walls by atypical lymphoid cells.

LEUKEMIA

Leukemic infiltrates can involve the lung and may be expected when the peripheral white blood cell counts exceed $10^5/mL^3$. In acute myelogenous leukemia, the malignant cells are "sticky," and they can adhere to the microvasculature and exit out of the vessels. When the leukocyte counts are high this produces a picture of exudative pulmonary edema associated with perivascular and alveolar leukemic cell infiltration (Fig. 11.11). However, even in the absence of elevated blood counts granulocytic sarcomas may develop in the lung and in other tissues.

In cases that present with fever and pulmonary infiltrates, the differential diagnosis includes infection and chemotherapy effects. Immunostains for CD34 and CD15 will demonstrate leukemic cells but the diagnosis is generally not difficult, if sampling is sufficient.

Chronic lymphocytic is a common disease in the elderly and attention should be oriented toward its presence either in perivascular infiltrates in the lung or in excised lymph nodes for other reasons. The lung may be infiltrated by monomorphic small CD20+ lymphocytes (Fig. 11.12).

Patients with myelodysplasia can develop pulmonary infiltrates. On biopsy, these show the findings of NSIP with prominent DIP-like reactions filling alveoli with CD68+ macrophages (Fig. 11.13). A high percentage of these patients have progressed to acute leukemia within weeks of the pulmonary diagnosis. It has been postulated that the lung reaction may represent an immune response to new antigens released by incipient malignant transformation.

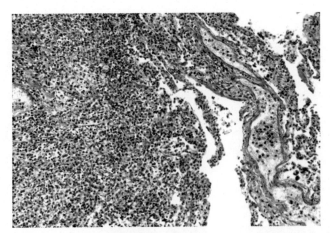

FIGURE 11.11   Infiltration of lung in patient with acute myelogenous leukemia.

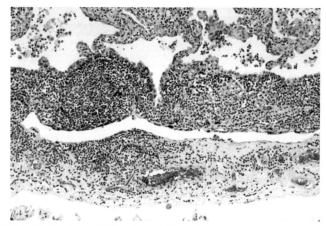

FIGURE 11.12   Interstitial lymphoid infiltrates in chronic lymphocytic leukemia.

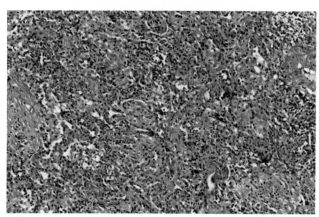

FIGURE 11.13   DIP-like changes in patient with myelodysplasia. Patient developed acute leukemia shortly after lung lesions appeared.

FURTHER READING

Isaacson, P.G., Du, M.Q., 2004. MALT lymphoma: from morphology to molecules. Nat Rev Cancer 4 (8), 644−653.
An interesting review of the immunology of these unusual lymphomas.
Swigris, J.J., Berry, G.J., Raffin, T.A., et al., 2002. Lymphoid interstitial pneumonia: a narrative review. Chest 122 (6), 2150−2164.
A review of a difficult topic in lung disease.
Colby, T.V., 2012. Current histological diagnosis of lymphomatoid granulomatosis. Mod Pathol 25, S39−S42.
A recent review of the features that characterize the diagnosis of this aggressive lesion.

Chapter 12

Iatrogenic Lung Diseases

A variety of therapeutic interventions can cause lung disease and this should be included in most differential diagnoses.

DRUG-INDUCED PNEUMONITIS

The diagnosis of drug-induced pneumonitis is virtually always one of exclusion. Perusal of the literature indicates that virtually all drugs have been implicated as causes of lung injury (Table 12.1). But in many instances, definitive proof is lacking. Yet there are a variety of drugs that reproducibly produce lung pathologies that are sufficiently distinctive to suggest the etiology (Table 12.2).

The patterns of injury caused by drugs include pulmonary edema, interstitial pneumonitis, organizing pneumonitis, eosinophilic pneumonia, granulomas, alveolar proteinosis, and hypersensitivity granulomatous inflammation. In some cases, the diagnosis is suggested by the administration of the drug with new symptoms and radiographic changes. It is my practice to recommend a lung biopsy only where there is a reasonable chance of encountering a diagnostic pathological reaction or excluding another diagnosis. Otherwise, the yield on lung biopsy is low. In addition, some drugs produce pathognomonic changes in the lung but may not be the cause of disease. This is true, e.g., of amiodarone, where the biopsy will often show an "amiodarone effect" but another process may be the cause of the clinical presentation.

OPIATE-INDUCED CHANGES

The administration of opiates, especially intravenous use of heroin or morphine, is capable of producing rapid onset pulmonary edema. In most cases, the clinical scenario suggests that the diagnosis and biopsy is not required. The edema fluid appears to reflect a central nervous system-mediated reflex vasoconstriction of pulmonary veins leading to elevated venous pressures and the transudation of fluid from the capillary bed (Fig. 12.1). A similar phenomenon may develop in patients with subarachnoid hemorrhage and intracerebral bleeding.

AMIODARONE

Amiodarone is widely used for the control of atrial and ventricular arrhythmias. The drug produces a change in the metabolism of alveolar type II cells and macrophages that causes them to take on a vacuolated appearance (Fig. 12.2A). Ultrastructural analysis shows that the affected cells contain whorled lamellar inclusions that are electron dense and reflect the accumulation of iodine (Fig. 12.2B). Similar changes may be observed in the thyroid gland. However, these changes occur independently of clinical or radiographic disease. When they are clinically significant there are usually fibrotic changes in the lung interstitium and a peculiar thickening of interlobular septa. Other patterns of lung pathology include nodules, organizing pneumonia, and DAD.

METHOTREXATE

Methotrexate (MTX) is a commonly used anti-inflammatory medication. As such, it may be difficult to distinguish the changes due to the underlying inflammatory disease from the pathological effects induced by MTX. The reaction to MTX is a hypersensitivity reaction and the pathology often shows nonnecrotizing granulomatous inflammation with giant cells and eosinophilia (Fig. 12.3). However, there may be substantial variations in the appearance of the lesions caused by MTX and only the first described changes are sufficiently characteristic to establish the diagnosis.

Understanding Pulmonary Pathology. http://dx.doi.org/10.1016/B978-0-12-801304-5.00012-5

TABLE 12.1 Drug-Induced Pulmonary Disease

Alveolar Hemorrhage	Obliterative Bronchiolitis
Acicumab	Busulfan
Cocaine	Penicillamine
Heparin	DAD
Penicillamine	Amiodarone
Sirolimus	Bleomycin
Statins	Cocaine
Urokinase	Cyclophosphamide
Organizing pneumonia	Interferon
Amiodarone	Methotrexate
Bleomycin	Nitrofurantoin
Busulfan	Procarbazine
Chloroquine	Vinblastine
Cyclophosphamide	
Gold	
Herceptin	
Interferon-alpha	
Mesalamine	
Nitrofurantoin	
Penicillamine	
Sirolimus	
Sulfasalazine	
DIP-Like Reactions	**Eosinophilic Pneumonia**
Amiodarone	Acetaminophen
Hydroxyurea	Amiodarone
Sulfasalazine	Ampicillin
Pulmonary edema	Bleomycin
Cocaine	Captopril
Haloperidol	Clarithromycine
Hydrochlorothiazide	Cocaine
Opiates	Chromalyn sodium
Penicillin	Dapsone
Propanolol	Ethambutol
Salicylates	Infliximab
	Interferon-alpha
Granulomas	Mesalazine
Cromolyn sodium	Montelukast
Etanercept	Naproxen
Interferon-alpha	Phenylbutazone
Methotrexate	Procarbazine
Sirolimus	Sulfonamides

Continued

TABLE 12.1 Drug-Induced Pulmonary Disease—cont'd

Hypersensitivity Pneumonitis	Trazadone
Hydroxyurea	Venlafaxin
Methotrexate	Alveolar proteinosis
Nitrofurantoin	Busulfan
Statins	Fentanyl
Sulfasalazine	Imatinib
Pulmonary Fibrosis	
Amiodarone	
BCNU	
Busulfan	
Chlorambucil	
Cocaine	
Cyclophosphamide	
Fluoxetine	
Gold salts	
Hydroxyurea	
Imatinib	
Interferon-alpha	
Melphalan	
Methotrexate	
Nitrofurantoin	
Phenytoin	
Procarbazine	
Sirolimus	
Statins	
Sulfasalazine	
Tryptophan	

MESALAMINE

Mesalamine is commonly used for the management of inflammatory bowel disease (IBD). It is well recognized that IBD can be associated primarily with a variety of lung pathologies including granulomatous lung disease and ILD. Drug-induced reactions to mesalamine may show a slow progression of dyspnea with interstitial infiltrates. Like MTX, the reactions appear to

TABLE 12.2 Drug-Induced Diseases with Characteristic Biopsy Pathologies

Amiodarone	Vacuolated macrophages and epithelial cells
Cytotoxic drugs (radiation)	Severe alveolar type II cell atypia
Sulfonamides	Eosinophilic pneumonia
Methotrexate	Granulomatous pneumonia with eosinophils

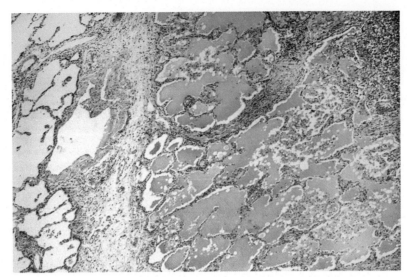

FIGURE 12.1   Alveolar flooding in patient with heroin-induced pulmonary edema.

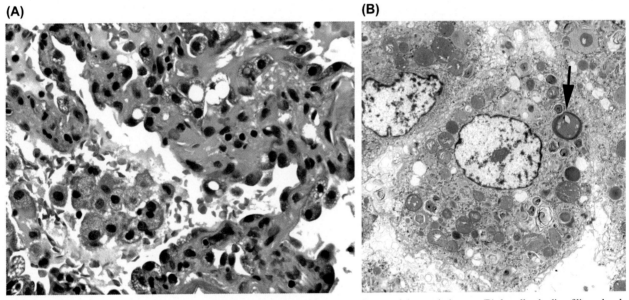

FIGURE 12.2   (A) Vacuolated pulmonary macrophages and epithelial cells in patient receiving amiodarone, (B) lamellar bodies filing alveolar macrophage (*arrow*).

include a degree of cell-mediated hypersensitivity and show nonnecrotizing granulomatous inflammation and intra-alveolar foamy macrophages. Clinical discernment may be necessary when distinguishing these changes from those caused by ILD. Clinical improvement following the withdrawal of the drug is generally accepted as "proof" of etiology.

SULFONAMIDES

Bactrim and other sulfonamide drugs are recognized to produce eosinophilic reactions in lung and other tissues. The diagnosis is usually made by the constellation of timing of administration, peripheral eosinophilia, and pulmonary infiltrates (Fig. 12.4). The lung is rarely biopsied but if tissue is obtained it will show an eosinophilic pneumonitis with alveolar and interstitial infiltration by eosinophils and macrophages. An eosinophilic angiitis may be present. BAL may be helpful in establishing the diagnosis if biopsy is contraindicated.

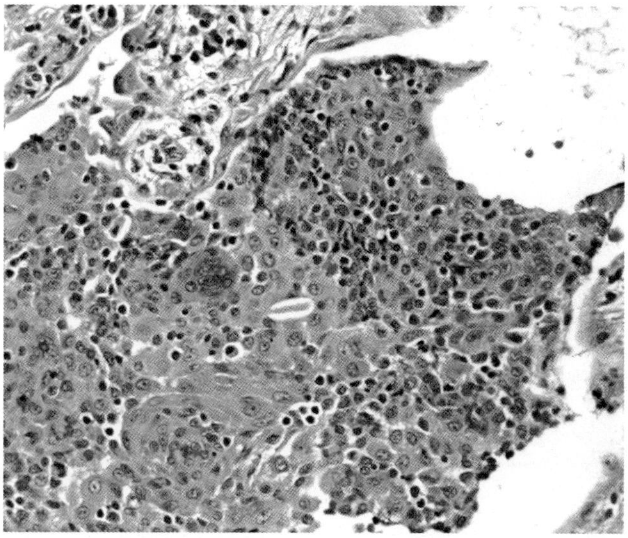

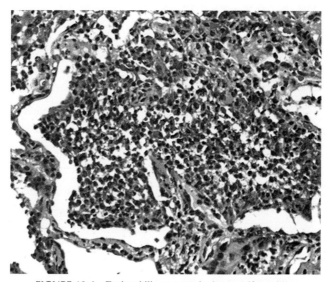

FIGURE 12.3 Granulomatous changes in patient receiving MTX for rheumatoid arthritis.

FIGURE 12.4 Eosinophilic pneumonia due to sulfonamides.

CYTOTOXIC DRUGS

Cytotoxic drugs including cyclophosphamide, busulfan, and bleomycin may cause acute and chronic lung disease. Establishing the diagnosis of cytotoxic drug injury is important in the setting of cancer treatment, as it may require a reformulation of treatment. The characteristic finding in these disorders is the presence of either DAD or nonspecific interstitial pneumonia associated with marked alveolar type II cell reactive atypia (Fig. 12.5). The use of cyclophosphamide as a second-line treatment for ILD should be tempered by the recognized capacity of this drug to cause lung injury, and one might want to seek alternative treatment options in this setting. Bleomycin (BLM) produces changes that are in part due to dose-responsive disease and a component of hypersensitivity. The changes are often irreversible and progressive. BLM has a propensity to induce DAD with fibrosis and squamous metaplasia of the lining alveolar cells.

RADIATION

Strictly speaking radiation is not a "drug"; radiation has the ability to produce lung disease via several pathways. The cytotoxic effects of radiation are well recognized. The changes in tissue include marked nuclear atypia of the type II pneumocytes (Fig. 12.6A) and other cells, dense fibrosis that tends to hyalinize (Fig. 12.6B) and that is usually limited to the portal of treatment, and radiation arteritis which show a characteristic intimal atheromatous change. In addition to

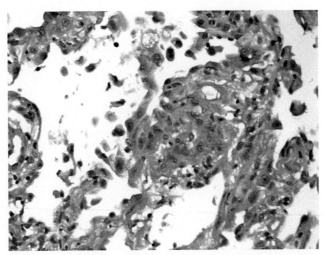

FIGURE 12.5   Atypical alveolar type II lining cells and fibrosis in patient treated with cyclophosphamide.

(A)   **(B)**

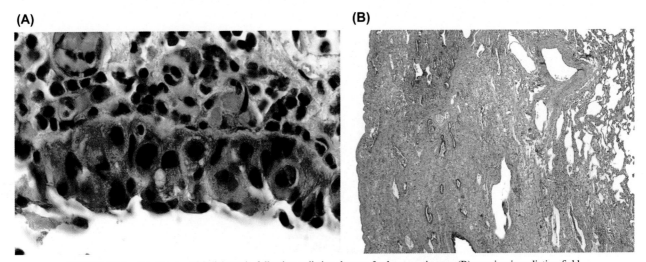

FIGURE 12.6   (A) Airway epithelial atypia following radiation therapy for lung carcinoma, (B) scarring in radiation field.

direct radiation-induced injuries, some patients develop a hypersensitivity response driven by T lymphocytes that can produce pulmonary infiltrates both within and outside of the portal of treatment. This radiation pneumonitis is typically a response to corticosteroid treatment and tends to occur several weeks to month following the beginning of radiation.

NITROFURANTOIN

This drug can cause serious pulmonary complications and fell into disfavor for many years for this reason, but it has re-emerged as it is an effective and generally well-tolerated chronic urinary bacterial suppressant. Nitrofurantoin produces a variety of lung pathologies including a lymphocytic interstitial pneumonitis and organizing pneumonia (Fig. 12.7). Some patients develop asthma with pulmonary eosinophilic infiltrates. Hemoptysis has been reported. The treatment requires permanent withdrawal of the drug and a trial of corticosteroids if function has been compromised.

TRANSFUSION-RELATED ACUTE LUNG INJURY

Transfusion-related acute lung injury (TRALI) may be evoked by the infusion of donor antibodies directed against recipient leukocytes. The infusion of donor anti-HLA (human leukocyte antigens) or anti-HNA (human neutrophil antigens) antibodies may cause complement activation, with the influx of neutrophils into the lung, resulting in subsequent

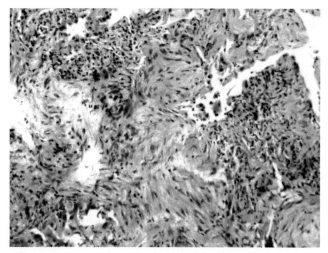

FIGURE 12.7   Inflammation and organizing pneumonia in patient treated with nitrofurantoin.

endothelial damage and capillary leak. Donor-derived antibodies to HLA class I antigens and neutrophils have been demonstrated in up to 89% of TRALI cases examined in the literature.

An alternate hypothesis argues that TRALI is the result of at least two independent clinical events: the first is related to the clinical condition of the patient (infection, cytokine administration, recent surgery, or massive transfusion) that causes activation of the pulmonary endothelium. This then leads to the sequestration of primed neutrophils to the activated pulmonary endothelium. The second event is the infusion of donor-derived anti-HLA or anti-HNA antibodies directed against antigens on the neutrophil surface and/or biological response modifiers (e.g., lipids) in the stored blood component that activate these adherent, functionally hyperactive neutrophils, causing neutrophil-mediated endothelial damage and capillary leak.

In most cases, the lung is not biopsied and the diagnosis is established by the timing of the response and the presence of donor antibodies to HLA class I major histocompatibility complex (MHC) antigens. In cases where the lung has been biopsied the findings have shown hyaline membranes and lymphocytic and neutrophilic infiltrates.

NEWER DRUGS

A variety of new drugs used in cancer and transplantation have been shown to have pulmonary toxicity and must be considered when pulmonary infiltrates develop in these settings. Tyrosine kinase inhibitors can produce organizing pneumonia and ILD without pathognomonic changes in the biopsy (Fig. 12.8). The toxicities caused by the family of tyrosine kinase inhibitors may be progressive and fatal.

Tacrolimus pulmonary toxicity almost always responds within several weeks to the withdrawal of the drug. Rituximab has also produced ILD with various patterns including organizing pneumonia (Fig. 12.9).

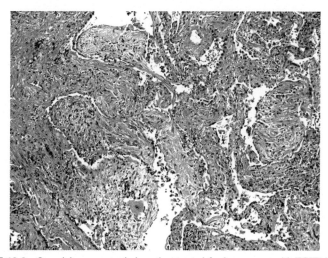

FIGURE 12.8   Organizing pneumonia in patient treated for lung cancer with EGFR inhibitor.

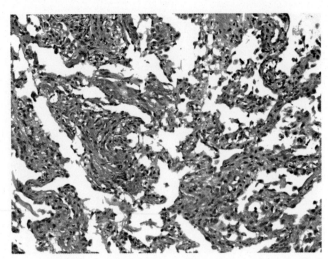

FIGURE 12.9   NSIP pattern with organizing pneumonia in patient treated with rituximab.

FURTHER READING

Camus, P.H., Foucher, P., Bonniaud, P.H., et al., 2001. Drug-induced infiltrative lung disease. Eur Respir J 32, 93s–100s.

A good review of a difficult topic.

Pneumotox Online, (pneumotox.com).

A website devoted to providing information on drug-induced pulmonary toxicities.

Chapter 13

Toxic Exposures

The lung is the target of a number of inhaled toxic materials that are encountered in the workplace. Occupational lung disease is almost certainly underdiagnosed, as the diagnostic work-up rarely includes sufficient detail to exclude the contribution of all potentially injurious inhaled materials. Furthermore, many inhaled toxins leave no discernible trace in the lung after causing injury so that one must rely upon the exposure history.

Detailed occupational histories are essential in triggering the search for a potential toxin. In exposures to asbestos that have prolonged latency periods, an exposed individual may no longer recall an exposure or know that a material or product that he was exposed to actually was composed in part of asbestos.

Surgical pathologists are generally only able to diagnose diseases caused by visible particulates. Ancillary methods including tissue digestion, energy dispersive X-ray analysis, and microprobe analysis are required to identify the offending agents with specificity, but these methodologies are only available only at research centers or nonmedical commercial laboratories.

The situation is further complicated by the fact that there is currently no accurate way to determine whether the presence of, e.g., foreign dust particles, is the cause of disease or simply a marker of exposure. For some disorders, e.g., exposure to asbestos, detailed diagnostic criteria have been established. But this is not the case for many other occupational exposures, and there is no uniform agreement concerning how best to interpret the significance of inhaled particulates in the lung.

TOXIC GASES

Inhaled phosphates, chlorine gases, chemical warfare "mustard" gas, and soot and chemicals inhaled at the scene of fires can produce rapid injury to the airways leading to chemical bronchitis, obliterative airway disease, and DAD. The history of exposure is usually evident. For burn victims, soot in the nares and oropharynx correlates well with the lower airway injury as does the level of CO gas in the blood. These patients must be carefully monitored and provided ventilator support and supplemental O_2 if necessary. The long-term prognosis will depend on the extent of fibrosis in the airways.

ASBESTOS

Asbestos is a generic commercial term for a number of fibrous mineral silicates used extensively in the 20th century in the building trades. Asbestiform minerals occur naturally and five forms have been commercially used. These differ in size, shape, and chemical constituents (Fig. 13.1). Amphibole asbestos (crocidolite, amosite, anthophyllite, and tremolite) shows a needle-like morphology characterized by a high aspect ratio (length: width). Serpentine (chrysotile) asbestos shows curly cue fibers that are particularly useful for weaving into textiles. Despite their physical differences, all forms of asbestos have been demonstrated to cause all of the diseases attributable to asbestos, although on a per fiber basis the amphiboles appear to be more potent in this regard.

Asbestos causes both benign and malignant diseases. For that reason, its use has been carefully regulated in recent years in the United States, and it has been banned altogether by European Union countries. However, it is still being widely being used in developing countries and will almost certainly result in a global health problem in the future.

The most common disorder caused by asbestos is a hyaline pleural plaque (benign asbestos-related pleural disease) (Fig. 13.2A and B). These are characteristically located along the lateral parietal pleura but most characteristically along the diaphragmatic leaflets. Plaques can occur along the visceral pleura, pericardium, mediastinal pleura, and even the spleen. They are usually calcified and visible on routine chest radiographs. Discrete pleural plaques rarely cause pulmonary dysfunction, although elements of restriction that are usually subclinical may be seen on detailed pulmonary function tests. Less commonly, diffuse pleural fibrosis is caused by asbestos and this can produce fibrothorax with pulmonary restriction (*plaque en cuirasse*).

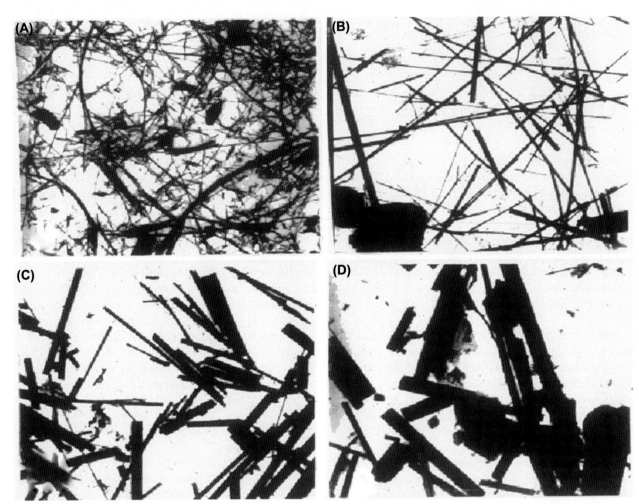

FIGURE 13.1 Ultrastructural appearance of (A) serpiginous chrysotile and needle-like fibers of amphibole, (B) crocidolite, (C) amosite, (D) anthophyllite.

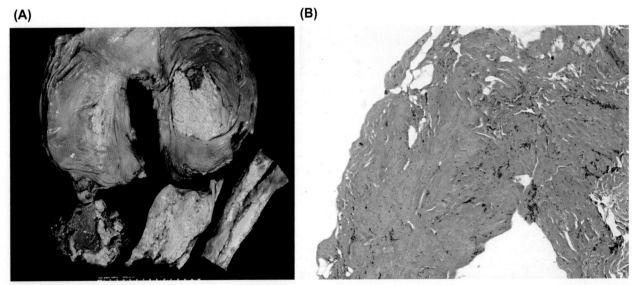

FIGURE 13.2 (A) "Candle-wax dripping" appearance of calcified asbestos-related pleural plaques on diaphragmatic surfaces and parietal pleura. (B) Hyaline pleural plaque.

(A)   **(B)**

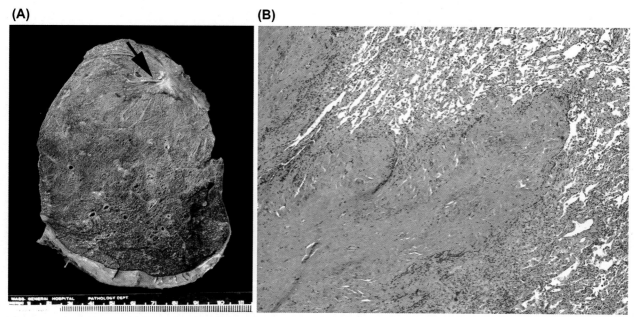

FIGURE 13.3 (A) Rounded atelectasis due to entrapment of lung beneath a benign pleural plaque (*arrow*). (B) Histology in right panel shows entrapped atelectatic lung beneath a fibrous pleural plaque.

Other benign pleural-related changes include rounded atelectasis and pleural effusions. Rounded atelectasis is the result of entrapment of lung tissue under a visceral pleural plaque (Fig. 13.3A and B). The appearance is characteristic on computed tomography images which show "rounded" collapsed lung tissue with a "comet's tail" of collapsed bronchovascular structures extending toward the hilum. When this appearance is present radiographically, it is sufficiently diagnostic that additional studies to exclude malignancy are not required. Multiple areas of rounded atelectasis may be seen in a given patient, and these should be followed radiographically and with pulmonary function studies to exclude progressive loss of pulmonary function and the superimposition of a malignant process.

Asbestos-related pleural effusion is essentially a diagnosis of exclusion as there is nothing pathognomonic concerning its presentation. The effusions are exudative and may recur. They can occur early or late following exposures. It is questionable whether they progress to produce diffuse pleural plaques.

Pulmonary asbestosis is caused by substantial occupational exposure to asbestos, in the range of 2−25+ fiber-cc-years, depending on individual susceptibility. The disease is detected radiographically as interstitial fibrosis in the lower lung zones often showing reticular "vertical lines." Honeycomb features may or may not be present. Asbestosis is generally diagnosed clinically based on the history of exposure, latency, and abnormal chest radiographs. Crackles may be present on lung examination as well as digital clubbing, but these changes are nonspecific. Low titer ANAs are also common.

The histological diagnosis requires the presence of interstitial fibrosis, which may range from mild to severe, and the presence of asbestos bodies (Fig. 13.4A), which are dumbbell shaped iron-encrusted asbestos fibers, showing a clear core (Fig. 13.4B). In the absence of asbestos bodies, the diagnosis cannot be established histologically and alternative diagnoses must be considered. However, vast majority of asbestos bodies appear to form on amphibole fibers, so that significant exposures to chrysotile may be underestimated. Currently, there is no clear solution to this diagnostic dilemma, as even the examination of the lung for fibers of chrysotile vastly underrepresents exposures due to the short half-life of chrysotile fibers in vivo. The presence of pleural plaques suggests the diagnosis but it is absent in ~15% of cases with asbestosis. The asbestos-related pleural pathologies are discussed in the chapter devoted to pleural diseases.

ASBESTOS-RELATED CANCER

The most common type of malignancy associated with asbestos exposure is lung carcinoma. All of the major types of lung cancer have been shown to occur in patients exposed to asbestos. There is also no specific site in the lung, e.g., lower versus upper lobes that show predominance. Many workers who were exposed to asbestos were also cigarette smokers. The studies by Selikoff et al. and recent follow-ups by Markowitz et al. have demonstrated an independent risk for developing

(A)

(B)

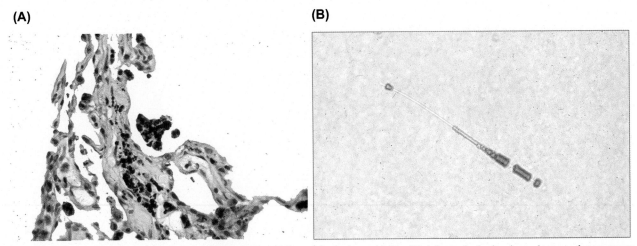

FIGURE 13.4 (A) Asbestosis with a focus on scarring and heavily ferruginized asbestos bodies. (B) Asbestos body showing a clear core that represents a thinly coated asbestos fiber.

lung cancer in asbestos-exposed nonsmokers, as well as an additive and supra-additive risk in those who were exposed and smoked and in smokers with asbestosis, respectively. From a medical legal perspective, there is little rationale for insisting on the presence of asbestosis before attributing a lung cancer to asbestos, especially if there has been a history of associated cigarette smoking.

I have personally seen three cases of lung cancer (adenocarcinoma) in occupationally asbestos-exposed nonsmokers who have had genetic abnormalities in EGFR exons 19, 20, and 21. Whether these genes are directly damaged by asbestos exposure or represent independently acquired somatic mutations is an area of active research.

Malignant mesothelioma, the most characteristic malignancy that is caused in most cases by asbestos is discussed in the chapter on pleural disease.

TALCOSIS

Workers exposed to industrial talc may develop changes in the chest that mimic those seen with asbestos exposure, including pleural plaques, interstitial fibrosis, and malignancies. This appears to reflect contamination of some sources of talc with the asbestiform amphibole tremolite (Fig. 13.5). Some brands of commercial talcum powders have been implicated as the cause of peritoneal mesotheliomas in women.

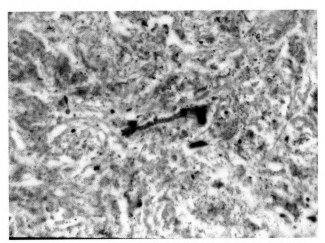

FIGURE 13.5 An asbestos body is seen in the lung of a talc miner. Talcosis shares features with asbestosis likely due to contamination of the talc by tremolite asbestos.

SILICOSIS

Silica is a mineral that is naturally found in quartz rock formations. Disease develops in coal and other miners, quarry workers, sandblasters, stone dressers, and glass workers. Like asbestosis, it takes decades for the disease to develop. The disease tends to primarily affect the upper lung zones (Fig. 13.6A). The pathology is due to presence of whorled nodules of scar tissue (Fig. 13.6B) associated with the presence of deposited crystals that can be identified with polarized light as weakly birefringent needle-like structures (Fig. 13.6C). Early nodules show anthraco-silicotic change. The disease has a lymphangitic distribution and often involves the draining of hilar lymph nodes that tend to show "egg-shell" calcifications that can be detected radiographically (Fig. 13.7). Confluent silicotic nodules produce progressive massive fibrosis (area of fibrosis >1 cm) and these may undergo cavitary degeneration (Fig. 13.8A and B). Patients with silicosis are at increased risk of developing mycobacterial infections, including tuberculous and nontuberculous infections. Caplan's syndrome is silicotic change in the wall of a rheumatoid nodule. Patients with silicosis are at an increased risk for the development of lung carcinoma. In addition, heavy exposures to silica can produce acute pulmonary alveolar proteinosis.

SILICATOSIS

Silicates are also capable of causing pulmonary fibrosis. Silicates are found in talc, mica, kaolin, and clays. People who mine these substances, work in foundries, or with building products, ceramics, pottery, and cosmetics, and can be

FIGURE 13.6 (A) Silicosis causes nodular scarring predominantly in the upper lung lobes, (B) the scarring shows nodular whorls of collagen and fibroblasts and anthracotic pigment, (C) silica crystals are weakly birefringent when examined by polarized light.

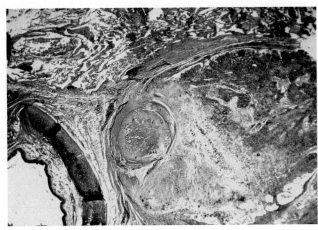

FIGURE 13.7 Lymph nodes in silicosis develop "egg shell calcifications" that can be seen radiographically.

(A)

(B)

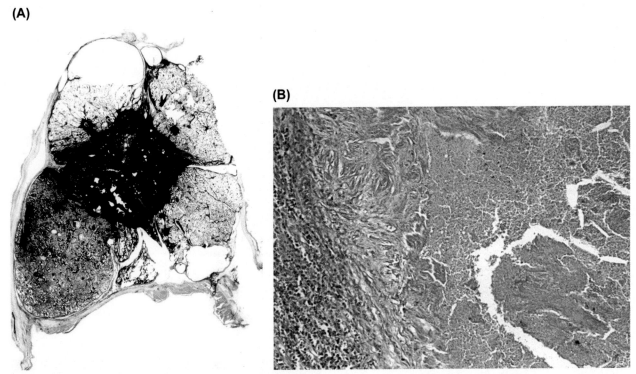

FIGURE 13.8 (A) Pneumoconiotic fibrotic nodules larger than 1 cm are referred to as progressive massive fibrosis. (B) These may undergo cavitary changes. The incidence of mycobacterial infection is increased in such cases.

occupationally exposed. The findings include peribronchiolar and interstitial fibrosis associated with intensely birefringent silicates (Fig. 13.9A and B). Areas of whorled fibrosis may result from silica contamination. The overall findings including progressive massive fibrosis mimic those of silicosis.

COAL WORKER'S PNEUMOCONIOSIS

Coal worker's pneumoconiosis is a disease of those who mine anthracite or hard coal. The toxic effects result from the codeposition of carbon together with silica and silicates. Focal upper lobe emphysema is a characteristic change and can occur in the absence of a history of cigarette smoking but is more severe in smokers. The extent of scarring within the lung primarily results from the silica content of the coal dust. The primary lesion is the carbon dust macule with larger nodules

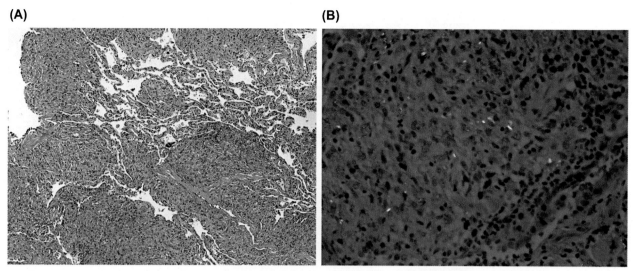

FIGURE 13.9 (A) Granulomatous change in silicatosis, (B) strongly birefringent silicates seen under polarized light with associated eosinophilic reaction.

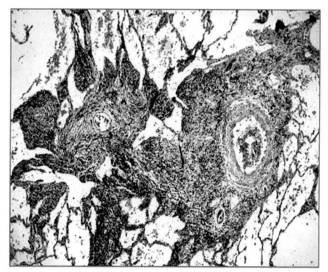

FIGURE 13.10 Coal miners pneumoconiosis depends largely on the degree of silica exposure as the carbon itself appears to cause little injury except mild centrolobular dilatation.

reflecting a mixed dust pneumoconiosis potentially leading to progressive massive fibrosis (Fig. 13.10). Simple coal worker pneumoconiosis (CWP) is confined to patients with macules and nodules, whereas as the presence of progressive massive fibrosis (PMF) suggests complicated disease.

HARD METAL PNEUMOCONIOSIS

"Hard" metals including cobalt and tungsten carbide can produce an unusual cellular pneumonitis characterized by the presence of bizarre multinucleated macrophages within alveolar spaces referred to as giant cell interstitial pneumonitis. These giant cells can exhibit ingestion of other inflammatory cells (emperipolesis) (Fig. 13.11). The surrounding alveolar walls show a lymphohistiocytic response without hyaline membranes, which helps to distinguish this lesion histologically from measles pneumonia, although the clinical presentations are sufficiently distinct that confusion should not occur.

The diagnosis requires confirming the presence of elevated levels of heavy metals in the lung tissue, although a good occupational history of exposure will suffice. Heavy metals can also cause other lung reactions, including hypersensitivity pneumonitis and asthma.

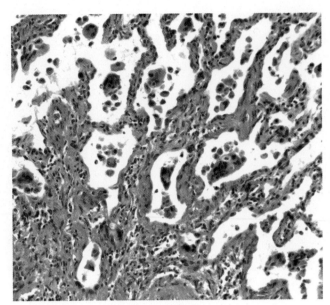

FIGURE 13.11   Giant cell pneumonitis due to cobalt exposure.

FIGURE 13.12   Beryllium produces a granulomatous hypersensitivity response in lung and regional lymph nodes that cannot be distinguished from sarcoidosis.

BERYLLIUM DISEASE

Pulmonary beryllium disease is histologically indistinguishable from sarcoidosis (Fig. 13.12). Patients with pulmonary beryllium disease appear to have been immunologically sensitized to a beryllium haptenated protein. Coincubation of peripheral blood lymphocytes or lymphocytes harvested from the BAL can be used to reveal sensitization via mitogenic proliferation in the presence of beryllium salts. Treatment of pulmonary beryllium disease consists of removing the sensitive patient from the work place. Patients with sarcoidosis should routinely be queried concerning possible occupational exposures to beryllium.

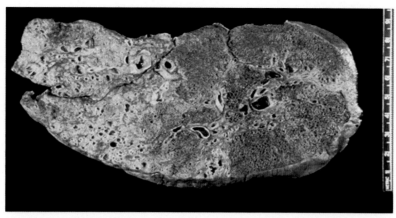

FIGURE 13.13   Iron workers may be exposed to iron causing a rust-like discoloration of the lung but is otherwise apparently benign.

SIDEROSIS

Individuals exposed to iron particles may develop siderosis. Pure iron generally produces radiographic abnormalities without pulmonary function abnormalities unless the exposures are heavy (Fig. 13.13). Exposures may occur at mines, steel mills, foundries, and in welders. However, contamination with silica can result in considerable pulmonary scarring. The diagnosis is generally made clinically based on the exposure history.

ALUMINOSIS

Occupational level of exposure to aluminum generally does not cause disease. But aluminum hydroxide that is generated in the presence of nonpolar lubricants can result in fibrotic nodules and symptoms. Aluminum particles are ingested by macrophages but can be seen within the lung interstitium. They are refractile but they are not birefringent. Aluminum has been implicated in some cases of alveolar proteinosis.

FURTHER READING

Currie, G.P., Rossiter, C., Dempsey, O.J., et al., 2005. Pulmonary amyloid and PET scanning. Respir Med 99 (11), 1463−1464.
Trapnell, B.C., Whitsett, J.A., Nakata, K., 2003. Pulmonary alveolar proteinosis. N Engl J Med 349 (26), 2527−2539.
Interesting reviews on these unusual disorders.

Chapter 14

Sundry Disorders

The lung shows a variety of pathologies that are not readily classified under any of the standard rubrics and these are addressed in this chapter.

PULMONARY EDEMA

Pulmonary edema is a common disorder of the lung. It is due to an increase in pulmonary vascular permeability that can reflect (1) increased transpulmonary hydrostatic pressure due to elevated pulmonary venous pressure, (2) changes in pulmonary vascular permeability, (3) endothelial injury, and (4) decreased oncotic pressure.

These factors are determined by the Starling equation, $Q = k(P_{cap} - P_{int}) - \sigma(p_{cap} - p_{int})$, in which the movement of edema fluid into the lung depends on the hydrostatic and oncotic pressure differences between the pulmonary capillary and the lung interstitium. Transudative pulmonary edema is rarely purposefully biopsied and the diagnosis rests on evidence of congestive heart failure on physical examination, radiographic evidence of lung edema, a history of cardiac systolic or diastolic dysfunction, and most definitively on right heart catheterization showing an elevated pulmonary capillary wedge pressure. Interstitial edema precedes alveolar flooding and can produce bronchospasm in some patients (cardiac asthma). Severe acute pulmonary edema due to cardiogenic shock can be hemorrhagic due to pulmonary venous hemorrhage (Table 14.1).

Pathologists tend to underestimate pulmonary edema, which may be subtle on lung biopsies showing only minimal alveolar fluid and proteinaceous accumulation. Chronic congestive heart failure may show "heart failure cells," i.e., hemosiderin-laden macrophages in alveolar spaces. Long-standing venous hypertension seen in the past from chronic rheumatic mitral valvular disease produced prominent siderotic nodules in the lung interstitium and cardiac fibrosis with diffuse mild interstitial scarring.

Exudative edema accompanies most forms of pulmonary injury. In its most severe form, it is seen in diffuse alveolar damage. However, even mild pulmonary capillary leak can promote interstitial fibrosis and most forms of interstitial lung disease are accompanied by an increase in pulmonary water.

DENDRIFORM OSSIFICATION

This is a common histological finding in patients with chronic interstitial disorders. The lesions are mostly a pathological curiosity but occasionally they can produce cysts and pneumothoraces (Fig. 14.1). This occurs when an ossified structure obstructs a small subpleural airway leading to a ball-valve effect. The ossifications tend to branch within the distal airways

TABLE 14.1 Causes of Pulmonary Edema

Left heart systolic and diastolic dysfunction

Left heart valvular stenosis and insufficiency

Congenital heart disease

Opiate and opioid induced edema

Other drugs (cocaine, salicylates, etc.)

Central nervous system hemorrhage (subarachnoid, intraventricular)

Acute lung injury

Chronic interstitial pneumonitis (subclinical)

Understanding Pulmonary Pathology. http://dx.doi.org/10.1016/B978-0-12-801304-5.00014-9

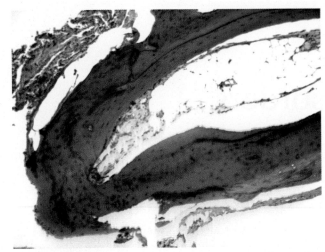

FIGURE 14.1   Branching metaplastic bone is frequently seen in fibrotic lungs and is termed a dendriform ossification.

leading to the term "dendriform." They must be distinguished from deposits of metastatic osteosarcoma, and this is easily done due to the paucity of cellularity in metaplastic ossifications.

ALVEOLAR MICROLITHIASIS

Alveolar microlithiasis is a rare disorder. It leads to progressive dyspnea and restrictive lung disease. Radiographically, the lung shows innumerable small nodular calcifications. The lung biopsy reveals large numbers of concentric lamellated calcifications associated with a matrix of fibrous tissue, although the underlying lung may be minimally altered in some cases (Fig. 14.2). Respiratory failure may develop in advanced cases and there may be a familial history. The differential diagnosis includes the presence of another microscopic curiosity, the *corpora amylacea*, but these are carbohydrate-rich PAS+ and noncalcified. They tend to be seen in lungs that are edematous or chronically fibrotic. Their etiology is unknown.

DYSTROPHIC CALCIFICATION

Calcification is a marker of chronic pathology. It is commonly seen in old scars or areas of remote infection (calcified granuloma). However, in patients with hypercalcemia from chronic renal disease, hyperparathyroidism, or metastatic disease to bone, calcium phosphate can rapidly precipitate out of the plasma into the pulmonary alveolar walls. These

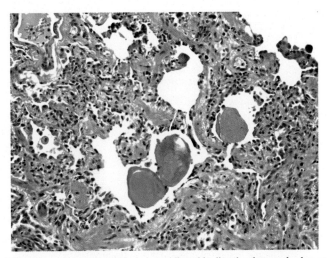

FIGURE 14.2   Pulmonary alveolar microlithiasis is an idiopathic disorder that can lead to respiratory failure.

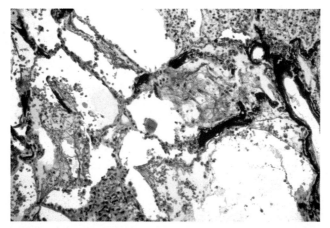

FIGURE 14.3   Elevated plasma calcium levels can result in metastatic calcifications around the microvasculature of the lung and other organs.

"metastatic" calcifications are usually asymptomatic but can cause pulmonary dysfunction. Treatment involves correction of hypercalcemia and control of phosphate levels (Fig. 14.3).

ASPIRATION PNEUMONIA

Microaspiration occurs for virtually everyone at night. In young healthy individuals, the various airway defense measures are sufficient to clear the lung of particulates or potentially injurious gastric contents. However, the defenses become less efficient with age leading to airway and lung parenchymal inflammation. Conditions, both anatomic and physiological, that decrease consciousness, interfere with normal swallowing mechanisms, or alter the normal anatomy of the upper airways promote the possibility of aspiration. The size of the aspirated substance will in part determine at what level the substance becomes entrapped. As most aspiration occurs at night, the dependent segments of the lung are predominantly involved, but this will depend on whether the individual sleeps on his or her back, side, or stomach.

FOOD

Aspiration of food particles can occur at any age including childhood. Eating habits may promote aspiration in addition to the factors that have been noted. The aspiration of particulate foods (Fig. 14.4A), particular lentils, and peanuts is frequently observed (Fig. 14.4B).

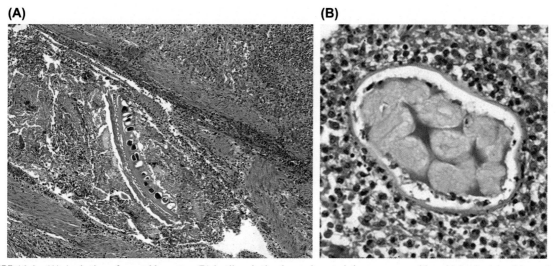

FIGURE 14.4   (A) Aspiration of vegetable matter, (B) lentil aspiration has a characteristic appearance with multiple glycogen-rich starch cells.

LIPID ASPIRATION

Patients who inadvertently aspirate lipid-rich preparations, often as vehicles in oil-based preparations such as nose drops, may develop localized or multifocal processes in the lung that lead to chronic scarring. *Exogenous* lipid pneumonia is recognized by the presence of globules of fat in the lung surrounded by multinucleated giant cells and associated with dense fibrosis (Fig. 14.5A). In recent years, there has been a change in the appearance of these lesions with increased amounts of microvesicular fat in macrophages and diminished amounts of fibrosis (Fig. 14.5B), and this may reflect a change in the manufacturing processes of the products. This finding must be distinguished from *endogenous* lipid pneumonia in which the lung is filled with macrophages that contain microvesicular fat. The aspiration of drug (Fig. 14.6A), feeding tube filler materials, and suicide attempts (Fig. 14.6B) can occur both at home and in the hospital setting.

Likely the most commonly encountered form of aspiration includes both macroaspiration of stomach acid leading to diffuse alveolar injury (Mendelsohn's syndrome) (Fig. 14.7A) or microaspiration that leads to chronic airway injury. These changes are often encountered in the elderly patient with GERD and diminished swallowing capacity. The spectrum of histological changes seen in acute gastric acid aspiration is large and nonspecific. One may see peribronchiolar inflammation with foreign body multinucleated giant cells, microgranulomas, fibrinous pneumonitis, organizing pneumonia, or DAD (Fig. 14.7B). When aspiration is chronic, airways with Lambertosis (Fig. 14.7C) and basilar bronchiectasis are common. These changes can be difficult to distinguish histologically from end-stage lung seen in UIP/IPF but the radiographic and clinical presentation is generally helpful in distinguishing these diagnoses.

COCAINE INHALATION ("CRACK")

Patients who use crack cocaine may develop a variety of pathological changes in the lung and the combination of findings should raise concern for inhaled drug use when encountered by the pathologist. These include focal areas of pulmonary hemorrhage, organizing pneumonia, alveolar edema, pigmented macrophages, organizing pneumonia (Fig. 14.8), and bullous emphysema. These findings suggest smoke-induced injury coupled with adrenergic effects. The diagnosis can be confirmed by insistent history taking or by toxicology screens that provide evidence of drug abuse.

AMYLOIDOSIS

Amyloid represents a prion disease showing the folding of 9 nm fibrils of acellular beta-pleated sheets (Fig. 14.9A). In the airways, amyloid can produce cough, dyspnea, and airways obstruction. Its appearance in the lung is stereotypic with sheets of amorphous eosinophilic material surrounded by a giant foreign body cell reaction and plasmacytic inflammation. It is important not to confuse amyloid with necrosis on frozen section interpretations.

In the lung, amyloid can involve virtually any compartment. Nodular densities of amyloid may be solitary or multiple. Amyloid deposition virtually always occurs around small blood vessels (Fig. 14.9B). In some cases, the disease is more diffuse and involves the alveolar walls, and these patients show the most pulmonary dysfunction with restrictive physiology and gas-exchange defects. Pulmonary and cardiac amyloid frequently coexist.

Amyloid stains intensely with Congo Red (Fig. 14.9C) and is birefringent under polarized light where it emits an apple green color (Fig. 14.9D) compared to collagens, which are weakly birefringent and white in color.

Treatment of tracheobronchial amyloid involves excision of obstructing lesions. Diffuse disease has been treated with moderate success by high-dose melphalan and bone marrow transplantation. Patients with cardiac dysfunction may require heart transplantation if their lung function can tolerate it.

An essentially identical histological appearance is produced by *light chain disease*. However, here the "amyloid-like" material fails to stain with Congo Red and ultrastructural examination shows electron-dense deposits rather than fibrillar sheets (Fig. 14.10).

HYALINIZING GRANULOMA

Hyalinizing granuloma is a rare disorder that can mimic amyloidosis histologically. However, in these cases the acellular matrix is collagenous rather than amyloid, as evidenced by staining with trichrome preparations (Fig. 14.11A–C). The lesions may be single or multiple, large or small in size. Some of these cases have been associated with histoplasma infection and the disorder may be a variant of mediastinal fibrosis that may be seen with this infection.

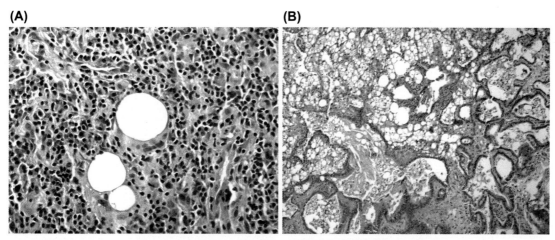

FIGURE 14.5   (A) Lipid aspiration can show large globules of fat and dense fibrosis that can be radiographically mistaken for malignancy. (B) Some cases show microvesicular fat globules and less scarring.

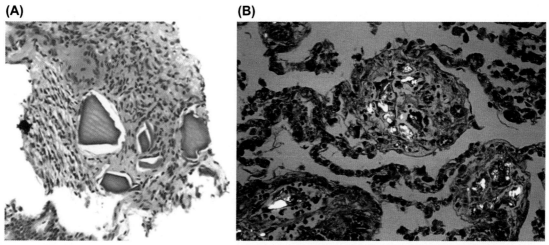

FIGURE 14.6   (A) Aspiration of various drugs and pill contents can cause pulmonary injury. This patient with hyperkalemia aspirated crystals of kayexalate. (*Courtesy of D. Flieder.*) (B) A successful suicide by ingestion of antifreeze (ethylene glycol) shows birefringent crystals in the lung.

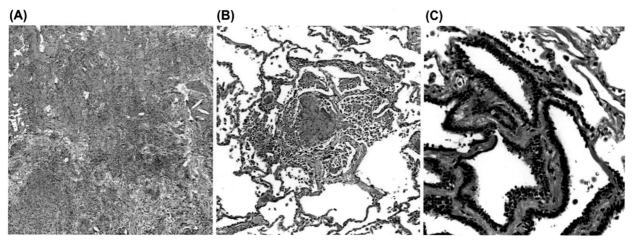

FIGURE 14.7   (A) Microaspiration of gastric contents produces peribronchiolar inflammation and scarring or (B) microgranulomas from aspiration, (C) chronic airways irritation from aspiration, smoking, or allergens produce bronchiolar metaplasia with respiratory epithelium growing along peri-bronchiolar surfaces, i.e., lambertosis.

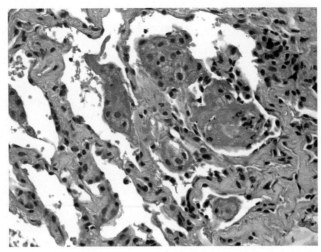

FIGURE 14.8   Inhalation of "crack cocaine" causes focal hemorrhage and pigmented smoker's macrophages.

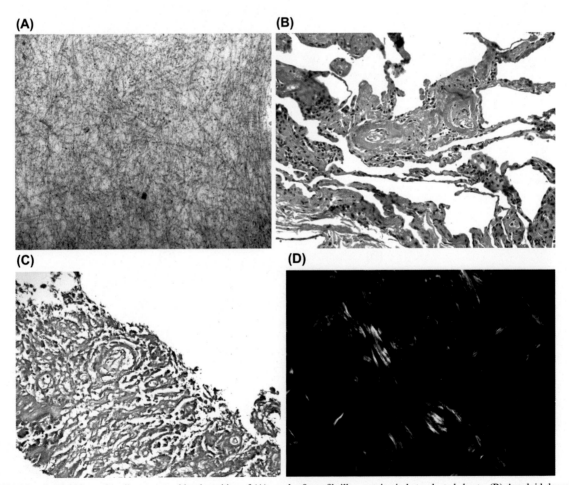

FIGURE 14.9   Amyloid is a prion disease caused by deposition of (A) regular 9 nm fibrillar proteins in beta-pleated sheets. (B) Amyloid deposition in the lung is invariably perivascular but can extend along alveolar septae or form localized nodules (amyloidomas), (C) amyloid stains with Congo Red, and (D) shows apple green birefringence when examined under the polarized light.

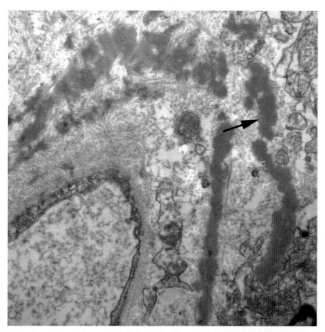

FIGURE 14.10   Amyloid-like deposits can occur in patients with light chain disease, which is distinguished by the failure to stain with Congo Red and electron dense rather than fibrillar deposits (*arrow*).

(A)   **(B)**

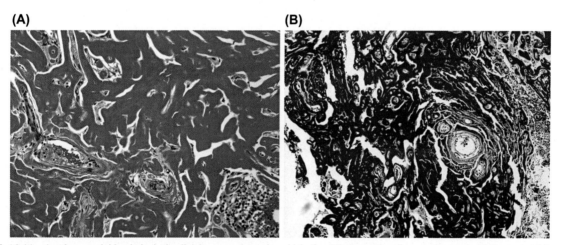

FIGURE 14.11   Another amyloid mimic is hyalinizing granuloma in which the (A) "amorphous" appearing eosinophilic material is actually; (B) paucicellular type I collagen. This disorder may be related to histoplasma infection in some cases.

PULMONARY ALVEOLAR LIPOPROTEINOSIS

Pulmonary alveolar lipoproteinosis (PAP) is an unusual disorder of macrophage metabolism. Patients present with dyspnea and cough and diffuse pulmonary infiltrates. Computed tomography (CT) scans show a characteristic mosaic pattern referred to as "crazy-paving." The serum lactate dehydrogenase is markedly elevated.

The primary form of disease is produced by a deficit of GM−CSF, which may occur either as a genetic defect in production or as the result of auto-antibodies directed at the moiety. This results in the accumulation of surfactant lipoproteins in macrophages that become engorged and finally burst filling the alveolar spaces with PAS-positive lipoproteinaceous membranous material inhibiting gas exchange.

The diagnosis is suggested by a BAL lavagate that shows milky opaque fluid. The diagnosis may be confirmed by examining cytologic smears or ultrastructurally by the presence of whorled lamellar bodies of surfactant material

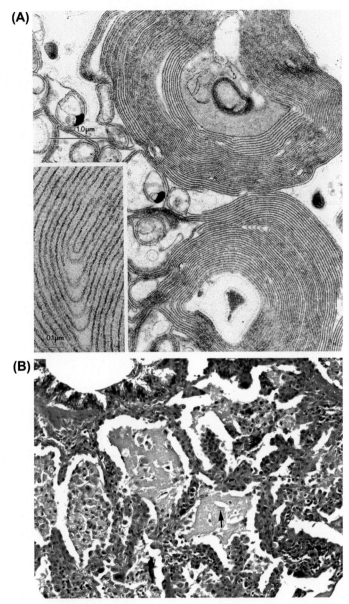

FIGURE 14.12 Pulmonary alveolar proteinosis results from the (A) accumulation of lamella surfactant in alveolar spaces (inset high power). (B) The lung shows amorphous eosinophilic, PAS-positive, intra-alveolar material with acicular clefts (*arrow*) and should not be confused with pulmonary edema.

(Fig. 14.12A). The appearance histologically is pathognomonic. The alveolar spaces are filled with proteinaceous material with small acicular clefts (Fig. 14.12B, arrow). Intense PAS staining of the exudates after diastase digestion helps to establish PAP and must not be confused with either proteinaceous pulmonary edema or pneumocystis infections both of which produce alveolar "flooding."

The treatment of this disorder is whole lung lavage that successfully removes the material, although it tends to reaccumulate. GM−CSF repletion has been tried with limited success and some patients have been treated with lung transplantation, although this is one of the diseases that may eventually recur in the graft. As a result of the macrophage defect, these patients are prone to opportunistic infections with mycobacteria and *nocardia*.

A variant of lipoproteinosis can be seen in patients who are immunosuppressed. This again suggests macrophage dysfunction but is not associated with anti-GM−CSF antibodies. The changes are rarely extensive and respond to a decrease in immunosuppression.

An acute variant of PAP has been observed in individuals exposed to large amounts of silica dust and to heavy metals.

PULMONARY SCARS

The lung may sustain a variety of insults during life that lead to scar formation. At times, these scars can raise concern for malignancy and may be resected for that reason. Old fibromuscular scars rarely show uptake of radiolabeled deoxyglucose uptake on positron emission tomography (PET) scans, however, low-grade adenocarcinomas can also fail to be PET avid. One may follow the progress of these scars on serial CT scans with a low threshold for resection if there is any change, especially in a smoker.

Large subpleural elastotic scars are seen frequently in resected lungs (Fig. 14.13). They may be due to past infection, pneumoconiosis, or remote thromboembolic disease. In the latter, a recanalized pulmonary artery confirms the etiology. The elastic network of the involved lung tissue can be discerned with elastic stains. In some cases, large amount soft lung tissue can be incorporated into a small scar and this may account for some of the lung volume reduction and hypoxemia that is occasionally observed. Carcinomas, both adenocarcinoma and squamous cell carcinoma, show a propensity to arise in preexisting scars that appear to entrap carbonaceous tars.

Apical pleural caps are dense, subpleural, and often hyalinized scars. They may show entrapped areas of traction bronchiolectasis and traction emphysema. In the past, when mycobacterial infection was common, these were attributed to "old granulomatous" infections, but currently it is more likely that they are the result of chronically ischemic regions at the apices. They are often seen as irregular areas of consolidation at the apices on CT scans.

Focal scars may occur as fibromuscular proliferations (Fig. 14.14) or as stellate fibrous scars. As previously noted, some of these have been attributed to burnt-out eosinophilic granuloma, and indeed one may see such scars in patients with active disease. In most cases the attribution is likely faulty, although these scars are often seen in cigarette smokers.

LYMPHANGIOLEIOMYOMATOSIS

In most textbooks, this disorder is classified as an "interstitial lung disease." However, it is in fact a proliferative disorder associated with abnormalities of the tuberose sclerosis (TSC)-1 and TSC-2 genes. Unlike tuberose sclerosis, LAM occurs almost exclusively in young women. They may present with dyspnea, hemoptysis, or chylothorax. About half of patients will have associated angiomyolipomas of the kidney or elsewhere in the retroperitoneum. Lymph node involvement in the mediastinum and retroperitoneum occurs and most have adenoma sebaceum. The pleural lymphatics may be involved leading to lymphatic obstruction and chylothorax. The chest radiographic is characteristic and shows both interstitial opacities and cystic changes. The latter may reflect emphysema or air-trapping behind a small airway occluded by the growth of what appears to be aberrant smooth muscle. The pulmonary function tests show a mixed picture of restriction and air-trapping with low DLCO.

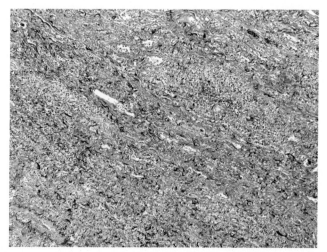

FIGURE 14.13 Elastotic subpleural scars are common. Large amount of lung can be condensed in these scars. Some are related to old pulmonary infarctions but most are idiopathic. They may become a focus for the development of malignant lung carcinomas.

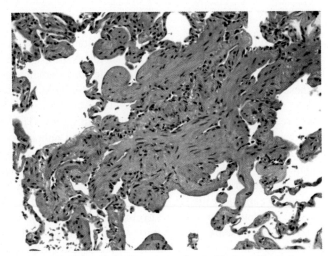

FIGURE 14.14 Fibromuscular scars are also idiopathic and should not be confused with the stellate lesions seen in old Langerhans' cell histiocytosis.

Microscopically one sees the proliferation of "smooth muscle" elements in the lung interstitium and within bronchovascular septa (Fig. 14.15A). However, the proliferating cell expresses HMB-45 and is closely related to a vascular pericyte (Fig. 14.15B). Foci of infiltration often contain slit-like spaces that mimic lymphatics. Intrapulmonary hemosiderosis may be present if there has been intrapulmonary hemorrhage (Fig. 14.15C) and clinically may be expressed as hemoptysis or be asymptomatic. Small pulmonary cysts are invariably present (Fig. 14.16).

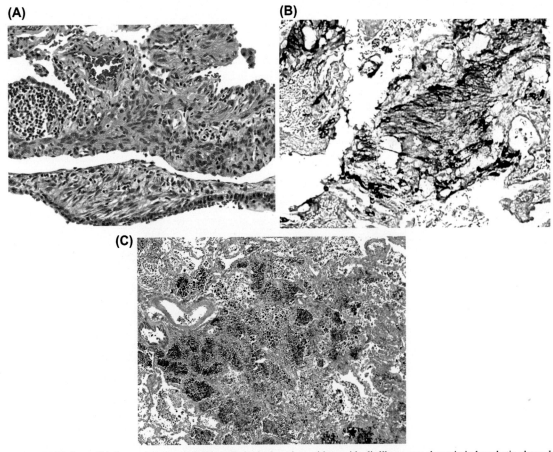

FIGURE 14.15 LAM shows (A) the growth of muscle-like cells in the lung interstitium with slit-like spaces that mimic lymphatic channels. (B) The proliferating cells express HMB-45. (C) Intraparenchymal areas of hemorrhage are common. Air-trapping produces emphysematous cysts, which are characteristic on CT scans.

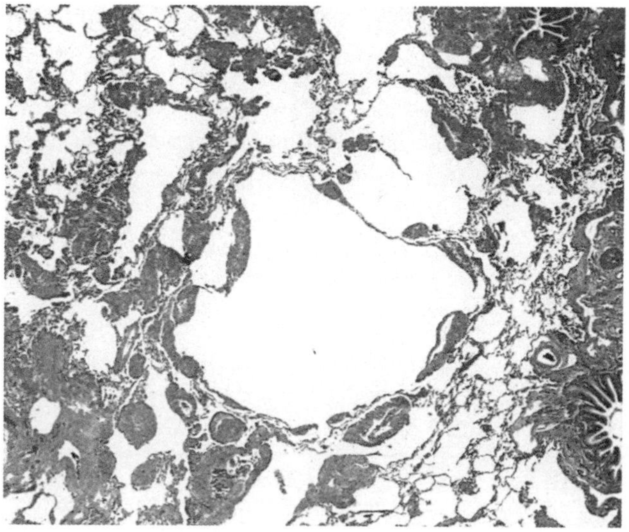

FIGURE 14.16 Emphysematous "cysts" are generally seen in LAM.

There is currently no definite medical cure for this disorder, although hormonal manipulation has been successful in anecdotal cases. Sirolimus (rapamycin or Rapamune) is the first US Food and Drug Administration approved treatment for lymphangioleiomyomatosis (LAM) and has been demonstrated to slow the progression of the disease. Definitive treatment includes orthotopic lung transplantation but LAM is one of the disorders that may recur in the allograft.

CHRONIC EOSINOPHILIC PNEUMONIA

Eosinophilic pneumonia may be caused by allergy, drugs, parasitic infections, or be idiopathic. Patients present with cough, dyspnea, chronic pulmonary consolidations that may wax and wane. Asthma may be a feature. They also develop constitutional features including fever, sweats, and weight loss that raise concern for chronic infection or malignancy. Radiographically, the lung shows multifocal consolidations that tend to be located peripherally.

Lung biopsies show an admixture of intra-alveolar macrophages and eosinophils. Disintegration of eosinophils can produce what appear to be *eosinophilic abscesses* (Fig. 14.17) in which Charcot—Leyden crystals may be present. When strongly suspected the diagnosis may be conformed by sputum induction or by bronchoalveolar lavage.

There is a spectral response between organizing pneumonia with large numbers of eosinophils and chronic eosinophilic pneumonia. Both diseases respond well to relatively low dosages of corticosteroids but may recur. Chronic eosinophilic

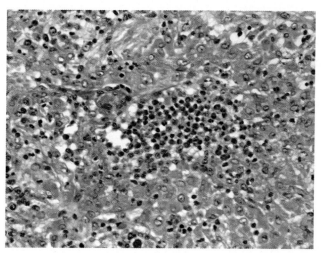

FIGURE 14.17 Chronic eosinophilic pneumonia shows intra-alveolar filling macrophages and eosinophils that may undergo necrosis to form "eosinophilic abscesses" that are sterile.

pneumonia should not be confused with acute eosinophilic pneumonia which represents DAD with eosinophils and is related to first-time smoke inhalation.

Other disorders that can show relatively large numbers of eosinophils include GPA, Churg—Strauss disease, methotrexate toxicity, drug hypersensitivities, and Langerhans' cell histiocytosis. The fibroproliferative phase of DAD may also show prominent tissue eosinophilia at 10—14 days postinsult. In most instances, diagnostic confusion does not occur for either the clinician or pathologist in distinguishing these disorders.

Chapter 15

Pulmonary Transplant Pathology

Orthotopic lung transplantation (OLT) is an accepted and potentially curative modality for the treatment of various inexorable chronic lung diseases. Surgical mortalities are currently low and long-term prognoses are overall favorable (Fig. 15.1). For diseases such as IPF, CF, end-stage COPD, and certain forms of PAH, OLT represents the best available therapeutic option for prolonged survival. Unfortunately, the problem of allograft rejection persists, although substantial progress in immunosuppressive regimens has occurred over the last two decades.

The interpretation of the transbronchial lung biopsy, which is currently used at most transplant centers to monitor the state of the allograft, is fraught with difficulties, as immunological graft rejection can be closely mimicked by infection and noninfectious causes of inflammation.

ACUTE ALLOGRAFT REJECTION

Acute allograft rejection is a disease of the pulmonary microvasculature and may occur at any time during the lifetime of the graft. In many cases, especially with mild graft rejection, patients are asymptomatic. Clinical expertise is required to determine when to intervene by increasing immunosuppression, as the risk of infection and posttransplant lymphoproliferative disorder (PTLD) are directly related to the level of immunosuppression. The problem in ascertaining graft rejection is the fact that the postcapillary venules are often involved in nonimmunological forms of graft inflammation, and judgment is required in distinguishing these possibilities.

Acute graft rejection (A1) reflects the infiltration of T lymphocytes cuffing small vessels (Fig. 15.2). Allograft rejection is not a homogeneous response and it is not uncommon to have small vessels selectively affected. Grade A1 rejection is diagnosed when the perivascular lymphocytic cuffing is just apparent at low-power inspection, in the absence of an alternative explanation.

Grade A2 rejection shows a more apparent level of perivascular infiltration at low power (Fig. 15.3A) and endothelialitis may be present (Fig. 15.3B). The latter, although nonspecific, supports the diagnosis of acute rejection, especially if there is no other apparent cause. The transplant clinicians will continue to observe these patients without changing the immunosuppressive regimen, if they are asymptomatic.

Grade A3 rejection constitutes more severe disease and requires treatment. One sees evidence of vascular and adjacent perivascular alveolar inflammation (Fig. 15.4). However, if the interstitium appears inflamed in the presence of adjacent

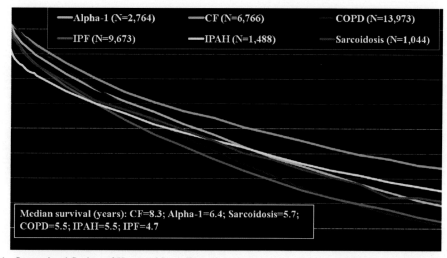

FIGURE 15.1 International Society of Heart and Lung Transplantation outcome experience with adult lung allograft transplantation.

Understanding Pulmonary Pathology. http://dx.doi.org/10.1016/B978-0-12-801304-5.00015-0

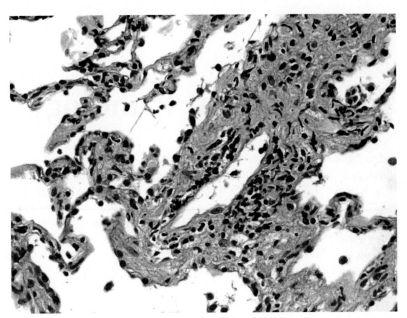

FIGURE 15.2   Grade A1 acute rejection shows barely perceptible lymphoid aggregates cuffing small pulmonary vessels.

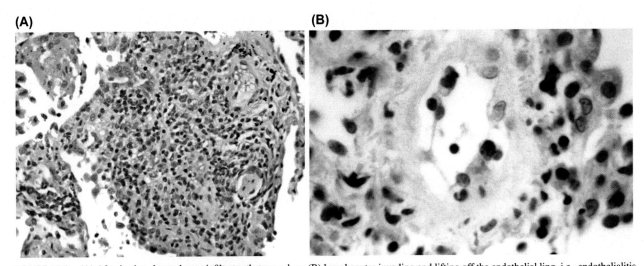

FIGURE 15.3   (A) A2 rejection shows denser infiltrates that may show (B) lymphocytes invading and lifting off the endothelial ling, i.e., endothelialitis.

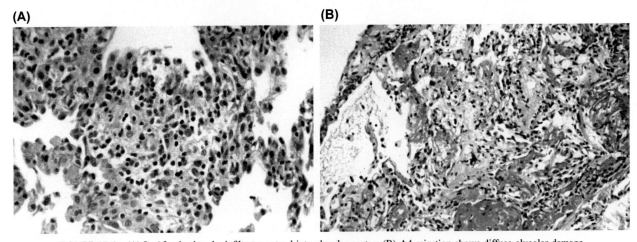

FIGURE 15.4   (A) In A3 rejection the infiltrates extend into alveolar septae. (B) A4 rejection shows diffuse alveolar damage.

normal appearing vessels, an alternative diagnosis should be considered. Grade A4 rejection portends irreversible graft rejection if not rapidly treated. In my experience, this is rarely seen except when patients have been totally noncompliant with their medications. In A4 rejection (Fig. 15.4B), the lung shows diffuse lymphocytic inflammation with alveolar wall damage, fibrinous infiltrates, and hyaline membranes.

AIRWAYS INFLAMMATION

Nonobliterative airways inflammation may be a precursor to chronic rejection. It is recommended that the airways be assessed and reported on for evidence of mild (B1) or more severe (B2) acute and chronic inflammation. B2 disease should show infiltration of the bronchiolar epithelium by neutrophils with epithelial injury or ulceration.

CHRONIC GRAFT AIRWAY REJECTION

Chronic rejection (C1) is defined as fibrous obliteration of the small airways, and chronic vascular rejection (D1) may also be present. However, the latter is difficult to assess on TBB and is underestimated by this approach. A variety of insults can damage the small airways of the lung. These include immunological rejection, infection, and aspiration. The latter is increasingly well recognized as a factor in chronic graft rejection and it may appear early in the course of the posttransplant period as opposed to immunological chronic rejection, which rarely develops in the first 6 months following graft implantation. Treatment for GERD-related disease may require intensive protein pump inhibitors and surgical fundoplication.

As noted, the diagnosis of C1 chronic rejection requires identifying the obliteration of small airways. Obliteration of the airways may be due to intraluminal fibrosis or constriction due to mural fibrosis (Fig. 15.5). Elastic stains are invaluable for identifying obliterated airways, as they show dense bundles of elastica located where the obliterated airway once was. Trichrome stains can also be helpful in this determination. The presence of dense peribronchiolar hyaline alveolar fibrosis in the biopsy although nondiagnostic suggests chronic rejection. Organizing pneumonia may be a variant of chronic rejection in some cases.

There is a tendency over time for the alveolar walls to become thickened posttransplantation, although the clinical significance of fibrous allelopathy is currently uncertain (Fig. 15.6). The normal alveolar wall is nonhomogeneous with respect to gas exchange, as one side of the alveolus consists of a fused endothelial and alveolar type I cell unit via which gas exchange normally occurs by diffusion, whereas the other side includes a loose matrix and rare mesenchymal cells. However, it is possible for the alveolar wall to thicken due to fibrosis in the non-gas-exchanging side of the alveolar wall without affecting gas exchange and this is observed in the first year following lung transplantation. However, after a period the fibrosis may become extensive enough to obliterate the capillary lumen leading to diminished gas exchange.

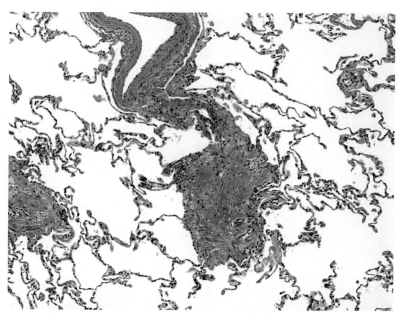

FIGURE 15.5 C1 rejection may be multifactorial. The end stage is obliterative bronchiolitis that may be intraluminal or constrictive. Scar is seen adjacent to the pulmonary vessel.

FIGURE 15.6   D1 rejection shows intimal proliferation and obliteration of small pulmonary vessels. These are rarely adequately sampled by transbronchial biopsies.

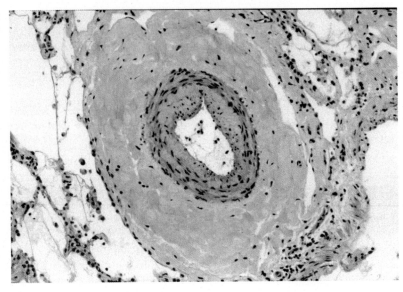

CHRONIC VASCULAR REJECTION

Chronic vascular rejection is infrequently diagnosed on TBBs. The vessels undergo an unusual intimal proliferative reaction that leads to luminal occlusion (Fig. 15.7). Vascular rejection virtually never occurs in the absence of chronic airways rejection. The diagnosis is suggested by elevated pulmonary artery pressures and may be confirmed by computed tomography (CT) angiogram, pulmonary angiography, or VATS biopsy.

FIGURE 15.7   Graft alveolopathy reflects the progressive thickening of the airway wall (F) as shown in this electron micrograph. Its implications are currently uncertain.

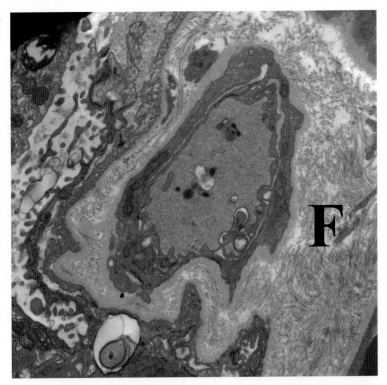

HUMORAL REJECTION

Humoral rejection remains a controversial topic in lung transplantation. It appears to be reported relatively frequent from some centers and not by others. One looks for evidence of endothelial injury and the deposition of C4d, which has been shown to be a reliable marker of humoral rejection in kidney allografts. However, C4d deposition in the lung is both nonspecific and often patchy. The diagnosis of humoral rejection should only be suggested when there is diffuse expression of C4d along the alveolar microvasculature (Fig. 15.8), the presence of anti-HLA antibodies against the donor, and an unexplained decrement in graft function.

OTHER FORMS OF GRAFT INJURY

A variety of other patterns of injury may be seen in this setting. Opportunistic infection is well recognized and is discussed in the chapter on lung infections. Infections tend to occur at different times within the life of the allograft. Most of the early viral and fungal infections, e.g., cytomegalovirus and pneumocystis carinii pneumonia, are treated prophylactically in transplant regimens. Anastomotic site infections are often problematic and may be due to bacterial infection or opportunistic fungal infections. Surgical intervention is generally required.

POSTIMPLANTATION RESPONSE

Immediately following implantation, reversible graft ischemia results in a postimplantation response. The lung shows alveolar wall edema with mild acute inflammation (Fig. 15.9). This picture may persist for some time but generally resolves to a near normal appearing lung parenchyma. Hemosiderin-laden macrophages are frequently seen in these grafts. It is supposed that they accumulate as the result of microhemorrhages but their etiology is uncertain.

ASPIRATION

Due to the damage to vagal pathways, most patients will show evidence of gastroesophageal reflux with acid aspiration. The lung in this setting generally shows fibrinous exudates and organizing pneumonia (Fig. 15.10). The picture may be difficult to separate from rejection and a consideration of the timing of the response is helpful in distinguishing these possibilities.

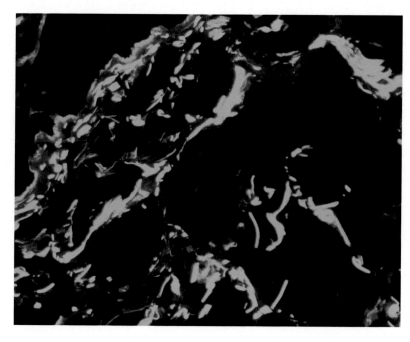

FIGURE 15.8 Deposition of C4d diffusely in the lung suggests humoral rejection but the diagnosis must be confirmed by the presence of circulating anti-HLA antibodies and graft dysfunction (immunofluorescence).

FIGURE 15.9 Reimplantation responses occur early following allograft placement and likely reflect ischemic changes in the graft. The lungs are diffusely edematous and there may be subsequent mild interstitial fibrosis.

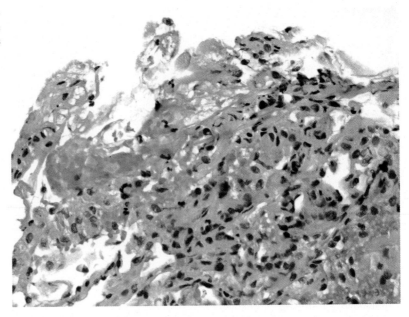

FIGURE 15.10 The presence of fibrinous infiltrates in the graft is often attributable to aspiration following the surgical severing of the vagus nerves. Chronic aspiration may contribute to the development of chronic allograft rejection along with infection and immune rejection.

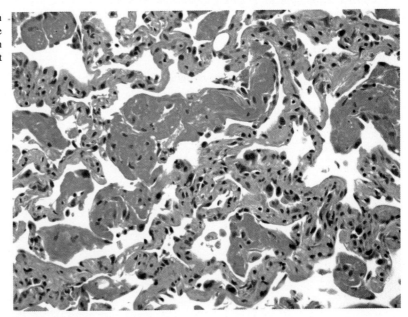

POSTTRANSPLANT LYMPHOPROLIFERATIVE DISORDER

PTLD is a serious complication of transplantation. The vast majority of cases of PTLD are related to EBV infections. The disorder can occur at virtually any time after graft implantation and is related to the degree of immunosuppression. The disease may be localized in the lung or multifocal. The target cell in most cases of PTLD complicating solid grafts appears to be of host recipient origin.

The disorder varies in severity. Polymorphic PTLD retains a degree of immune organization and shows dense infiltrates of plasmacytoid B cells as well as immunoblasts but is not frankly malignant in appearance. Monomorphic PTLD as the name implies is frankly malignant in its appearance and consists primarily of large lymphocyte with immunoblastic features (Fig. 15.11).

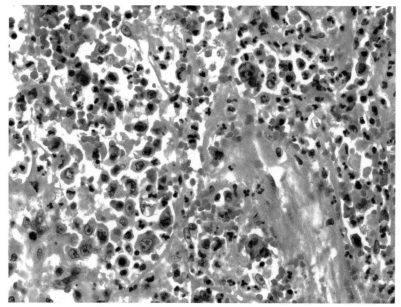

FIGURE 15.11 Posttransplant lymphoproliferative disease can occur anytime in the life of the grant and results from immunosuppression and reactivation of Epstein-Barr virus within B cells. Polymorphic diseases can be controlled with reduction of immunosuppression but frank lymphomatous transformation carries a poor prognosis.

TABLE 15.1 Disorders that May Recur in Lung Allografts

Sarcoidosis

Lymphangioleiomyomatosis

Langerhans' cell histiocytosis

Talc granulomatosis

Diffuse panbronchiolitis

Pulmonary alveolar proteinosis

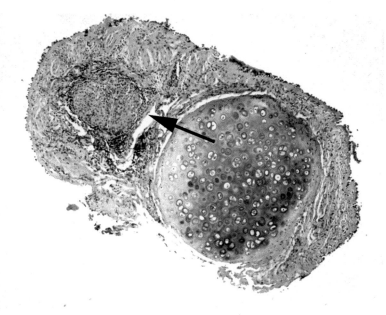

FIGURE 15.12 Recurrent sarcoidosis in allograft (*arrow*).

The initial treatment is to reduce immunosuppression. This may be sufficient to control patients, especially those with polytypic disease. Early intervention may be beneficial, although this is uncertain. Overall, the prognosis is guarded. Chemotherapeutic regimens have had marginal success. Good short-term responses have been seen with anti-B cell regimens (rituximab) but detailed long-term follow-up is not yet available.

RECURRENCE OF DISEASE IN AN ALLOGRAFT

It has been recognized that certain diseases have the capacity to recur in the transplanted lung within months to years following implantation (Table 15.1). The most common disease to recur is sarcoidosis (Fig. 15.12). However, a rather lengthy list of disorders can recur. In such cases, the options include medical treatment as the risk of recurrence in a second allograft is generally too great to contemplate a second graft.

FURTHER READING

Christie, J.D., Edwards, L.B., Kucheryavaya, A.Y., et al., 2011. The registry of the International Society for Heart and Lung Transplantation: twenty-eighth adult lung and heart-lung transplant report—2011. J Heart Lung Transplant 30 (10), 1104—1122.
Recent in-depth review of the state of the art with respect to transplantation.
Stewart, S., Fishbein, M.C., Snell, G.I., et al., 2004. Revision of the 1996 working formulation for the standardization of nomenclature in the diagnosis of lung rejection. J Heart Lung Transplant 26 (12), 1229—1242.
Current criteria for grading lung allograft rejection.

Chapter 16

Pediatric Lung Disease

For years, pulmonary pediatric pathology was a specialized field with its own unusual jargon. Increasingly, it has become part of the purview of the specialist in pulmonary pathology. However, clinicians should be aware that some of the older literature tends to reflect a classification system that may be unfamiliar. It is a good idea for adult and pediatric pulmonologists and pathologists to attend each other's conferences to benefit from insights gained from each group.

HYALINE MEMBRANE DISEASE

This is a disease of immaturity at gestation before normal surfactant production has been achieved by the embryonic lung. Neonates develop acute respiratory distress and require mechanical ventilation and supplemental oxygenation. Surfactant replacement and corticosteroids may speed recovery. The pathology shows diffuse alveolar injury with hyaline membranes akin to DAD in adults.

Mechanical ventilation may produce *pulmonary interstitial emphysema* in these infants with an attendant risk of tension pneumothorax and pneumomediastinum. Infants who survive hyaline membrane disease can develop *bronchopulmonary dysplasia*, in which the newborn lung shows interstitial fibrosis with anatomic distortion, bronchiolar obliteration, cystic changes, and pulmonary hypertensive remodeling (Fig. 16.1).

SEQUESTRATIONS

Sequestrations by definition lack communication with the airway of the normal lung. Extralobar sequestrations present as well-defined areas of consolidation adjacent to the diaphragm or in the mediastinum and almost always on the left side. The extralobar sequestered lung is invested in its own pleura. Patients may be asymptomatic or have nonspecific pulmonary

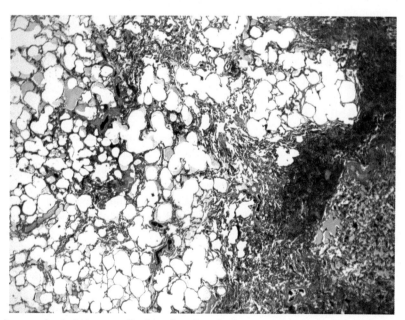

FIGURE 16.1 Bronchopulmonary dysplasia is a complication of newborn hyaline membrane disease. The lung shows scarring and architectural distortion.

(A)   **(B)**

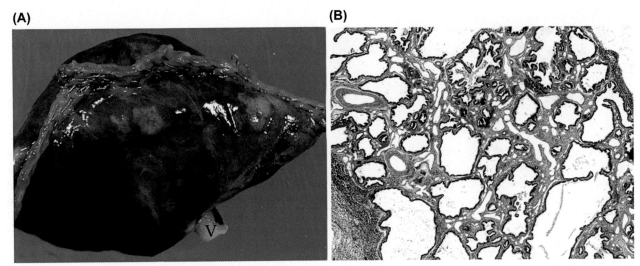

FIGURE 16.2   (A) Extralobar sequestrations show a systemic arterial feeding vessel (V) and are invested in their own pleura. (B) The sequestration shows cystic changes lined by bronchiolar epithelium. *Courtesy S. Vargas.*

symptoms. A critical aspect of this diagnosis is recognizing the sequestered lung's blood supply, which is generally from the descending aorta or other regional systemic arteries prior to resection. If a subdiaphragmatic arterial feeder vessel is not first located and clamped prior to resecting the sequestration, the feeding artery may retract into the abdomen with lethal hemorrhagic complications. Unlike intralobar sequestrations, extralobar sequestrations are associated with a wide range of congenital abnormalities.

Microscopically, the extralobar sequestration shows a large systemic arterial feeder vessel (Fig. 16.2A). There may be evidence of the atretic airway with mucocele formation and maldevelopment of the lung with microcystic changes (Fig. 16.2B).

INTRALOBAR SEQUESTRATION

It is not entirely certain whether intralobar sequestration is a congenital malformation or an acquired form of bronchiectasis, although the abnormality has been increasingly noted with prenatal ultrasound examination suggesting its congenital origin. However, it is rarely accompanied by other congenital abnormalities. It tends to present in late childhood or early adulthood. The sequestration is most often seen in the left lower lobe associated with an atretic airway and mucocele formation (Fig. 16.3). The distal lung shows cystic dilatation of airways, chronic inflammation, and may be infected. If symptomatic, resection is required.

BRONCHIAL ATRESIA AND CONGENITAL CYSTIC ADENOMATOID MALFORMATION

The common element in sequestration, as noted, is bronchial atresia but this change is also responsible for a variety of other congenital pulmonary abnormalities, especially congenital adenomatoid malformation. Isolated atresia is most commonly seen in the upper lung zones. Patients present with dyspnea and chest infections in late childhood or in adulthood.

The Stocker classification divided these into a macrocystic variant (type I) (Fig. 16.4), a microcytic variant [type II congenital cystic adenomatoid malformation (CCAM)] (Fig. 16.4B), and a solid adenomatoid variant (type III) (Fig. 16.4C). The type I CCAM has been associated with increased incidence of mucin-producing adenocarcinomas with *KRAS* mutations.

The type II variant appears to be a part of the spectrum of bronchial atresia. The large cyst variant (>2.0 cm) presents in infancy with respiratory distress. The abnormality is usually limited to a single lobe that shows extensive air trapping. The airways lack cartilaginous support but communicate with surrounding normal airways and are not true cysts. The adenomatoid variant represents a form of pulmonary hyperplasia and the lungs are large and show immature differentiation. It may be clinically confused with congenital lobar overinflation which is likely attributable to a ball-valve affect with the narrowing of a lobar bronchus. It may be related to polyalveolar lobe syndrome.

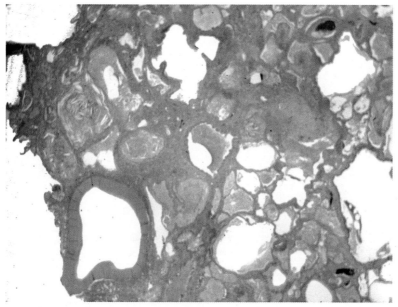

FIGURE 16.3 Intralobar sequestration results from an atretic airway. The lung is supplied by a large bronchial artery and shows cystic changes. The surrounding lung is normal.

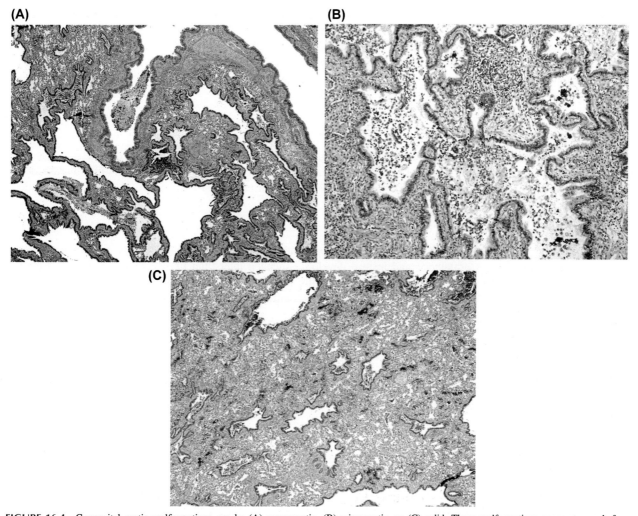

FIGURE 16.4 Congenital cystic malformations can be (A) macrocystic, (B) microcystic, or (C) solid. These malformations appear to result from bronchial atresia.

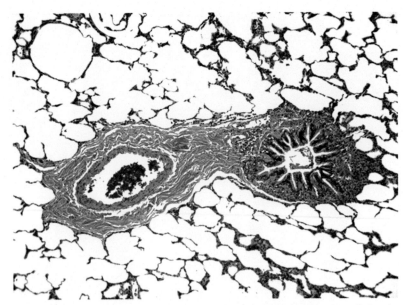

FIGURE 16.5   Chronic tachypnea of childhood is related to the secretion of bombesin by neuroendocrine cells in otherwise normal small airways.

NEUROENDOCRINE HYPERPLASIA OF CHILDHOOD

This disorder occurs in infants who develop persistent tachypnea after birth. They may show wheezing and respiratory distress. Chest radiographs reveal hyperinflation. The lung biopsy shows minimal abnormalities other than an increase in small airway neuroendocrine cells that secrete bombesin, as well as parenchymal clusters of neuroendocrine cells with multinucleated cells (Fig. 16.5). It has been postulated that bombesin plays a role in airway narrowing and leads to persistent obstructive disease in these patients. Whether this is related to the adult variant of neuroendocrine hyperplasia that also causes air trapping is currently uncertain.

INTERSTITIAL LUNG DISEASES

There has been an increasing recognition of infants and young children who have interstitial lung disease associated with abnormalities in surfactant genes. Some of these patients have survived into adulthood with interstitial disease or first present at later ages. A recent classification of these disorders is shown in Table 16.1.

SURFACTANT PROTEIN DEFICIENCY

Surfactant deficiency is associated with alveolar lipoproteinosis but many patients with surfactant protein abnormalities do not show classical features of proteinosis and instead develop associated interstitial fibrosis that is pauci-inflammatory. Lipoproteinaceous material may be scant (Fig. 16.6) and the pulmonary alveolar macrophages may show a foamy appearance with evidence of disintegration. The diagnoses are confirmed by genetic testing of the patients, parents, and siblings.

ABCA3 DEFICIENCY

This is caused by an autosomal recessive mutation in the ABCA3 gene on chromosome 16. It may present at birth, in older children, or in adults. Prognosis is variable depending on the severity of disease.

SURFACTANT-PROTEIN C DEFICIENCY

This may present as a chronic pneumonitis in infancy or may present in older children and adults as histologically heterogenous interstitial lung diseases. It is an autosomal dominant disorder due to a mutation of surfactant-protein C

TABLE 16.1 Proposed Classification Scheme for Pediatric Diffuse Lung Disease

I. Disorders More Prevalent in Infancy

A. Diffuse Developmental Disorders

1. Acinar dysplasia
2. Congenital alveolar dysplasia
3. Alveolar–capillary dysplasia with pulmonary vein misalignment

B. Growth Abnormalities

1. Pulmonary hypoplasia
2. Chronic neonatal lung disease
   A. Prematurity-related chronic lung disease (bronchopulmonary dysplasia)
   B. Acquired chronic lung disease in term infants
3. Structural pulmonary changes with chromosomal abnormalities

A. Trisomy 21

B. Others

4. Associated with congenital heart disease in chromosomally normal children

C. Specific Conditions of Undefined Etiology

1. Pulmonary interstitial glycogenosis
2. Neuroendocrine cell hyperplasia of infancy

D. Surfactant Dysfunction Mutations and Related Disorders

1. SPFTB genetic mutations—PAP and variant dominant histologic pattern
2. SPFTC genetic mutations—Chronic pneumonitis of infancy (CPI) dominant histologic pattern; also DIP and NSIP
3. ABCA3 genetic mutations—PAP variant dominant pattern; also CPI, DIP, NSIP
4. Others with histology consistent with surfactant dysfunction disorder without a yet recognized genetic disorder

II. Disorders not Specific to Infancy

A. Disorders of the Normal Host

1. Infectious and postinfectious processes
2. Disorders related to environmental agents: hypersensitivity pneumonia, toxic inhalation.
3. Aspiration syndromes
4. Eosinophilic pneumonia

B. Disorders Related to Systemic Disease Processes

1. Immune-related disorders
2. Storage disease
3. Sarcoidosis
4. Langerhans' cell histiocytosis
5. Malignant infiltrates

C. Disorders of the Immunocompromised Host

1. Opportunistic infection
2. Disorders related to therapeutic intervention
3. Disorders related to transplantation and rejection syndromes
4. Diffuse alveolar damage of unknown etiology

D. Disorders Masquerading as Interstitial Disease

1. Arterial hypertensive vasculopathy
2. Congestive vasculopathy, including veno-occlusive disease
3. Lymphatic disorders
4. Congestive changes related to cardiac dysfunction

Adopted from ATS official guideline. Am J Respir Crit Care Med 2013;188:376–94.

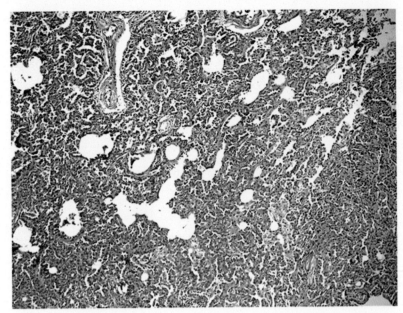

FIGURE 16.6   Abnormalities in surfactant genes can produce lipoproteinaceous accumulation in the lung.

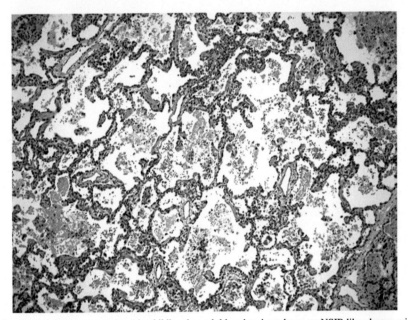

FIGURE 16.7   SP-C deficiency can present in childhood or adulthood and produces an NSIP-like change within families.

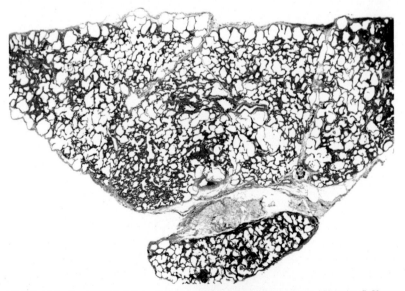

FIGURE 16.8   Rare TTF-1 deficiencies cause diffuse microcystic changes. *Courtesy S. Vargas.*

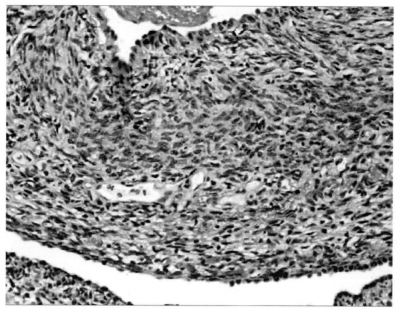

FIGURE 16.9 Pleuropulmonary blastoma type I involves the lung and surrounding pleura and has a relatively good prognosis. *Courtesy S. Vargas.*

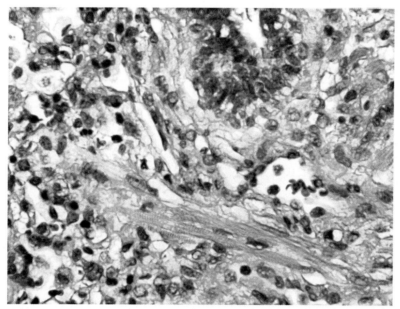

FIGURE 16.10 Type 2 pleuropulmonary blastoma shows primitive differentiation including skeletal muscle and has a poor prognosis. These disorders are to be differentiated from the adult disease in which multiple mesenchymal lines of differentiation are seen. *Courtesy S. Vargas.*

(SP-C) on chromosome 8. Some patients have shown a picture of cholesterol pneumonitis, whereas others have diffuse interstitial disease (Fig. 16.7). The alveolar lining cells may become severely atypical and one case ultimately developed pulmonary adenocarcinoma. Other rare forms of congenital cystic lung disease have been observed in families with TTF-1 deficiency (Fig. 16.8).

MALIGNANT TUMORS

Primary pulmonary epithelial malignancies are rare in childhood. However, sarcomatous malignancies do occur. Pleuropulmonary blastoma is an unusual lesion that involves the lung and adjacent pleura in patients before the third year. These tumors may be cystic or solid. Low-grade cystic lesions can be cured surgically in ∼50% of cases but local recurrences do occur (Fig. 16.9). The high-grade lesions are treated with chemotherapy followed by surgery but recurrences are frequent.

These lesions in the past were diagnosed as rhabdomyosarcoma or mesenchymal hamartomas. The tumor appears to arise from the embryonic mesenchyme or *blastema*. The cystic spaces are lined by a *cambium layer* of mesenchyme. There may be differentiation toward skeletal muscle (Fig. 16.10), and cartilaginous elements may be present as well as bizarre histiocytoid giant cells. Unlike pulmonary blastomas in adults, these lesions lack an epithelial component and there is no association with cigarette smoking.

FURTHER READING

Deutsch, G.H., Young, L.R., Deterding, R.R., et al., 2007. Diffuse lung disease in young children: application of a novel classification scheme. Am J Respir Crit Care Med 176 (11), 1120–1128.
The current classification schema for pediatric medical lung disease.
MacSweeney, F., Papagiannopoulos, K., Goldstraw, P., et al., 2003. An assessment of the expanded classification of congenital cystic adenomatoid malformations and their relationship to malignant transformation. Am J Surg Pathol 27 (8), 1139–1146.
A critical examination of the classification of congenital cystic disease.

Chapter 17

Pleural Diseases

The pleura normally is a microscopic mesothelial lined surface that envelops the lung (visceral pleura) and reflects onto the chest wall (parietal pleura). It can be a site of primary disease or secondarily involved by underlying lung inflammation. Malignancies can arise primarily in the pleura, e.g., malignant mesothelioma, but more commonly the pleura is the site of metastatic tumors from the lung and other sites.

REACTIVE FIBRINOUS PLEURITIS

Pleural reactions were discussed in Chapter 2. Organizing fibrinous pleuritis is the most common finding in patients with exudative pleural effusions due to any number of causes including infectious pleuritis, parasympathetic effusion, pulmonary embolic disease, drug reactions, etc. The pleura show exudative fibrin, granulation tissue, fibrosis, and mesothelial proliferation, which may be exuberant and show cytological atypia. These changes can also be present when there is either a primary or secondary malignancy and may delay the diagnosis of malignancy.

Some cases of pleuritis show features that suggest their etiology. Eosinophilic pleuritis is evoked in reaction to recent pneumothorax, but parasitic infection, tuberculosis, and drugs can also yield eosinophilic pleuritis.

Granulomatous pleuritis is most often due to mycobacterial tuberculosis infection or fungal infections, but may be seen with sarcoidosis and Wegener's granulomatosis, or in response to drugs, e.g., methotrexate. If the patient had been previously surgically treated or received pleurodesis with talc, one sees a foreign body giant cell reaction that includes strongly birefringent crystals. As talc can be contaminated with tremolite asbestos, its use is best avoided in young patients. Giant cells in granulomas of sarcoid or infections can produce oxalate crystals that should not be confused with a foreign body reaction.

Catamenial hemoptysis and hemothorax occurs in women during their premenopausal cycles may be due to endometriosis involving the lung or pleura, respectively. Rarely one can see ectopic splenic tissue in the pleura due to post-traumatic laceration of the spleen.

PLEURAL PLAQUES

Most pleural plaques are caused by exposure to asbestos or talc. They are characteristically seen on chest radiographs along the chest wall parietal surfaces parallel to the ribs and most characteristically along the diaphragmatic leaflets. They show a distinctive histology with a paucicellular hyalinized basket-weave collagen (Fig. 17.1), which may show dystrophic

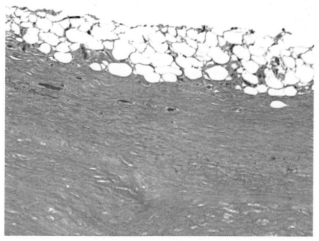

FIGURE 17.1   Benign asbestos-related pleural disease (hyaline fibrous plaque).

calcification. Rarely, this appearance may see similar changes along the visceral pleural adjacent to an underlying malignancy. Old empyemas and hemothoraces can produce comparable findings but are distinguishable histologically from asbestos plaques by the presence of persistent chronic inflammation or evidence of hemosiderin. Asbestos can also uncommonly cause diffuse pleural fibrosis that causes fibrothorax. This may result from repeated asbestos-related pleural effusions.

Asbestos pleural plaques are responsible at times for the infolding of the underlying lung to produce rounded atelectasis that can mimic malignancy. However, the CT appearance of a band of atelectatic lung coursing toward the hilum to produce a "comet's tail" is characteristic and should suffice in making the correct diagnosis. It is worth noting that asbestos bodies are virtually never seen in pleural plaques or for that matter in the substance of malignant mesothelioma and a search for asbestos bodies should be limited to the lung parenchyma. It has been shown by ultrastructural analysis that chrysotile fibers have a predilection to translocate from the lung and to deposit as short fibers in the pleura, which may play a role in both plaque formation and the pathogenesis of malignant mesothelioma.

BENIGN TUMORS OF THE PLEURA

Solitary fibrous tumors (SFT)—referred to as benign mesothelioma in the older literature—are the most common benign lesion of the pleura (Fig. 17.2). They are usually solitary but small numbers of tumors can occur. They arise from the visceral or parietal pleura and may grow to large sizes if undetected. Occasional tumors are hormonally active and produce an insulin-like substance that leads to hypoglycemia. They are also a known cause of digital clubbing.

This tumor, which can also occur in the lung parenchyma, is composed of bland spindle cells in an edematous and hyalinized matrix. They exhibit a prominent staghorn vascular pattern. The lesion must be distinguished from malignant SFT that is cellular and shows increased mitotic activity, necrosis, and the potential to recur and metastasize. It must also be distinguished from localized sarcomatoid malignant mesothelioma. Immunostains are helpful, as SFT is strongly CD34 and Stat-6 positive, which is not the case for malignant mesothelioma.

DESMOID TUMORS

These are seen in the chest wall and are identical to desmoid tumors in the soft tissues. They are associated with either previous chest surgery or trauma and essentially are exuberant scars. Their appearance is one of bland fibrosis with a hyaline collagenous matrix (Fig. 17.3A and B). The tumor cells immunostain for beta-catenin and the lesion must be distinguished from SFT and benign neural tumors.

NEURAL TUMORS

Schwannomas and neurofibromas generally occur in the posterior mediastinum but can be seen in the pleura as a result of growth off of the intercostal nerves. They are histologically identical to neural sheath tumors at other sites and immunostain

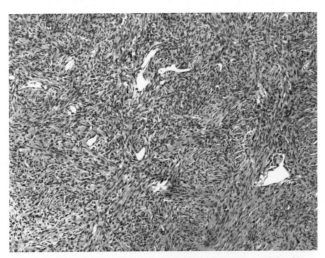

FIGURE 17.2 Solitary fibrous tumor with staghorn vessels and a cellular fibrous matrix.

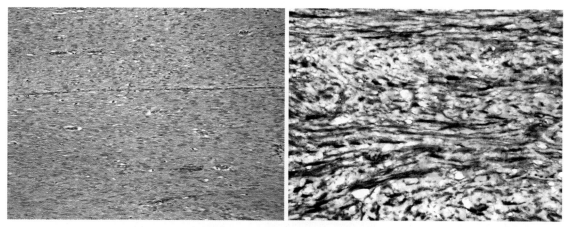

FIGURE 17.3   Paucicellular desmoid tumor.

(A)   **(B)**

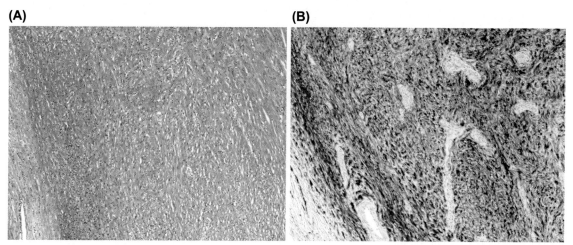

FIGURE 17.4   (A) Schwannoma of the pleura arises from sensory nerve twigs and are (B) strongly S-100 immunopositive.

strongly for S-100 (Fig. 17.4A and B). Rare malignant peripheral neural sheath tumors can occur and these show features of malignancy with sarcomatous spindle cells, mitoses, and necrosis.

METASTATIC PLEURAL TUMORS

The most common malignancy of the pleura is secondary metastatic carcinoma from the lung. However, malignancies from other primary sites, including esophagus, stomach, pancreatobiliary tree, colon, liver, prostate, breast, uterine cavity, etc., can all metastasize to the pleura and present with pain and effusions. The diagnosis may be made cytologically from the effusion but can require a pleural biopsy if the effusion cytology is nondiagnostic and there remains a high index of suspicion for malignancy.

MALIGNANT MESOTHELIOMA

The most common primary malignancy of the pleura is diffuse malignant mesothelioma. This tumor was virtually never reported prior to the introduction of asbestos into the workplace and it is considered a "signal" tumor for asbestos exposure (Fig. 17.5). Other accepted causes are far less common and include therapeutic radiation and erionite, an asbestiform mineral that is present in the Anatolia region of Turkey.

Malignant mesothelioma is caused by all types of asbestos, although the amphiboles appear to be more potent carcinogens on a per fiber basis. However, chrysotile was the most commonly used asbestos fiber in the United States and it tends to concentrate in the pleura following inhalation.

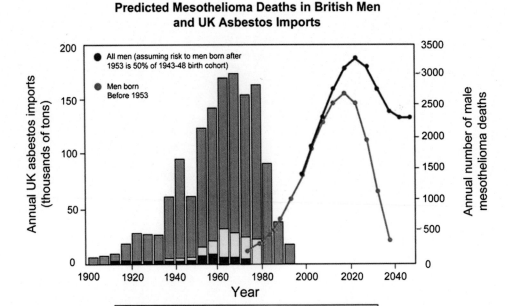

**Predicted Mesothelioma Deaths in British Men
and UK Asbestos Imports**

☐ Chrysotile   ☐ Amosite   ■ Crocidolite

FIGURE 17.5   Mesothelioma was virtually unknown before the introduction of asbestos into the workplace as seen in this epidemiological study from the UK. *(Reproduced by kind permission of Professor J. Peto and the Lancet 1995;345:535, with permission from Elsevier.) Courtesy of G. Mark.*

The latency period for the development of malignant mesothelioma after asbestos exposure is measured in decades, and it may take 40 or more years to develop. The disease is dose responsive but there is currently no accepted safe level of occupational or para-occupational (secondary domestic) exposure to asbestos. Housewives who have laundered the clothing of their asbestos-exposed husbands are a well-recognized target for the development of this disease.

The patients present with dyspnea, cough, chest pain, weight loss, and malaise. Chest imaging shows a pleural effusion, often associated with nodular pleural thickening which may encase the lung (Fig. 17.6). The tumors are invariably positron emission tomography (PET) avid. Radiographic staging may require both PET scans to assess regional node involvement and magnetic resonance imaging scans to determine chest wall and diaphragmatic invasion.

The diagnosis is frequently delayed. The yield by cytology is roughly 50%. The diagnosis is complicated by associated areas of reactive fibrinous pleuritis, which may be sampled in small biopsies. VATS biopsies are the optimal approach as they also allow the surgeon to assess the gross appearance of the pleura.

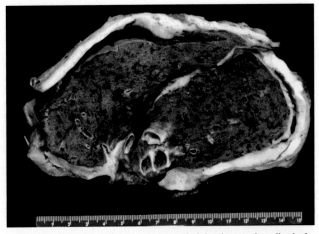

FIGURE 17.6   Diffuse malignant mesotheliomas form a rind that entraps the underlying lung and mediastinal vessels and extends into the chest wall.

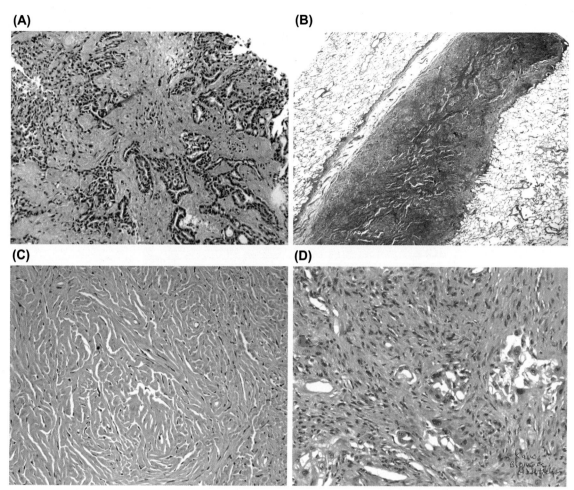

FIGURE 17.7   Mesotheliomas may be (A) epithelioid, (B) sarcomatoid (spindle celled) mesothelioma with extension into pulmonary septum, (C) desmoplastic mesothelioma that may difficult to distinguish from a benign fibrosis, (D) mixed epithelioid and spindle cell features, and (D) may show dense.

The histological appearance is multiform. Roughly one-third show a proliferation of epithelioid cells with solid, tubular, or papillary growth (Fig. 17.7A). Another one-third show sarcomatoid growth (Fig. 17.7B), which may include marked desmoplasia that can at times mimic a benign pleural plaque (Fig. 17.7C). Another one-third show biphasic growth with both epithelioid and sarcomatous elements (Fig. 17.7D). Histologic subsets of mesotheliomas include small-cell mesotheliomas, in which cells vaguely resemble small-cell carcinoma, inflammatory mesothelioma, in which there are large numbers of lymphocytes and macrophages, adenomatoid variants, and sarcomas producing heterologous elements including bone and cartilage. None of these variants has a substantial impact on clinical treatment or the uniformly poor outcome.

The major differential diagnosis especially for the epithelioid type is metastatic pulmonary adenocarcinoma. Metastatic sarcomatoid carcinomas, usually squamous cell carcinoma of the lung, can at times mimic sarcomatoid mesothelioma. The distinction is made by immunostaining and electron microscopy. Epithelioid malignant mesothelioma immunostains positive in most cases for pancytokeratins, calretinin, WT-1, CK7, CK5/6, D2-40, mesothelin, and glut-1, and does not stain for TTF-1, Napsin A, p40, p63, CEA, Ber-Ep4, or CD15.

There is a pseudomesotheliomatous adenocarcinoma that appears to arise in the lung periphery and closely mimics diffuse malignant mesothelioma, both in its growth pattern and histological appearance, but immunostains positively for markers of adenocarcinoma. It has been suggested that this malignancy may also be asbestos related (Fig. 17.8A−C).

The immunostaining pattern of sarcomatoid malignant mesotheliomas is less consistent and in some cases these tumors may not immunostain for any of the aforementioned antigens. In such cases, the diagnosis is one of exclusion. At times ultrastructural analysis, which characteristically shows long microvilli, a feature of mesotheliomatous differentiation, can be of assistance, but this may not apply to very poorly differentiated sarcomatoid mesothelial malignancies.

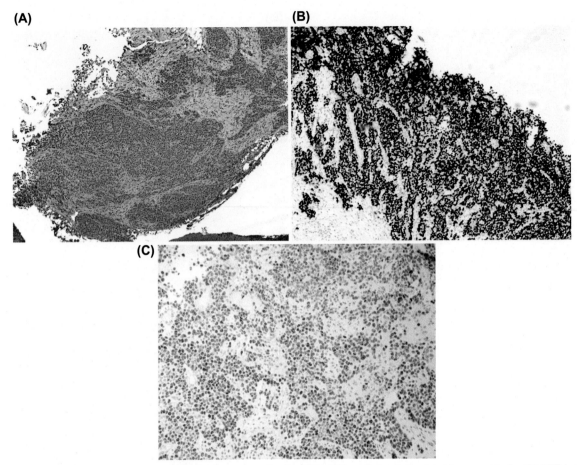

FIGURE 17.8 (A) Pseudomesotheliomatous adenocarcinoma may closely mimic malignant mesothelioma but immunostain for (B) TTF-1, (C) this tumor also showed a *NUT* midline carcinoma mutation.

Localized malignant mesothelioma is rare but it otherwise shares clinical and histological features with the diffuse disease and appears to be caused by asbestos in many cases.

The prognosis of this disease is dismal. Aggressive surgeries produce severe complications and most patients cannot tolerate them. Currently, optimal therapy for those who can tolerate surgery is surgical decortication and combination chemotherapy with pemetrexed and carboplatinum, but survival rates are generally less than 2 years.

RARE PLEURAL MALIGNANCIES

Epithelioid hemangioendothelioma can occur primarily in the pleura and mimic a poorly differentiated malignant mesothelioma (Fig. 17.9). The tumor cells show small vascular spaces that are immunopositive for CD31, factor VIII, and fli-1 and may be cytokeratin positive. Several of the originally reported cases had a history of asbestos exposure but not all.

Synovial sarcoma of the pleura can closely mimic the biphasic growth of malignant mesothelioma (Fig. 17.10). The tumor shows spindle cells, a prominent vascular pattern, and well-differentiated epithelial tubules. The tumor does not immunostain for the usual mesothelial antigens other than calretinin but is positive for CD99 and the epithelial component is positive for CD56, cytokeratin, and EMA. Most, but not all, show a characteristic gene translocation t (X; 18). There has been no definite link to asbestos exposure.

Primary effusion lymphoma is an unusual large B-cell lymphoma that specifically involves the pleural, pericardial, and peritoneal cavities (Fig. 17.11). It tends to occur in immunosuppressed patients with acquired immunodeficiency syndrome. Patients present with effusions and the diagnosis can be made by thoracentesis with cytofluorometry. The malignant cells show plasmacytic features and are immunopositive for CD30, CD138, and CD38. There is a strong association with EBV infection as evidenced by in situ hybridization and human herpesvirus-8 infection that is ascertained by the polymerase chain reaction. The malignancy is recalcitrant to most therapies and has a poor prognosis.

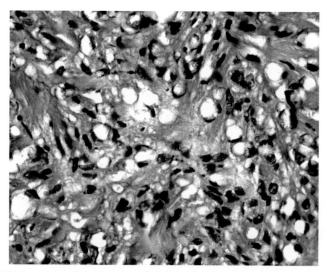

FIGURE 17.9 Epithelioid hemangioendothelioma can mimic mesothelioma and may be associated with asbestos exposure.

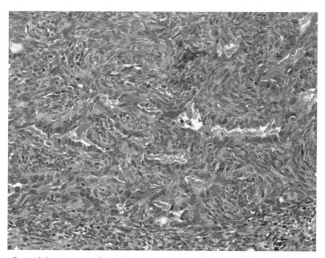

FIGURE 17.10 Synovial sarcomas of the pleura are rare but like mesothelioma are often biphasic tumors.

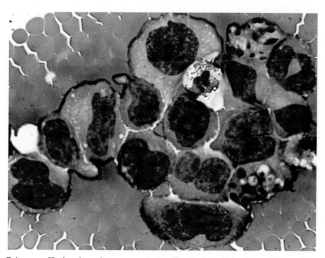

FIGURE 17.11 Primary effusion lymphomas are generally diagnosed by examining cells in the pleural fluid.

FURTHER READING

Cardillo, G., Carbone, L., Carleo, F., et al., 2009. Solitary fibrous tumors of the pleura: an analysis of 110 patients treated in a single institution. Ann Thorac Surg 88 (5), 1632–1637.

Good reviews of the most common malignant and benign tumors of the pleura.

Markowitz, S., 2015. Asbestos-related lung cancer and malignant mesothelioma of the pleura: selected current issues. Semin Respir Crit Care Med 36 (3), 334–346.

Glossary of Terms

Abscess A collection of neutrophils with necrotic tissue forming a mass. These can be small microabscesses or large cavitary abscesses.

Acinar A mode of tumor growth for adenocarcinoma that shows gland formation.

Acinus An idealized abstraction of the gas-exchange unit. It includes all gas-exchanging structures distal to the terminal bronchiole. The acinus cannot be visualized radiographically or by light microscopy, although 3–5 "acini" are generally found within a pulmonary lobule.

Acute allograft rejection Host T-lymphocytic response directed against the microvasculature of the graft. Acute rejection may occur at a point in the life of the graft if protective immunosuppression is decreased.

Acute inflammation A term used to describe the presence of neutrophils in tissues by pathologists.

Adenocarcinoma A malignant gland forming tumor. Although defined by gland formation, in some cases, this cannot be identified. Mucin production and the expression of certain nuclear and cytoplasmic antigens define pulmonary adenocarcinomas at the immunohistochemical level.

Adenoma A benign glandular tumor.

Alveolus The terminal gas exchanging sac. It consists of a gas-exchanging surface with pulmonary capillary, and a nongas potential space and alveolar epithelial cells (type I and II). The type I cell and the capillary endothelium share a fused basement membrane and this unit represents the shortest distance for gas exchange.

Amyloid A prion disease due to proteins deposited in the lung that form beta-pleated sheets of fibers 9 nm in size. Amyloid stains with Congo red, iodine, and cresyl violet. Congophilic amyloid shows apple-green birefringence when examined with polarized light.

Angiitis Vascular inflammation that may be characterized by neutrophilic or lymphocytic inflammation. Angiitis is a very nonspecific finding and can be primary or part of an inflammatory process that secondarily involves vessels.

Angiomatoid lesion Seen in severe pulmonary hypertension, it is the result of thrombosis and recanalization of a small pulmonary arteriole, the *plexiform lesion* resulting in poststenotic capillary dilatation.

Apoptosis A universal mechanism of programmed cell death. In tissue it is recognized as condensed "apoptotic" nuclei and must be distinguished from mitotic figures.

Asbestos A family of fibrous mineral silicates that were used commercially and have the capacity to cause both benign and malignant diseases.

Asbestosis An interstitial lung disease caused by occupational levels of exposure to asbestos. It is to be distinguished from *asbestos pleural plaques*, which are limited to the pleura and rarely cause clinical symptoms.

Aspiration The inspiration of foreign material into the lung. Aspiration can be due to food, liquids, oropharyngeal secretions, or gastric acid. This can produce either a bacterial or chemical pneumonitis or occlude the airways.

Atresia A segment of airway that has not developed properly leading to a focus of narrowing.

Basement membrane The secreted matrix of epithelial and endothelial cells. It consists of glycosaminoglycans and specialized proteins including laminin and type IV collagen. Its thickening in the airways generally reflect collagen deposition and is a marker of chronic inflammation.

Basophilia The dark blue staining seen as a result of tissues taking up hematoxylin. Nuclear chromatin stain with hematoxylin.

Blasteme Undifferentiated embryonic mesenchyme.

Bronchiectasis A structurally distorted airway with loss of cartilaginous support, mural fibrosis, and dilation of the airway lumen. Three types of bronchiectasis have been described: cylindrical, varicose, and saccular, reflecting increasing severity of airway distortion and loss of distal lung tissue.

Bronchiole A small airway that lacks cartilage.

Bronchitis/iolitis Pathologically defined as an inflamed airway.

Bronchopneumonia An infection of the lung that begins in small airways and then extends to the peribronchiolar lung tissue. It may be patch or diffuse and at times distinct areas may coalesce to produce changes that mimic a lobar process.

Bronchopulmonary dysplasia A complication of infantile respiratory distress syndrome, which results in scarring with anatomic distortion in the developing lung.

Bronchovascular bundle The term that refers to the coursing of a pulmonary airway together with its accompanying pulmonary artery within a fibrous septum.

Bronchus A cartilaginous airway.

Brown–Hopps stain A tissue gram stain that detects most gram-positive and -negative bacteria as well as fungal microconidia and *candida* yeast.

Cambium layer A term used to describe a dense area of linear growth of cells mimicking a tree ring.

Capillaritis Acute microvascular inflammation characterized by vascular necrosis, fibrin, hemorrhage, neutrophils, and nuclear dust (karyorrhexis).

Carcinoma A malignant tumor derived from epithelial cells. It is not synonymous with "cancer," which generically refers to all malignancies.

Cavity A necrotic area of lung that communicates with an airway to produce an air-filled "cavity."

Charcot–Leyden crystals Crystalized product of eosinophil protein due to the breakdown of eosinophils.

Cholesterol clefts A marker of cellular lysis. Cholesterol in cell membranes may crystallize to form characteristic clefts that are often engulfed by macrophages. The implications of the change are nonspecific but it is often seen in lipoid pneumonias and in hypersensitivity pneumonitis.

Chronic allograft rejection Defined as obliterative bronchiolitis. Chronic vascular rejection often accompanies chronic airways rejection but may be difficult to establish in a small biopsy. Nonimmune factors, e.g., aspiration, may contribute to chronic graft rejection.

Chronic inflammation Inflammation that is rich in lymphocytes, macrophages, and plasma cells. It often follows an acute inflammatory response and tends to persist and lead to scarring.

Cilia Specialized microtubules in a characteristic $9 + 2$ configuration that are present in airway epithelial lining that beat adorally and assist in the clearance of inhaled particulates and microbes.

Curschmann's spirals Small mucus plugs expectorated by patients with asthma or chronic bronchitis.

Dead space A physiological term that refers to air-filled structures that do not engage in gas exchange. The normal anatomic dead space includes the conducting airways that account for 150 mL of air. Physiological dead space refers to areas of lung that are ventilated but not perfused as in pulmonary thromboembolic disease. It is the opposite of shunt in which blood is not oxygenated due to collapse of ventilated areas (atelectasis), intrapulmonary, or intracardiac communications between the pulmonary and systemic circulation (right to left shunt).

Diapedesis The movement of a cell out of the circulation and into the adjacent tissues.

Diffuse alveolar damage The underlying pathology seen in either acute respiratory distress syndrome (ARDS) or acute interstitial pneumonitis (AIP) characterized by alveolar injury with hyaline membranes.

Diffusing capacity The amount of O_2 that diffuses across the alveolar gas-exchange as a function of time. The DLCO uses inhaled carbon monoxide as a surrogate marker for O_2. The DLCO is dependent on the alveolar volume and may be corrected for it as the DLCO/VA (alveolar volume).

Driver mutations Mutations within the cell genome that play a key role in regulating the cell cycle. Several driver mutations that can be targeted by pharmacological agents have recently been identified in lung carcinomas.

Dysplasia A morphological change that signals a change along the pathway toward malignancy. Dysplasia is generally graded from mild to severe with the distinction between severe atypia and in situ carcinoma often subjective.

Elastic tissue A specialized mesenchymal element termed elastin allows tissue to expand and contract like an elastic band. The lung is generously invested with elastin fibers. Elastin fibers are highlighted by specialized elastic silver stains so that the fibers appear black under the microscope.

Emphysema A collection of air in cyst-like spaces in the lung due to destruction of the alveolar wall and to be differentiated from true cysts, which are epithelial lined structures, and areas of hyperinflation without alveolar wall destruction. Air that dissects into the pulmonary interstitium or soft tissues of the chest is also referred to as "emphysema."

Endarteritis obliterans An eccentric proliferative reaction of the intima of blood vessels adjacent to a focus of active inflammation.

Endothelialitis Vascular intimal inflammation that is nonspecific but a key feature seen in acute allograft rejection when T lymphocytes infiltrate the endothelial lining causing injury and separation of the endothelial cell from its basement membrane.

Eosinophilic Based on the appearance of tissue following staining with the standard hematoxylin and eosin (H&E) stain. Eosinophilia refers to the pink to red appearance as seen under the microscope. The cytoplasm of cells stains with eosin as do matrix proteins including fibrin and collagens.

Fibrin The result of either intravascular or extravascular thrombus formation. Fibrin is a marker of acute injury to the microvasculature. It is amorphous and eosinophilic.

Fibrinoid necrosis A misnomer, as the term *fibrinoid* refers to the presence of fibrin. The change is most often used to refer to vascular wall changes in a necrotizing angiitis.

Fibroblastic focus An active outgrowth of fibroplasia in response to microulceration of the alveolar wall and limited to it. It is a characteristic feature of usual interstitial pneumonitis (UIP). When the alveolar wall injury is diffuse confluent fibroblastic foci represent the fibroproliferative phase of diffuse alveolar damage.

Fibrosis An elemental response to injury that includes the recruitment of fibroblasts to the site of injury with the deposition of collagens and other matrix proteins. This process leads in some cases to irreversible scarring and anatomic distortion. Fibrosis is best highlighted with a histochemical trichrome or pentachrome stain in which it stains blue or green.

Fite stain A modified Ziehl–Neelsen stain that is decolorized with weak acid. It reacts with mycobacteria, *nocardia* spp., *Legionella micdadei*, *Rhodococcus*, the cortex of schistosome ova, and the hooklets of echinococcus.

Flow-volume loop The standard representation of the forced expiration of a patient from total lung capacity (TLC). The amount of air exhaled forcefully in 1 s is termed the *FEV1* and the total exhaled volume is the *FVC*. A ration of *FEV1/FVC* of <0.6 indicates an obstructive defect. A normal or increased *FEV1/FVC* ratio may reflect a restrictive defect when the FVC is reduced.

Follicular hyperplasia A term that relates to a hyperplastic change within lymphoid tissue. The "follicle" refers to the presence of germinal center formation. The change indicates an ongoing stimulus for chronic inflammation, often due either to infectious or autoimmune etiologies.

Fontana–Masson stain A histochemical stain that detects melanosomes present in pigmented tumors and pigmented fungi.

Foregut One of the precursor elements of embryonic growth. The lung and esophagus are derived from the foregut.

Foreign body (giant cell) reaction A host response to foreign particulates that consists of macrophages showing multinucleated giant cells (multinucleated macrophages). When localized this may also be termed a foreign body granuloma.

Gomori–methenamine silver This is one of a family of silver stains that reacts with glycosaminoglycans in the wall of fungi and gram-positive bacteria.

Granuloma Technically speaking the term means a nodular collection of epithelioid macrophages. However, the term is used in various ways in pathology. A solitary "granuloma" is an old term for any nodule in the lung.

Granulomatous inflammation When it includes nodular granulomas is generally modified as either necrotizing or nonnecrotizing. The former suggests an infectious etiology. However, the other usage of the term refers to any inflammatory reaction comprised of lymphocytes, macrophages, and plasma cells, with multinucleated macrophages, e.g., Wegener's granulomatosis. Each use of the term has different diagnostic implications and clinicians need to be clear as to how it is being applied.

Hamartoma A form of growth that represents the disordered but benign growth of normal mesenchymal and epithelial elements.

Helminth A parasite "worm." Helminthic infections are caused by round worms (*nematodes*), flukes (*trematodes*), and tapeworms (*cestodes*). All can infect the lung and cause allergic eosinophilic reactions.

Hematoxylin & eosin The standard histochemical stain applied to tissues for diagnosis in virtually all pathology laboratories.

Histiocytic An antiquated term for macrophages but one still widely used by pathologist.

Histochemistry The approach to nonimmunological chemical staining of cellular constituents.

Honeycomb lung A markers of end-stage changes in the lung. More accurately it refers to areas of bronchiolectasis with intralobular fibrosis.

Hyaline A peculiar appearance of type I collagen that shows little cellularity and intense eosinophilic staining.

Hyperplasia A *benign* proliferation of a cell type, e.g., alveolar type II cell hyperplasia.

Immunohistochemistry A histochemical technique based on the ability of antibodies to bind to proteins associated with cells and matrix. In the case of malignancies, immunohistochemistry can be extremely useful in differentiation cell lineage.

Iron stain A histochemical stain that detects iron moieties in tissues. It is used routinely to detect evidence of hemosiderin in macrophages and ferruginated asbestos fibers (asbestos bodies).

Karyorrhexis The fragmentation of cell nuclei in inflammation.

Lambertosis Bronchiolar metaplasia that surrounds an airway. The term implies extension of bronchiolar epithelium through the "canals of Lambert" that connect distal airways to adjacent alveoli. Lambertosis indicates chronic airway injury and is seen in smokers, hypersensivity pneumonitis, aspiration, and airway centered interstitial fibrosis.

Langerhans' cell A specialized antigen presenting cells found normally in epithelium and lymph nodes. In Langerhans' cell histiocytosis, these cells proliferate to form radiographically detectable nodules. They immunostains for S-100, CD1a, and Langerin.

Large cell carcinoma A variant of non-small-cell lung carcinoma that lacks any other specific mode of differentiation either histologically or immunohistochemically.

Lepidic A growth pattern that develops using the alveolar wall as a "scaffolding." Lepidic growth of an adenocarcinoma if it does not include any invasive element is currently considered adenocarcinoma in situ and replaces the older term of *bronchioloalveolar carcinoma*.

Lipoid pneumonia A misnomer for a reaction seen generally in response to proximal airway obstruction. The lesion shows macrophages that show small vacuoles of lipid. Lipid pneumonia is usually restricted to changes caused by the aspiration of exogenous lipids.

Lymphoma A monoclonal proliferation of either B or T lymphocytes.

Lymphoproliferative disorder A poorly defined term that implies either a benign or questionably malignant proliferation of lymphocytes.

Lysosomes Intracytoplasmic membrane-bound organelles.

Masson bodies First described in uremic lung, these are tufts of active fibrosis that characterize organizing pneumonia.

Mesothelioma A malignant tumor of the pleura that is strongly associated with asbestos exposure.

Metaplasia Transformation of a normally found cell type in an anatomic location to a different histological cell type that while benign is not normally seen in that location.

Micropapillary A growth pattern seen in adenocarcinomas that show small papillary excrescences.

Mucin A family of acid, neutral, and basic mucopolysaccharides secreted by specialized cells. Mucins play a role in host defense.

Mucociliary escalator A physiological term that suggests the combined activities of secreted mucus and respiratory cilia in ridding the airways of inhaled particulates and microbes.

Neuroendocrine carcinoma A high-grade malignant tumor that shows evidence of neuroendocrine differentiation. High-grade neuroendocrine carcinomas are divided into small and large cell variants somewhat subjectively based on cell size.

Non-small-cell carcinoma A generic term used to describe a lung carcinoma that is not "small cell carcinoma." The diagnosis should be limited to small biopsies where the entire pattern of differentiation cannot be determined due to sampling. Current techniques allow pathologists in most cases to further denote non-small-cell carcinoma as either squamous cell or adenocarcinoma. However, as complex patterns of carcinomatous growth can occur, e.g., adenosquamous carcinoma, the term is appropriate for small biopsies.

Nonspecific interstitial pneumonitis (NSIP) A common syndromic form of lung injury often seen with autoimmune disorders but may be due to other etiologies or be idiopathic.

Organizing pneumonia A term used to describe an airway response to injury. It can be either idiopathic or secondary to other etiologies. In addition, the term is sometimes used very nonspecifically to describe changes in small biopsies that show an admixture of macrophages, lymphocytes, and fibrosis.

Organizing thrombus This term is a source of substantial confusion as it is used in two distinct ways. Organization of a thrombus can refer to the internal architecture of the thrombus to distinguish it from a post-mortem blood clot. Organizing thrombi show "lines of Zahn," which is the admixture of fibrin, platelets, and red blood cells.

The second usage refers to the relationship of the thrombus to the vascular wall. In pulmonary thromboembolic disease, angioneogenesis arising from the vascular walls to penetrate the thrombus and eventually incorporate it as a remodeled neointima. From a medico-legal perspective, it is possible to roughly date the age of a thrombus based on the extent of organization into the vascular wall.

Papillary Any growth that shows long frond-like excrescences with a central feeding blood vessel.

Periodic acid Schiff stain (PAS) This histochemical stain highlights glycogen and glycosylated proteins including mucins. It stains most fungal species.

Pleomorphic carcinoma A poorly differentiated form of lung cancer that shows a significant degree of either bizarre cytological changes with giant cells or spindle cell "sarcomatoid" growth.

Pneumoconiosis A lung disease attributable to the inhalation, usually occupational, of various types of dust that can cause pulmonary fibrosis.

Pneumonitis A term that is interchangeable with pneumonia. It refers to lung inflammation. Most cases of infection are termed "pneumonias." However, inflammatory conditions like usual interstitial pneumonitis may also be referred to as usual interstitial pneumonia.

Pseudoamyloid The deposition of hyalinized collagen that mimics amyloid but does not stain with Congo red.

Pulmonary thrombo/embolus Pulmonary thromboemboli arise from peripheral veins and lodge in the pulmonary vasculature at various levels dependent on the size of the embolus. However, thrombi can also develop in situ when there is vascular intimal injury, or a procoagulant stimulus, as can occur, e.g., when there is severe lung injury with release of tissue thromboplastin.

Pyogenic A tissue response rich in neutrophils

Reactive atypia A benign response by any cell type to injury. It often includes nuclear enlargement but with a retained nuclear to cytoplasmic ratio. When the cytological changes are marked, there is a risk of the pathologist confusing it with malignancy.

Respiratory bronchiolitis A marker of inhaled dust, most often cigarette smoke that reflects uptake of pigmented particulates by macrophages (pigmented macrophages).

Sarcoma A form of malignancy that recapitulates mesenchymal elements. Most but not all lack expression of keratins. Sarcomas that arise in the lung may take on the appearance of any intrinsic mesenchymal cell. Sarcomas, unlike carcinomas that metastasize primarily through lymphatic channels, tend to spread via the peripheral blood.

Sarcomatoid A growth pattern with spindle cells that mimics malignant sarcoma.

Sequestration A congenital disorder in which the lung lacks an airway connection and derives its primary blood supply from a bronchial or systemic artery. Extralobar sequestration has its own pleural investment.

Squamous cell carcinoma A form of epithelial malignancy that recapitulates the lining of the skin. Histologically squamous cell carcinomas make keratin at times in rounded structures called "keratin pearls" and show intercellular cytoplasmic "bridges " between cells.

Synechiae Microscopic adhesions.

Tracheopathia osteoplastica An idiopathic disorder characterized by bony metaplasia in the tracheal wall.

Tumorlets A benign neoplastic growth that is rarely of clinical significance but is not uncommonly seen in lungs resected for other reasons. Tumorlets either show features of neuroendocrine carcinoids or meningothelial cells.

Usual interstitial pneumonitis (UIP) A pattern of chronic lung injury. It is always the pathology referred to clinically as idiopathic interstitial pneumonia (IPF) but may be seen with autoimmune disorders and other etiologies.

V/Q mismatch One the causes of hypoxemia. It occurs when there is a mismatch in the normal pairing of ventilation to perfusion. A classic example is pneumonia. V/Q of zero is termed *shunt* and means that there is no ventilation of an area that is perfused. A classic example is atelectasis. A V/Q ratio of ∞ is equal to *dead space* and means that there is ventilation in an area with no blood flow. A classic example is pulmonary vascular disease

Vasa vasorum Systemic vessels that provide nutrients to pulmonary vessels.

Vasculitis Like angiitis an inflammatory change in a vessels wall. However, the term usually reflects a situation where the vasculature is the primary target and there is fibrinoid necrosis of the vessel wall.

Warthin–Starry stain One of several silver impregnation stains (Steiner, Dieterle) that reacts with all eubacteria and is used primarily to detect spirochetes and gram negative bacteria that do not stain well with tissue grams stain, e.g., *Legionella* spp.

Ziehl–Neelsen stain Reacts with fatty acid moieties in mycobacteria tuberculosis (TB) and most atypical mycobacteria.

Index

'*Note*: Page numbers followed by "f" indicate figures and "t" indicate tables.'

A

ABCA3 deficiency, 320
Actinomycosis, 183–184, 186f
Acute allograft rejection, 309–311, 310f
Acute bronchopneumonia, 178, 180f
Acute fibrinous and organizing pneumonia
    (AFOP), 87, 87f
Acute inflammation
    acute neutrophilic pulmonary inflammation,
      22, 23f
    fibrin diffuse deposition, 21, 22f
    regulation, 21, 22f
Acute interstitial pneumonia, 114–120, 120f
Adenocarcinomas (ACAs), 251f
    atypical adenomatous hyperplasia, 249,
      250f
    atypical cells, small lepidic ACA, 249, 250f
    distal ACA, unfixed lung, 249, 249f
    lepidic growth insitu, 243, 244f
    micropapillary ACA, 249, 250f
    mucin-secreting ACA and multifocal
      mucinous ACA, 249, 250f
    in scar, 249, 249f
Adenoid cystic carcinoma, 261, 262f
Adenovirus, 167–168, 169f
Airway diseases
    amyloidosis, 91–93, 92f–93f
    asthma
      airway thickening and mucus plugging,
        66, 66f
      Charcot–Leyden crystals, 64, 65f
      Churg–Strauss syndrome (CSS), 66, 67f
      clinical triggers, 63, 65t
      curschmann spirals, 64, 65f
      defined, 63
      eosinophils, in airway wall, 64, 65f
      features, 63, 64t
      flow-volume loop, before and after
        bronchodilator use, 63, 64f
    bronchial epithelium, 59, 61f
    bronchiectasis, 37–38, 39f, 76–77
    bronchiolitis obliterans organizing
      pneumonia (BOOP), 84–87
    chronic bronchitis, 67–69, 69f
    chronic obstructive pulmonary disease
      (COPD), 67, 67t, 68f
      exacerbations, 75–76
    collapsibility, 37, 38f
    cystic fibrosis, 77–80
    dilated mucin-filled ducts, chronic
      bronchitis, 62, 63f

distal trachea long segment stricture and left
    main bronchus, 40f, 41
    emphysema, 69–75
    follicular bronchiolitis, 87–90
    hypertrophic smooth muscle, asthma,
      62–63, 63f
    increased seromucous glands, 62, 62f
    luminal diameter, 37, 38f
    neuroendocrine cell, 59, 61f
    normal flow-volume loop, 59, 60f
    obliterative bronchiolitis (OB)
      constrictive bronchiolitis, 83–84,
        84f
      elastic stain highlights airway obliteration,
        83–84, 83f
      focal peri-arterial scar, 84, 85f
      intralumenal fibrous obliteration,
        83–84, 83f
    pulmonary cilia ultrastructure, 59, 61f
    pulmonary obstruction, causes, 59, 59t
    Reid index, 62, 62f
    relapsing polychondritis, 93
    rightward displacement, 40, 40f
    small airways, 81–82, 82f
    spirometric analysis, 59, 59t
    squamous cell carcinoma, 41, 41f
    stenosis, 38–40, 39f
    thickened airway basement membrane,
      asthma, 59, 62f
    thickened airways, chronic bronchitis,
      62, 63f
    tracheal stenosis, 90, 91f
    tracheobronchomalacia, 93
    tracheopathia osteoplastica, 93
    upper airway diseases, 90, 91f
    wall thickness, 37
Airways inflammation, 311
Airway wall
    granulation tissue, 28, 28f
    lipoid pneumonia, 28, 28f
    mucinous metaplasia, small airway,
      27–28, 27f
    squamous metaplasia, chronically inflamed
      airway, 27–28, 27f
Allergic bronchopulmonary aspergillosis,
    212–213, 218f
Aluminosis, 295
Alveolar interstitium
    corpora amylacea, 30–31, 31f
    fibroblastic focus, 29, 29f
    hyaline membrane, 29, 29f

    metaplastic smooth muscle extension,
      30–31, 30f
    reactive atypia, 29–30, 30f
Alveolar microlithiasis, 298, 298f
Amiodarone, 279, 282f
Amyloidosis, 91–93, 92f–93f, 300,
    302f–303f
Angiitis, 33
Angioinvasive aspergillosis, 216–217,
    223f–224f
Ankylosing spondylitis, 110, 111f
Anthrax, 236–238, 238f
Antialveolar basement membrane disease
    (ABM), 153–155, 153f–155f
Arteriovenous malformations, 156, 156f
Asbestos, 287–289, 288f–290f
    cancer, 289–290
Aspergillus spp., 2–3, 209, 216f–217f
    A. bronchitis, 215, 220f
    A. nidulans, 219, 226f
    A. terreus, 219, 225f
Aspiration pneumonia, 299
Asthma
    airway thickening and mucus plugging, 66,
      66f
    Charcot–Leyden crystals, 64, 65f
    Churg–Strauss syndrome (CSS), 66, 67f
    clinical triggers, 63, 65t
    curschmann spirals, 64, 65f
    defined, 63
    eosinophils, in airway wall, 64, 65f
    features, 63, 64t
    flow-volume loop, before and after
      bronchodilator use, 63, 64f
    hypertrophic smooth muscle, 62–63, 63f
    thickened airway basement membrane, 59,
      62f
Atypical mycobacteria, 196–199, 199t,
    200f–203f
Azzopardi effect, 247

B

Bacillus anthracis, 236–237
BCG. *See* Bronchocentric granulomatosis
    (BCG)
Benign tumorlets, 268f–269f, 269
Benign tumors, 263, 264f
Beryllium disease, 294, 294f
Bioterrorism, 235–236
    anthrax, 236–238, 238f
    plague pneumonia, 238, 238f

Bioterrorism (*Continued*)
 pleural infection, 239f—241f, 242
 tularemia pneumonia, 239f, 241
Blastomyces, 204—205, 208f
Bone marrow embolus postresuscitation,
 145, 146f
Bronchial atresia, 318
Bronchial circulation, 13, 13f
Bronchiectasis, 76f, 76t, 176—177,
 177f—179f
 bronchial artery (BA) hypertrophy,
 79, 81f
 narrowed and scarred airways, 77, 78f
 peribronchiolar fibrosis, 77, 78f
 saccular bronchiectasis, 77, 77f
Bronchioles, 7
Bronchiolitis, 81—82, 82f—83f, 160,
 162f
Bronchiolitis obliterans organizing
 pneumonia (BOOP), 84, 88t
 acute fibrinous and organizing pneumonia
 (AFOP), 87, 87f
 bronchioloalveolar metaplasia, 87, 89f
 eosinophils, 84—85, 86f
 foreign body giant cells, 87, 88f
 hyaline membranes, 86, 86f
 Masson bodies, 84—85, 85f
Bronchoalveolar lavage, 1, 1t
Bronchocentric granulomatosis (BCG),
 213—214, 219f
Bullous emphysema, 72, 73f

C

Candida spp., 208, 214f—215f
Cardiopulmonary exercise testing, 15
Centrilobular emphysema, 69, 70f
Cestodes, 234—235, 236f—237f
Charcot—Leyden crystals, 64, 65f
Chronic bronchitis, 67—69, 69f
Chronic eosinophilic pneumonia, 307—308,
 307f
Chronic graft airway rejection, 311,
 311f—312f
Chronic inflammation, 23, 24f
Chronic interstitial diseases, classification,
 95, 97t
Chronic necrotizing aspergillosis, 215
Chronic obstructive pulmonary disease
 (COPD), 67, 67t, 68f
 exacerbations, 75—76
Chronic thromboembolic pulmonary
 hypertension (CTEPH), 144—145,
 145f
Chronic vascular rejection, 312, 312f
Churg—Strauss syndrome (CSS), 66
Cigarette smoking. *See also* Non-neoplastic
 smoking related disorders
 chronic bronchitis, 69
 emphysema, 72
 Lambertosis, 87
Coal worker's pneumoconiosis (CWP),
 292—293, 293f
Cocaine inhalation, 300, 302f
Coccidioides immitis, 206—207, 212f—213f

Common lung diseases, 37
 airway diseases
 bronchiectasis, 37—38, 39f
 collapsibility, 37, 38f
 distal trachea long segment stricture
 and left main bronchus,
 40f, 41
 luminal diameter, 37, 38f
 rightward displacement, 40, 40f
 squamous cell carcinoma, 41, 41f
 stenosis, 38—40, 39f
 wall thickness, 37
 diffuse lung disease. *See* Diffuse lung
 disease
 direct signs
 bronchiolectasis, 42
 centrilobular opacities, 41, 42f
 tree-in-bud nodules, 41, 42f
 indirect signs, 42, 43f
 infection, 42
 bronchopneumonia, 43, 44f
 cavitation, 45, 45f
 ground glass halo, 45, 46f
 hematogenous infection, 43—45, 45f
 interstitial pattern, 43, 44f
 lobar pneumonia, 43, 44f
 location, 45, 46f
 lung neoplasia, 54—57
 small airways disease, 41
COPD. *See* Chronic obstructive pulmonary
 disease (COPD)
Corpora amylacea, 30—31, 31f
Cryobiopsies, 2
Cryptococcus neoformans, 205
Cryptogenic organizing pneumonia (COP),
 84—87
Cryptosporidium, 226, 231f
CSS. *See* Churg-Strauss syndrome
 (CSS)
Cuboidal alveolar type II epithelial cell, 8,
 12f
Cystic fibrosis
 acute inflammation, polypoid respiratory
 epithelium, 78—79, 80f
 atelectatic lung, right middle lobe syndrome,
 79—80, 81f
 bronchial artery hypertrophy, in
 bronchiectasis, 79, 81f
 endogenous lipoid pneumonia,
 80, 82f
 gram-negative mucoid *pseudomonas*,
 78—79, 80f
 lung cross-section, 77, 79f
Cytomegalovirus (CMV), 168, 169f—170f
Cytotoxic drugs, 283, 284f

D

Decreased diffusing capacity, 95, 97t
Dematiaceous fungi, 222—223, 228f
Dendriform ossification, 30—31, 297—298,
 298f
Desmoid tumors, 326, 327f
Desquamative interstitial pneumonitis (DIP),
 130—131, 130f

Diffuse alveolar damage (DAD), 86,
 114—119, 116f—119f, 160—161,
 162f—163f, 163t, 164f, 164t, 165f
 delayed alveolar re-epithelialization, 120,
 121f
Diffuse lung disease, 46
 calcified nodules, 49, 49f
 centrilobular nodules, 48
 eosinophilic diseases, 53, 53f
 hexagon-shaped secondary pulmonary
 lobules, 46, 47f
 honeycomb, 47, 48f
 hypersensitivity pneumonitis, 52, 52f
 lung cysts, 49, 50f
 lung volumes, 49, 50f
 nonspecific interstitial pneumonia, 51, 51f
 organizing pneumonia (OP), 52, 52f
 perilymphatic nodules, 48, 48f
 peripheral consolidation, 49, 50f
 pulmonary hemorrhage, 53, 54f
 random nodules, 49, 49f
 septal lines and reticulation, 47, 47f
 traction bronchiectasis and bronchiolectasis,
 48
 usual interstitial pneumonia (UIP), 51, 51f
Dirofilaria, 229—230, 233f
DNA viruses
 adenovirus, 167—168, 169f
 cytomegalovirus (CMV), 168, 169f—170f
 hantavirus, 171, 174f
 herpesvirus, 168, 171f—173f
 varicella zoster, 170—171, 173f
Drug-induced angiitis, 155
Drug-induced pneumonitis, 279, 280t—281t
Dyspnea, 14—15
Dystrophic calcification, 298—299, 299f

E

Ebstein—Barr virus (EBV), 174, 175f
Emphysema, 13—14, 70t
 centrilobular emphysema, 69, 70f
 computed tomography (CT) scans, 70
 emphysematous lung, 72, 73f
 fibrobullous disease, upper lobes, 73—75,
 75f
 floating fragments, lung, 71, 71f
 hyperinflation and atelectasis, bullous
 emphysema, 72, 73f
 infected bullae, 73, 75f
 lung hyperinflation, 71, 71f
 panlobular emphysema, 72, 74f
 paraseptal emphysema, 72, 74f
 pulmonary acinus, centriacinar emphysema,
 69, 69f
 upper lobe predominance and anthracotic
 pigment, 72, 72f
Endobronchial biopsies, 2, 3t
End-stage honeycomb lung, 99
Eosinophilic granuloma, 131—134. *See also*
 Langerhans cell histiocytosis (LCH)
Epithelioid hemangioendothelioma, 330,
 331f
Extranodal marginal zone B-cell lymphoma,
 273, 273f

F

Fetal lung, 5
Fibrinoid necrosis, 32
Fibroblastic focus, 29, 29f
Fibrosis, 69
    established hyaline fibrosis, pulmonary scar,
        23, 26f
    intraluminal fibrosis, organizing pneumonia,
        23, 25f
    pathways, 23, 25f
Fine needle aspiration biopsy, 157
Follicular bronchiolitis, 87–90, 89f–90f
Francisella tularensis, 239f, 241
Fungus balls, 216, 221f–222f
Fusarium spp., 221, 227f

G

Gastroesophageal reflux, 15
Glomus tumors, 263, 265f
Goodpasture's disease, 153–155, 153f–155f
Granulation tissue, 28, 28f
Granulolymphocytic interstitial lung disease,
    126, 126f
Granulomatosis with polyangiitis (GPA),
    150–151, 150f–151f
Granulomatous pneumonia, 190

H

Hamartoma, 263
Hantavirus, 171, 174f
Hard metal pneumoconiosis, 293, 294f
Herpesvirus, 168, 171f–173f
High-resolution computed tomography
    (HRCT), 37
Histiocytic reactions, 23–27, 26f
Histochemical stains, 4, 4t
Histoplasmosis, 201–202
    amphophilic, H&E-stained sections, 202, 206f
    Candida glabrata, 202, 206f
    debris within macrophage phagolysosomes,
        203, 207f
    intracytoplasmic *H. capsulatum*, 203, 207f
    irregular yeast forms and pseudohyphae,
        202, 205f
    mediastinal granuloma, 203, 207f
    microcalcifications, 203, 206f
    narrow necked budding 2-4 μM yeast, 202,
        205f
    necrotizing granulomatous infection,
        202–203, 205f, 207f
    nodular pneumonia, 202, 204f
    reticulin stains, 202, 204f
Hodgkin's lymphoma, 274, 274f–275f
HRCT. *See* High-resolution computed
    tomography (HRCT)
Humoral rejection, 313, 313f
Hyaline membrane, 29, 29f, 317, 317f
Hyalinizing granuloma, 300, 303f
Hypersensitivity pneumonitis (HP), 112t,
    115t, 214–215, 219f
    causes, 111–112, 112t
    chronic hypersensitivity pneumonitis, rare
        microgranuloma, 114, 115f

fibrinous pneumonia, 111–112, 113f
intra-alveolar foamy macrophages, 112, 114f
upper lobe predominance, 111–112, 113t
Hyphate fungi, 209
Hypoxemia, 10, 12t

I

Iatrogenic lung diseases
    amiodarone, 279, 282f
    cytotoxic drugs, 283, 284f
    drug-induced pneumonitis, 279, 280t–281t
    mesalamine, 281–282
    methotrexate (MTX), 279, 283f
    newer drugs, 285, 285f–286f
    nitrofurantoin, 284, 285f
    opiate-induced changes, 279, 282f
    radiation, 283–284, 284f
    sulfonamides, 282, 283f
    transfusion-related acute lung injury
        (TRALI), 284–285
Idiopathic pulmonary arterial hypertension
    (IPAH), 138–139, 140f
Idiopathic pulmonary fibrosis (IPF). *See*
    Usual interstitial pneumonitis (UIP)
Idiopathic pulmonary hemosiderosis (IPH),
    155
Idiopathic tracheal stenosis, 90
IgG4-mediated lung disease, 127–128, 128f
Immune reconstitution syndrome, 196
Immunostains, 4
Interstitial lung disease (ILD), 320, 321t
    acute interstitial pneumonia, 114–120, 120f
    diffuse alveolar damage (DAD), 114–119,
        116f–118f
        delayed alveolar re-epithelialization, 120,
            121f
        fibroblastic foci, interstitial pneumonitis,
            120, 122f
    granulolymphocytic interstitial lung disease,
        126, 126f
    hypersensitivity pneumonitis, 111–114
    IgG4-mediated lung disease, 127–128, 128f
    lymphocytic interstitial pneumonitis (LIP),
        110–111
    management, 102, 103f
    nonspecific interstitial pneumonitis (NSIP),
        102–110
    pleuro-pulmonary fibroelastosis, 127, 127f
    sarcoidosis, 122–125
    usual interstitial pneumonitis (UIP), 95
        cystic bronchiolectasis and intralobular
            scarring, 97–99, 99f
        cystic space lined, low cuboidal
            epithelium, 101, 101f
        features, 97, 98t
        honeycomb lung, 101, 101f
        morphologic mimics, 97, 98t
        multiple fibroblastic foci, 97–99,
            100f
        myositis syndrome, 97–99, 99f
        nonspecific interstitial pneumonitis
            (NSIP), 101
        patchy subpleural fibrosis and cysts,
            97–99, 99f

pulmonary cirrhosis, 97–99, 98f
        subpleural fatty metaplasia and smooth
            muscle metaplasia, 97–99, 100f
Intralobar sequestration, 318, 319f
Intravascular large B-cell lymphoma, 275,
    275f
IPAH. *See* Idiopathic pulmonary arterial
    hypertension (IPAH)

K

Klebsiella, 183, 185f

L

Langerhans cell histiocytosis (LCH)
    cavitary change, 131, 131f
    CD1a immunostaining, 132, 133f
    cleaved nuclei, 131, 132f
    DIP-like change, 131, 132f
    interlobular septum infiltration,
        Erdheim–Chester disease, 133f, 134
    low power of nodule, 131, 131f
    macrophage accumulation, Rosai–Dorfman
        disease, 133f, 134
    macrophages immunostain, S-100, 133f,
        134
    stellate scar, 131, 132f
    ultrastructural "tennis racket" appearance,
        132, 133f
Large-cell undifferentiated (LCC) carcinoma,
    243, 253–254
Legionella spp., 188, 188f
Leukemia, 276, 276f–277f
LIP. *See* Lymphoid interstitial
    pneumonitis (LIP)
Lipid aspiration, 300,
    301f
Lipoid pneumonia, 28, 28f
Lung anatomy
    airway anatomy
        branching segmental anatomy, 5, 8f
        cuboidal alveolar type II epithelial cell, 8,
            12f
        essential aspects, 5, 7t
        gas exchange, 9
        normal pseudostratified ciliated
            respiratory epithelium, 5–7, 9f
        pulmonary acinus, 7, 11f
        pulmonary airway and artery, 7, 10f
        respiratory epithelium, 5–7, 9f
        secondary pulmonary lobule, 7, 10f
        squamous alveolar type I cell, alveolar
            wall, 8, 11f
    alveolar phase, 5, 5f
    bronchial circulation, 13, 13f
    canalicular phase, 5, 5f
    elastic structure, 13–15, 14f, 14t
    gross pulmonary anatomy, 5, 6f, 7t
    immune anatomy, 17–18, 17f–18f
    pleura, 15–16, 16f
    pseudoglandular phase, 5, 5f
    pulmonary lymphatics, 13
    pulmonary vascular elastic fibers, 16–17,
        16f

Lung anatomy (*Continued*)
  pulmonary vessels, 9
    hypoxemia, causes, 10, 12t
    increased pCO₂ and respiratory acidosis,
      10—12, 12t
    physiological parameters, 10—12
Lung biopsy, 138—139
Lung cancer, 54, 243t—244t
  adenocarcinomas, 251f
    atypical adenomatous hyperplasia, 249,
      250f
    atypical cells, small lepidic ACA, 249,
      250f
    distal ACA, unfixed lung, 249, 249f
    lepidic growth insitu, 243, 244f
    micropapillary ACA, 249, 250f
    mucin-secreting ACA and multifocal
      mucinous ACA, 249, 250f
    in scar, 249, 249f
  adenoid cystic carcinoma, 261, 262f
  alveolar and papillary adenoma, 267, 267f
  benign tumorlets, 268f—269f, 269
  benign tumors, 263, 264f
  glomus tumors, 263, 265f
  immunohistochemical staining, 245, 245t
  immunohistology, 253
  intermediate malignant potential,
    259f—260f, 260—261
  large-cell carcinoma, 247—249, 248f
  large-cell undifferentiated carcinoma,
    253—254
  malignant melanoma, 254—255, 255f
  metastatic carcinoma, 256, 256f
  mixed tumors, 263, 263f
  molecular analysis, 252—253, 252f—254f
  mucoepidermoid tumors, 261
  multiple tumors, 251—252
  perivascular cell tumor, 263, 265f
  pleomorphic carcinomas, 254, 255f
  salivary gland tumors, 261
  sarcomas, 257—258, 257f—259f
  sclerosing hemangioma, 263, 265f
  small-cell carcinoma, 247—249, 248f
  squamous cell carcinoma (SCC), 245—247,
    246f—247f
  squamous papillomas, 266, 266f
  WHO pathology staging schema, 243,
    244t—245t
Lung hyperinflation, 71, 71f
Lung neoplasia, 57f
  adenocarcinoma, 54—55, 55f
  lung cancer, 54
    staging, 55—56, 56f
  necrotic lymph nodes, 56—57
  pulmonary metastases, 57, 57f
  small cell carcinoma, 54, 54f
  squamous cell carcinoma, 55, 56f
Lymphangioleiomyomatosis, 305—307,
  306f—307f
Lymphocytic interstitial pneumonitis (LIP),
  110—111, 111f, 271—272, 272f
Lymphoid lesions
  benign intrapulmonary lymph nodes, 271,
    271f

    extranodal marginal zone B-cell lymphoma,
      273, 273f
    Hodgkin's lymphoma, 274, 274f—275f
    intravascular large B-cell lymphoma, 275,
      275f
    leukemia, 276, 276f—277f
    lymphoid interstitial pneumonitis (LIP),
      271—272, 272f
    lymphomatoid granulomatosis, 272—273,
      273f
    nodular lymphoid hyperplasia, 271, 272f
    primary B-cell lymphoma, 274, 274f
    T-cell lymphomas, 275, 276f
Lymphomatoid granulomatosis, 272—273,
  273f

M

Macrophage reactions, 23—27
Malignant melanoma, 254—255, 255f
Malignant mesothelioma, 327—330,
  328f—329f
Malignant pulmonary nodules, 1
Malignant tumors, 323—324, 323f
Measles, 166, 168f
Melioidosis, 200, 203f
Mesalamine, 281—282
Metastatic carcinoma, 256, 256f
Metastatic pleural tumors, 327
Methotrexate (MTX), 279, 283f
Microsporidia, 227, 231f
Middle eastern respiratory syndrome
  (MERS), 164
Mucinous metaplasia, 27—28, 27f
Mucoepidermoid tumors, 261
Multiple tumors, 251—252
Mycobacterial infection, 191—194,
  192f—193f, 194t
Mycobacterium tuberculosis, 191
Mycoplasma pneumonia, 172, 174f—175f

N

Nematodes, 232f
  cestodes, 234—235, 236f—237f
  Dirofilaria, 229—230, 233f
  necrotizing granulomatous and
    eosinophilic inflammation, 228, 232f
  paragonimiasis, 233—234, 235f
  pulmonary larval phase, 228, 231t
  trematodes, 230—233, 234f
Neural tumors, 326—327, 327f
Neuroendocrine cell, 59, 61f
Neuroendocrine hyperplasia, 320,
  320f
Nitrofurantoin, 284, 285f
Nocardia spp., 184—188, 187f
Nodular lymphoid hyperplasia, 271, 272f
Non-neoplastic smoking related disorders
  desquamative interstitial pneumonia (DIP),
    130—131, 130f
  Langerhans cell histiocytosis (LCH)
    cavitary change, 131, 131f
    CD1a immunostaining, 132, 133f
    cleaved nuclei, 131, 132f

    DIP-like change, 131, 132f
    interlobular septum infiltration,
      Erdheim—Chester disease, 133f, 134
    low power of nodule, 131, 131f
    macrophage accumulation,
      Rosai—Dorfman disease, 133f, 134
    macrophages immunostain, S-100, 133f,
      134
    stellate scar, 131, 132f
    ultrastructural "tennis racket" appearance,
      132, 133f
  respiratory bronchiolitis (RB), 129—130,
    129t
  smoking-related interstitial fibrosis, 134,
    134f
Non-small cell lung cancer (NSCLC), 252
Nonspecific interstitial pneumonitis (NSIP),
  101, 108f
  ankylosing spondylitis, 110, 111f
  autoimmune interstitial lung disease, 102,
    104t
  cellular NSIP, 104, 105f
  CREST syndrome, 108—109, 110f
  features, 102, 104t
  lymphoid follicles, autoimmune lung
    disease, 104—105, 106f
  pulmonary arterial bed, 108—109
  rheumatoid lung, 107t
    plasma cells, 105, 106f
  scleroderma, 107, 109f
  Sjogren's syndrome, 109, 110f
  systemic lupus erythematosus (SLE), 107,
    108t
    pulmonary capillaritis, 109f
  usual interstitial pneumonitis (UIP), 104

O

Obliterative bronchiolitis (OB)
  constrictive bronchiolitis, 83—84, 84f
  elastic stain highlights airway obliteration,
    83—84, 83f
  focal peri-arterial scar, 84, 85f
  intralumenal fibrous obliteration, 83—84, 83f
Open thoracoscopic biopsy, 158
Orthotopic lung transplantation (OLT), 309

P

Panlobular emphysema, 72, 74f
Papillary thyroid carcinomas, 256
Paracoccidioides, 207, 214f
Paragonimiasis, 233—234, 235f
Parainfluenza, 166, 167f
Paraseptal emphysema, 72, 74f
Parasites, 224
PD1+ lymphocytes, 23
Pediatric lung disease, 317
  ABCA3 deficiency, 320
  bronchial atresia, 318
  cystic adenomatoid malformation, 318
  hyaline membrane disease, 317, 317f
  interstitial lung diseases, 320, 321t
  intralobar sequestration, 318, 319f
  malignant tumors, 323—324, 323f

neuroendocrine hyperplasia, 320, 320f
sequestrations, 317–318, 318f
surfactant-protein C deficiency, 320–323, 322f
surfactant protein deficiency, 320, 322f
Penicillium marneffei, 223
Peribronchiolar fibrosis, 77, 78f
Perivascular cell tumor, 263, 265f
Plague pneumonia, 238, 238f
Pleomorphic carcinomas, 254, 255f
Pleura, 34–36, 34f–35f
Pleural diseases, 325
    desmoid tumors, 326, 327f
    malignant mesothelioma, 327–330, 328f–329f
    metastatic pleural tumors, 327
    neural tumors, 326–327, 327f
    pleural plaques, 325–326, 325f
    rare pleural malignancies, 330, 331f
    reactive fibrinous pleuritis, 325
    solitary fibrous tumors (SFT), 326, 326f
Pleural infection, 239f–241f, 242
Pleural plaques, 325–326, 325f
Pleuro-pulmonary fibroelastosis, 127, 127f
Plexiform lesions, 32
Pneumocystis jirovecii, 174–175, 176f–177f
Positron emission tomography (PET), 17
Postimplantation response, 313, 314f
Posttransplant lymphoproliferative disorder (PTLD), 314–316, 315f
Primary B-cell lymphoma, 274, 274f
Primary effusion lymphoma, 330, 331f
Protozoa
    amebic abscess, "anchovy paste" gross appearance, 224–225, 229f
    Cryptosporidium, 226, 231f
    E. histolytica, erythrocyte ingestion, 224–225, 230f
    liquefactive necrosis, 224–225, 230f
    microsporidia, 227, 231f
    Toxoplasma gondii, 225, 230f
Pseudallescheria, 221, 227f
Pulmonary acinus, 7, 11f
Pulmonary alveolar lipoproteinosis, 303–304, 304f
Pulmonary angiitis, 149
Pulmonary arterial hypertension, 137–138, 137t, 138f
Pulmonary asbestosis, 289
Pulmonary capillary hemangiomatosis (PCH), 140, 141f
Pulmonary defenses, 159, 160f
Pulmonary edema, 297, 297t
Pulmonary host response, 158–159
Pulmonary infection
    actinomycosis, 183–184, 186f
    acute bronchopneumonia, 178, 180f
    allergic bronchopulmonary aspergillosis, 212–213, 218f
    angioinvasive aspergillosis, 216–217, 223f–224f
    approach to sampling, 157, 158t
    Aspergillus bronchitis, 215, 220f

Aspergillus nidulans, 219, 226f
Aspergillus spp., 209, 212, 216f–217f
Aspergillus terreus, 219, 225f
atypical mycobacteria, 196–199, 199t, 200f–203f
bacterial infections, 178, 180f
bronchiectasis, 176–177, 177f–179f
bronchiolitis, 160
bronchocentric granulomatosis (BCG), 213–214, 219f
Candida spp., 208, 214f–215f
chronic necrotizing aspergillosis, 215
Coccidioides immitis, 206–207, 212f–213f
Cryptococcus, 205–206, 210f–212f
dematiaceous fungi, 222–223, 228f
differential diagnosis, 221–222
diffuse alveolar damage (DAD), 160–161, 162f–163f, 163t, 164f, 164t, 165f
DNA viruses
    adenovirus, 167–168, 169f
    cytomegalovirus (CMV), 168, 169f–170f
    hantavirus, 171, 174f
    herpesvirus, 168, 171f–173f
    varicella zoster, 170–171, 173f
Ebstein–Barr virus (EBV), 174, 175f
fine needle aspiration biopsy, 157
fungal infection, yeasts, 200–201, 204t
fungus balls, 216, 221f–222f
Fusarium, 221, 227f
granulomatous pneumonia, 190
group A streptococci, 181, 182f
histoplasmosis, 201–202
    amphophilic, H&E-stained sections, 202, 206f
    Candida glabrata, 202, 206f
    debris within macrophage phagolysosomes, 203, 207f
    intracytoplasmic *H. capsulatum*, 203, 207f
    irregular yeast forms and pseudohyphae, 202, 205f
    mediastinal granuloma, 203, 207f
    microcalcifications, 203, 206f
    narrow necked budding 2-4 μM yeast, 202, 205f
    necrotizing granulomatous infection, 202–203, 205f, 207f
    nodular pneumonia, 202, 204f
    reticulin stains, 202, 204f
hypersensitivity pneumonitis, 214–215, 219f
hyphate fungi, 209, 220, 226f
Klebsiella, 183, 185f
Legionella, 188, 188f
lung abscess, oropharyngeal aspiration, 183, 186f
lung biopsy, 158, 159f
lung injury patterns, 160, 161t
melioidosis, 200, 203f
microbes, bioterrorism, 235–236
    anthrax, 236–238, 238f
    plague pneumonia, 238, 238f
    pleural infection, 239f–241f, 242
    tularemia pneumonia, 239f, 241

miliary infection, 160
mycobacterial infection, 191–194, 192f–193f, 194t
Mycoplasma pneumonia, 172, 174f–175f
nematodes, 232f
    cestodes, 234–235, 236f–237f
    Dirofilaria, 229–230, 233f
    necrotizing granulomatous and eosinophilic inflammation, 228, 232f
    paragonimiasis, 233–234, 235f
    pulmonary larval phase, 228, 231t
    trematodes, 230–233, 234f
Nocardia, 184–188, 187f
paracoccidioides, 207, 214f
parasites, 224
Penicillium marneffei, 223
pneumococcal pneumonia, 179, 181f
Pneumocystis jirovecii, 174–175, 176f–177f
protozoa
    amebic abscess, "anchovy paste" gross appearance, 224–225, 229f
    Cryptosporidium, 226, 231f
    E. histolytica, erythrocyte ingestion, 224–225, 230f
    liquefactive necrosis, 224–225, 230f
    microsporidia, 227, 231f
    Toxoplasma gondii, 225, 230f
Pseudallescheria, 221, 227f
pulmonary defenses, 159, 160f
pulmonary host response, 158–159
pulmonary tuberculous infection, 194–195, 194t, 195f–198f
reactivation tuberculosis, 195–196, 199f
Rhodococcus equi, 189, 189f–190f
RNA viruses
    influenza, 162–163, 165f–166f
    measles, 166, 168f
    middle eastern respiratory syndrome (MERS), 164
    parainfluenza, 166, 167f
    respiratory syncytial virus (RSV), 165, 167f
    serious acute respiratory syndrome (SARS), 163
Staphylococcal aureus, 181–182, 183f
tracheobronchitis, 160
transbronchial biopsy (TBB), 157
transbronchial needle aspiration biopsy, 157–158
Tropheryma whippelii, 189–190, 190f–191f
video-assisted and open thoracoscopic biopsy, 158
Pulmonary inflammation, 21t
    acute inflammation
        acute neutrophilic pulmonary inflammation, 22, 23f
        fibrin diffuse deposition, 21, 22f
        regulation, 21, 22f
    airway wall
        granulation tissue, 28, 28f
        lipoid pneumonia, 28, 28f

Pulmonary inflammation (*Continued*)
   mucinous metaplasia, small airway, 27–28, 27f
   squamous metaplasia, chronically inflamed airway, 27–28, 27f
   alveolar interstitium
     corpora amylacea, 30–31, 31f
     fibroblastic focus, 29, 29f
     hyaline membrane, 29, 29f
     metaplastic smooth muscle extension, 30–31, 30f
     reactive atypia, 29–30, 30f
   chronic inflammation, 23, 24f
   fibrosis, 23
   histiocytic reactions, 23–27, 26f
   pleura, 34–36, 34f–35f
   pulmonary vessels
     angiomatoid and dilatation lesions, 32, 32f
     inflamed pulmonary artery, focal necrosis, 32, 32f
     pulmonary vasculitis, 33, 33f
     recanalized vascular intima, 33, 33f
     vascular remodeling, in pulmonary hypertension, 31–32, 31f
Pulmonary lymphatics, 13
Pulmonary scars, 305, 305f–306f
Pulmonary thromboembolic disease
   arterial wall by endothelium ingrowth, 142, 144f
   hemorrhagic infarction and infarction, 142, 143f
   lines of Zahn, 142, 144f
   old recanalized thrombus, 142–143, 144f
   severe pulmonary vascular congestion, incipient infarction, 141–142, 142f
   subpleural pulmonary infarctions, 142, 143f
Pulmonary transplant pathology, 309f
   acute allograft rejection, 309–311, 310f
   airways inflammation, 311
   aspiration, 313, 314f
   chronic graft airway rejection, 311, 311f–312f
   chronic vascular rejection, 312, 312f
   graft injury, 313
   humoral rejection, 313, 313f
   lung allografts, 315t, 316
   postimplantation response, 313, 314f
   posttransplant lymphoproliferative disorder (PTLD), 314–316, 315f
   recurrent sarcoidosis in allograft, 315f, 316
Pulmonary tuberculous infection, 194–195, 194t, 195f–198f
Pulmonary vascular disorders
   amniotic fluid emboli, squames, 147, 147f
   antialveolar basement membrane disease (ABM), 153–155, 153f–155f
   arteriovenous malformations, 156, 156f
   bone marrow embolus postresuscitation, 145, 146f
   causes, 141
   chronic thromboembolic pulmonary hypertension (CTEPH), 144–145, 145f

dirofilaria, 148, 148f
drug-induced angiitis, 155
fat emboli, 145, 146f
granulomatosis with polyangiitis (GPA), 150–151, 150f–151f
granulomatous pulmonary angiitis, 148–149, 149f
idiopathic PAH (IPAH), 138–139, 140f
Oil Red O staining, 145, 146f
pulmonary alveolar hemorrhage syndromes, 152, 152t, 153f
pulmonary angiitis, 149
pulmonary arterial hypertension, 137–138, 137t, 138f
pulmonary capillary hemangiomatosis (PCH), 140, 141f
pulmonary thromboembolic disease
   arterial wall by endothelium ingrowth, 142, 144f
   hemorrhagic infarction and infarction, 142, 143f
   lines of Zahn, 142, 144f
   old recanalized thrombus, 142–143, 144f
   severe pulmonary vascular congestion, incipient infarction, 141–142, 142f
   subpleural pulmonary infarctions, 142, 143f
pulmonary veno-occlusive disease (PVOD), 139–140, 140f
silicone microemboli, 148, 148f
talc microemboli, 147–148, 147f
tumor emboli, 149, 149f
Pulmonary veins, 13, 13f
Pulmonary veno-occlusive disease (PVOD), 139–140, 140f
Pulmonary vessels, 9
   hypoxemia, causes, 10, 12t
   increased pCO2 and respiratory acidosis, 10–12, 12t
   physiological parameters, 10–12
Pulmonary volumes/capacities, 95, 96t

R

Radiographic patterns. *See* Common lung diseases
Reactive atypia, 29–30, 30f
Reactive fibrinous pleuritis, 325
Reid index, 62, 62f
Relapsing polychondritis, 93
Respiratory acidosis, 10–12, 12t
Respiratory bronchiolitis (RB), 129–130, 129t, 130f
Respiratory distress syndrome, 5
Respiratory syncytial virus (RSV), 165, 167f
Restrictive defect, 95, 95t
Rhodococcus equi, 189, 189f–190f
RNA viruses
   influenza, 162–163, 165f–166f
   measles, 166, 168f
   middle eastern respiratory syndrome (MERS), 164
   parainfluenza, 166, 167f

respiratory syncytial virus (RSV), 165, 167f
serious acute respiratory syndrome (SARS), 163
Round worms, 228–235

S

Saccular bronchiectasis, 77, 77f
Salivary gland tumors, 261
Samples handling, 3–4
Sarcoidosis, 122t
   airways, 123, 123f
   central fibrinoid necrosis, 124, 125f
   granulomas
     airways, 123, 123f
     lymphatic distribution, 123, 123f
   hyalinizing fibrosis, 125, 126f
   necrotizing sarcoidal granulomatosis, 123, 124f
   perivascular granulomas, 123, 124f
Sarcomas, 257–258, 257f–259f
SCC. *See* Squamous cell carcinoma (SCC)
Scleroderma, 107, 109f
Sclerosing hemangioma, 263, 265f
Sequestrations, 317–318, 318f
Serious acute respiratory syndrome (SARS), 163
Siderosis, 295, 295f
Silicatosis, 291–292, 293f
Silicone microemboli, 148, 148f
Silicosis, 291, 291f–292f
Simon's foci, 191
Sjogren's syndrome, 109, 110f
Small airways disease, 41
Smoking-related interstitial fibrosis, 134, 134f
Solitary fibrous tumors (SFT), 326, 326f
Spirometric analysis, 59, 59t
Squamous cell carcinoma (SCC), 245–247, 246f–247f
Squamous metaplasia, 27–28, 27f
Squamous papillomas, 266, 266f
Staphylococcal aureus, 181–182, 183f
Streptococcal pneumonia, 179
Sulfonamides, 282, 283f
Sundry disorders
   alveolar microlithiasis, 298, 298f
   amyloidosis, 300, 302f–303f
   aspiration pneumonia, 299
   chronic eosinophilic pneumonia, 307–308, 307f
   cocaine inhalation, 300, 302f
   dendriform ossification, 297–298, 298f
   dystrophic calcification, 298–299, 299f
   food, 299, 299f
   hyalinizing granuloma, 300, 303f
   lipid aspiration, 300, 301f
   lymphangioleiomyomatosis, 305–307, 306f–307f
   pulmonary alveolar lipoproteinosis, 303–304, 304f
   pulmonary edema, 297, 297t
   pulmonary scars, 305, 305f–306f

Surfactant-protein C deficiency, 320–323, 322f
Surfactant protein deficiency, 320, 322f
Synovial sarcoma, 330, 331f
Systemic lupus erythematosus (SLE), 107, 108t
    pulmonary capillaritis, 109f

T

Talc microemboli, 147–148, 147f
Talcosis, 290, 290f
Tapeworms, 234–235, 236f–237f
TBB. *See* Transbronchial biopsy (TBB)
T-cell lymphomas, 275, 276f
Thoracoscopic biopsies, 1
Toxic exposures
    aluminosis, 295
    asbestos, 287–289, 288f–290f
        cancer, 289–290
    beryllium disease, 294, 294f
    coal worker's pneumoconiosis (CWP), 292–293, 293f
    hard metal pneumoconiosis, 293, 294f
    siderosis, 295, 295f
    silicatosis, 291–292, 293f
    silicosis, 291, 291f–292f

talcosis, 290, 290f
toxic gases, 287
Toxoplasma gondii, 225, 230f
Tracheal stenosis, 90, 91f
Tracheobronchomalacia, 93
Tracheopathia osteoplastica, 93
Transbronchial biopsy (TBB), 2, 3t, 157
Transbronchial needle aspiration biopsy, 157–158
Transfusion-related acute lung injury (TRALI), 284–285
Trematodes, 230–233, 234f
Tropheryma whippelii, 189–190, 190f–191f
Tularemia pneumonia, 239f, 241
Tumor emboli, 149, 149f

U

Upper airway diseases, 90, 91f
Usual interstitial pneumonitis (UIP), 95
    cystic bronchiolectasis and intralobular scarring, 97–99, 99f
    cystic space lined, low cuboidal epithelium, 101, 101f
    features, 97, 98t
    honeycomb lung, 101, 101f
    morphologic mimics, 97, 98t

    multiple fibroblastic foci, 97–99, 100f
    myositis syndrome, 97–99, 99f
    nonspecific interstitial pneumonitis (NSIP), 101
    patchy subpleural fibrosis and cysts, 97–99, 99f
    pulmonary cirrhosis, 97–99, 98f
    subpleural fatty metaplasia and smooth muscle metaplasia, 97–99, 100f

V

Varicella zoster, 170–171, 173f
Video-assisted thoracoscopic (VATS) lung biopsy, 158

W

Wegener's granulomatosis, 150–151, 150f–151f
Wheezing, 63
Whipple's disease, 189–190

Y

Yersinia pestis, 238, 238f

Printed in the United States
By Bookmasters